Working Lives
c. 1900

Working Lives c. 1900

A PHOTOGRAPHIC ESSAY

Erik Olssen

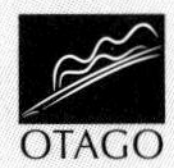
OTAGO

RIGHT The new chimney for Donaghy's Rope and Twine Co, erected in 1895.

Hocken Collections, A.G.-202/0425, S12-528g.

FRONT COVER DCC Archives, Drainage & Sewerage Board Photograph Album, DD & SC 157. (see page 92 for caption).

BACK COVER This photograph of the female staff at Irvine and Stevenson's, probably c. 1895, was taken by S. Collins. One suspects that Mrs Low, who stands to the left in the middle row, and the woman seated to her right, Mrs Bird, were forewomen, possibly but not necessarily widows.The person who donated this (and other) photographs of the staff also included the first and last names of almost all the workers in the picture. Their air of confidence no less than their dresses eloquently portrays something of the independent spirit of the 'New Women'.

Hocken Collections, S14-102a

First published 2014

ISBN 978-1-877578-51-9

A catalogue record for this book is available from the National Library of New Zealand.

Publisher: Rachel Scott
Editor: Wendy Harrex
Design/layout: Wendy Harrex & Fiona Moffat
Index: Diane Lowther

Printed in China through Asia Pacific Offset Ltd.

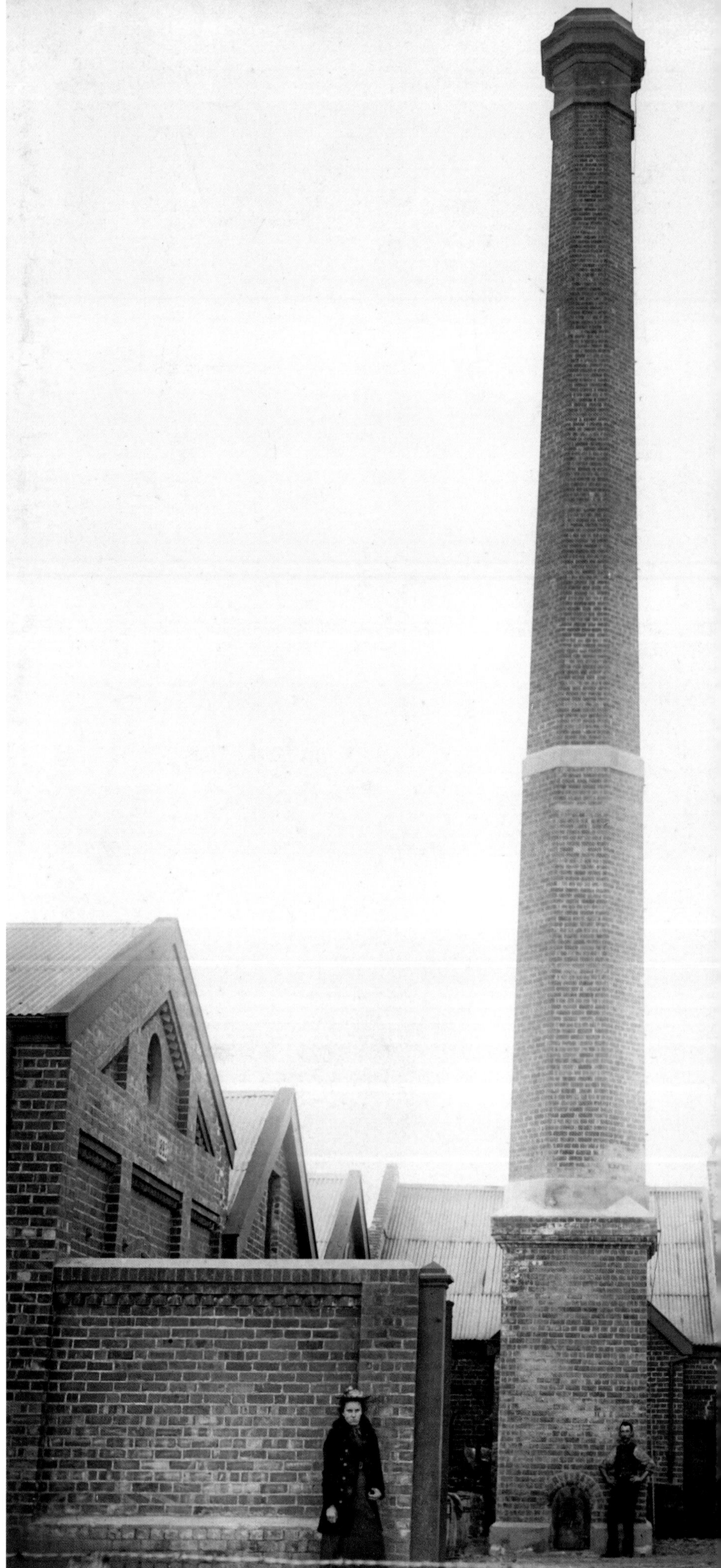

Contents

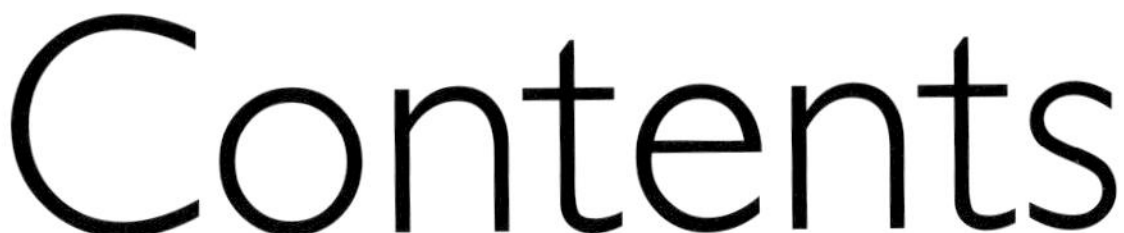

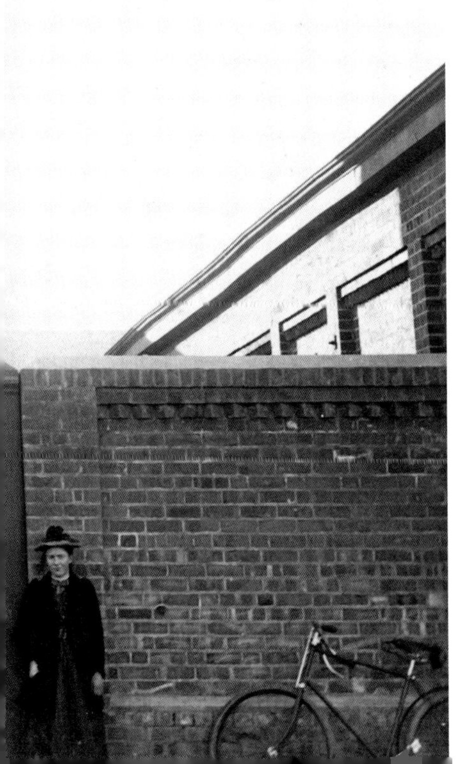

INTRODUCTION

Creating a New Society

FOR THE BEST PART OF THE LAST 30 YEARS I have studied the lives of workers in New Zealand. What began as a study of how capitalism operated here – appropriating the workers' 'surplus value' until the contradictions inherent ushered in a socialist millennium – became in time a study of how these workers used the state, their own organisations, and even the opportunities afforded by a new and lightly populated society, to modify capitalism. In a phrase commonly heard around 1900, men and women spoke of 'civilising capitalism'.

That story has been told in my *Building the New World* (Auckland University Press, 1995; now an ebook) and, with assistance from Clyde Griffen and Frank Jones, in *An Accidental Utopia?* (Otago University Press, 2011). Unpaid work, whether voluntary or domestic, sat in an uncomfortable relationship to paid work, an issue explored fully in *Sites of Gender*, edited by Barbara Brookes, Robin Laws and Annabel Cooper (Auckland University Press, 2003; now on Google Books).

This present book – a photographic essay – is not focused on unions, industrial relations, or the protracted gestation that produced the Labour Party, although those topics are not ignored. Instead I use the great wealth of photographs available from around 1900 to investigate the nature of work in New Zealand's first industrial suburbs. The book illustrates two processes fundamental to creating a new society: the transformation of a wild landscape into farmland and then industrial heartland; and the transplantation of the knowledge and skill acquired in the Old World that were essential to building a new world.

It is not often realised how recently the market for people's labour has emerged. Until the Industrial Revolution, most people doing manual work were either servants of some sort or worked for no pay within a household economy. Labour markets, in short, represent an invention. Although the Left in Britain and New Zealand has long depicted the Industrial Revolution as a social and cultural disaster, especially for working people, in the last 20 years historians have overturned this traditional view. During the Long Depression between 1878 and 1895, when many employers tried to reduce costs by subcontracting jobs to outworkers, the uproar over 'sweated' labour saw the practice ended first in Dunedin, by the intervention of an outraged middle class, and then more generally in New Zealand, by the state. Nor were the statutory restraints of the freedom of labour that were fundamental in England and Australia, such as a Masters and Servants Act, ever enforced here.

The Industrial Revolution (1750–1850), when capitalised, denotes the first instance of the breakthrough from an agrarian, handicraft economy to one dominated by the use of inanimate power, especially steam, to drive newly invented machines. At the same time, labour also became a commodity which could be bought and sold. New Zealand was one of the first countries in the world where almost all labour outside the home was done for pay. For the men and women of the skilled trades, the expertise and knowledge of their respective crafts were a source of identity and pride. The cultures of craft were much more important than the culture of class and were widely seen as underlying civilisation. Printers, bakers and brewers, tailors and tailoresses, smiths, fitters and turners, cabinet-makers and joiners – all had knowledge central to their craft, often secret, and all believed their craft skill fundamental to civilised life. And in a New World society such as New Zealand, the unskilled – regarded as almost a separate biological (and inferior) species in nineteenth-century England and Scotland – had skills and knowledges widely recognised as essential to the transformation of swamp and wilderness into farmland and cityscape. This grand task of colonisation enlisted the labour of all migrants, regardless of class or nationality.

The book's centre of gravity falls between, but is not confined by, Queen Victoria's Jubilee in 1897 and her death in 1903. The reader might be startled to find a book on working lives in New Zealand framed by two imperial events, but those photographed would not have been startled. These workers had usually begun their working lives in Britain. Their children, already a majority of the population (although most not yet 30 years old), often thought of themselves as having two homes, Britain and New Zealand. When the Duke and Duchess of Cornwall visited Dunedin in 1901, the entire populace rejoiced in their membership of the world's greatest empire (although some Ngāi Tahu, Chinese and Irish Nationalists doubtless dissented in private). Events elsewhere greatly strengthened that sense of shared membership; Queen Victoria's Jubilee and death, the Boer War and the departure for South Africa of New Zealand volunteers quickened this sense of dual identity (quite apart from any sense they had of themselves as Scottish, English, Irish, Scotch-Irish, Chinese, 'Syrian' or

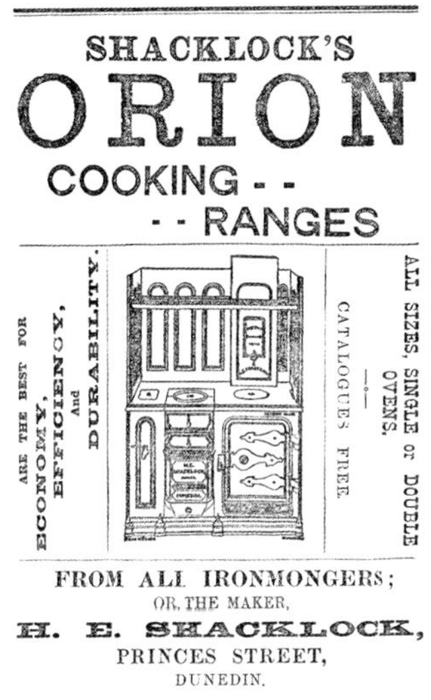

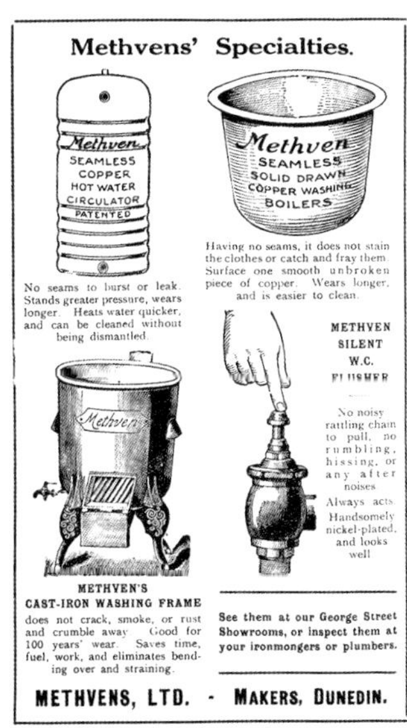

German). The great majority were proud of a British heritage, but confident that in this far-flung province of the Empire they would create a new and fairer society.

The outlook was promising. The selection of photographs focuses on Otago and the city of Dunedin for good reasons. Thanks to the discovery of gold, Dunedin had grown from village to small city and the entire province had boomed. Julius Vogel's public works and immigration schemes saw both Dunedin and Otago continue to prosper. In the 1870s, the city became the colony's leading commercial, financial and industrial capital. The presence of opportunities and the availability of work became major reasons why people came to Dunedin and stayed. Prosperity and the rapid growth in population created a market for a wide range of goods. In the early 1880s, poverty and unemployment stalked the land. Desperate men competed by under-selling each other. Wages fell. When the provincial economy crashed, more people left than arrived. A generation brought up on the Bible described this outflow as 'The Exodus' and it became a major political issue. By 1900, the city had survived both its first economic depression and 'The Exodus', and the province had just celebrated its fiftieth jubilee.

Working Lives illustrates the dramatic creation of an industrial and urban landscape in what had been a wilderness. (As far as we can tell, Kai Tahu Otakou traversed but otherwise did not much use this area, enjoying the riches of superior mudflats, wetlands and swamps in the lower Otago Harbour.) The chronological focus for the book captures Dunedin in its heyday as the industrial as well as commercial and financial capital of the country. In the 1900s, these were the world's best paid and freest manual workers. They had also played a critical role in making their new homeland 'the world's social laboratory'. They stood tall.

I

Set against this story of spectacular change, most of the images in *Working Lives* are of people at work, the places in which they lived and worked, and the way they went about protecting and extending the new opportunities and rights that migration had brought them. Each image, whether of a land- or townscape, a workplace or an event, is a snapshot, a moment in time. The impression of timelessness is the false beauty of the photograph. Houses, townscapes, and working lives change. Occupational labels may seem reassuringly solid: once a lawyer, always a lawyer, we assume, or once a labourer, always a labourer. In fact, the former was not always the case, and the latter was often not the case. In the 1890s and the 1900s, about 10 per cent of those men older than 21 had changed class a decade later. Roughly the same proportion had also changed occupation. Each class, like each street, neighbourhood and suburb, was like a tram: always full, but full of different people. On 'the Flat' of southern Dunedin there was a strong tendency for those in rental housing in South Dunedin, the main rental area, to move to the new suburban frontiers, first of St Kilda and Kew,

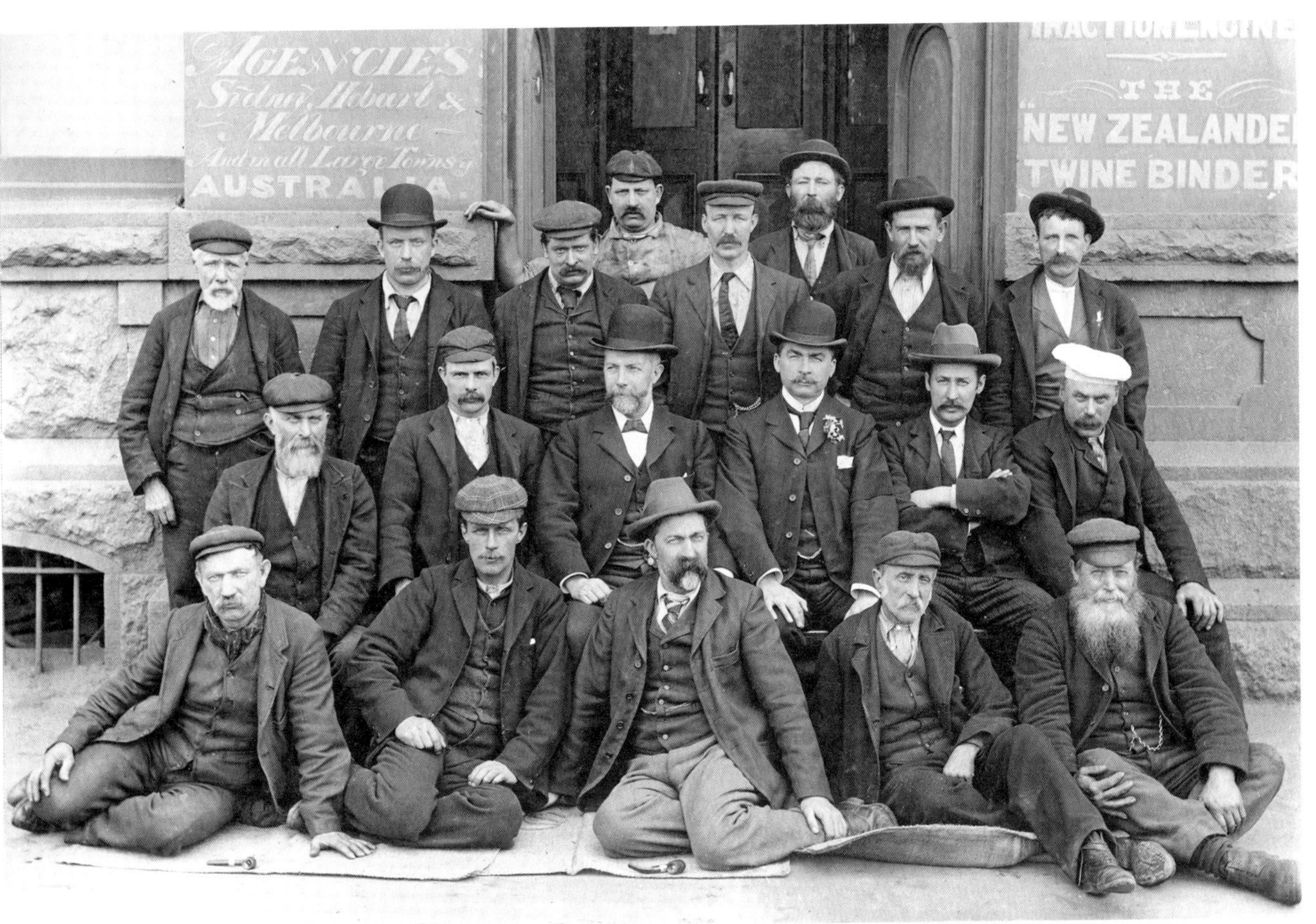

ABOVE **The men from Reid & Gray's smiths' shop and the machine shop** pose for Armstrong & Greer in 1899. Someone at Reid & Gray's wrote down the name, occupation, and in some cases length of service, of each person. The two clerks and the foreman fettler are better dressed, and the clerks sport fob watches and bowler hats, instead of cloth caps or trilby hats, but the men all sit together, their children went to school together, and they possibly all worshipped together.

Hocken Collections, Reid & Gray Photograph Album, Ms-1165-063; image S12-528b.

then later Musselburgh, Tainui and Andersons Bay to the east. Indeed, this propensity to move became a structural feature of New Zealand society; we move house more than any other people on the planet.

As well as obscuring fundamental processes of change and emphasising the appearance of continuity, each photograph also reflects what the photographer thought worthy of recording; many subjects were deemed either uninteresting or unacceptable. Grand houses and squalid slums, notable people, the transformation of the wilderness, the rise of new cities and the growth of industry: all had their photographers. So did many of the new society's leisure activities. However, many subjects that would interest us now do not figure in the photographic record. Housework and the home interested nobody, although as the market for home appliances grew we find advertisements for coal ranges and coppers. The heroic work of draining and waterproofing southern Dunedin is almost unrecorded, apart from one album commissioned by the Drainage and Sewerage Board. Many events that now intrigue us remain un-recorded. Imagine, for instance, hundreds paying to enter a large tent at the Oval to watch a 'scientist' demonstrate the potential applications of electrical currents to cure various physical ailments, including sexual impotence. That was a male crowd. Or another crowd of (mainly) men in the Exchange listening to soap-box orators: socialists, anarchists, salvationists, theosophists, rationalists, and so on. Because an abundance of leisure defined this new society, any number of photographers found a ready market for shots of the spectators and performers at the New Year games at the 'Cally', the biggest sporting event in the province between the 1880s and the 1900s. The crowds and horses at race meetings at Forbury Park also proved popular. So did the exhibits at the Winter Show. And the crowds enjoying St Clair beach, or promenading on the Esplanade, also attracted their photographers.

Some topics were not to be discussed in polite circles and that informal censorship operated to ensure no photographs were taken. Earth privies were almost universal until the start of the twentieth century, but nobody thought one worthy of recording. Nor were the nightsoil men with their night cart or even the nightsoil depot near the intersection of David Street and the Main South Road considered worthy of recording. If a genie transported us back, it is often said, the smell would shock us most, but smells could not be photographed, of course. There is an occasional shot of the open drains that carried waste and effluent across the Flat to the harbour, but the mud flats that became an unsightly and foetid rubbish dump and an embarrassment to the city were also deemed unworthy. Nor did the clouds of filthy dust from the unpaved streets in summer, or even the squalid mud after two or three days of drizzle, interest any photographer. As a result, of course, the men who had to deal with these waste products were never photographed.

Perhaps there was something about the craft of being a photographer that determined what was photographed and what was ignored. Most of the men whose work is to be seen in this book – for apart from the paintings by two local women, Ellen Valpy and Doris Lusk, every image was taken by a man – were professionals and had to sell what they photographed in order to keep their families. They also needed to be careful of their equipment. Like artists, whose work photography was transforming, photographers invariably had a studio and took a high proportion of all their shots there. Few studio shots appear here. While almost all photographers undertook their own version of the artist's *plein air* sketch, what they could achieve was limited. Around 1900, cameras of that time could not capture movement, or a moment in a movement. Nor, it seems, were they tolerant of dust, hence we have virtually no images of brickfields, quarries, or potteries. Given that the Flat boasted three large quarries and several smaller ones, two very large brickfields and the city's most important pottery, that is most unfortunate. Nor could they easily take photographs in dark places: most small workshops were dark.

As one wag said, it was less impressive that they could or would not do certain things than that they could capture any image at all. And that was then the verdict. This was the age of the postcard; Muir & Moodie, who took over Burton Brothers in 1898, reissued many of Alfred Burton's

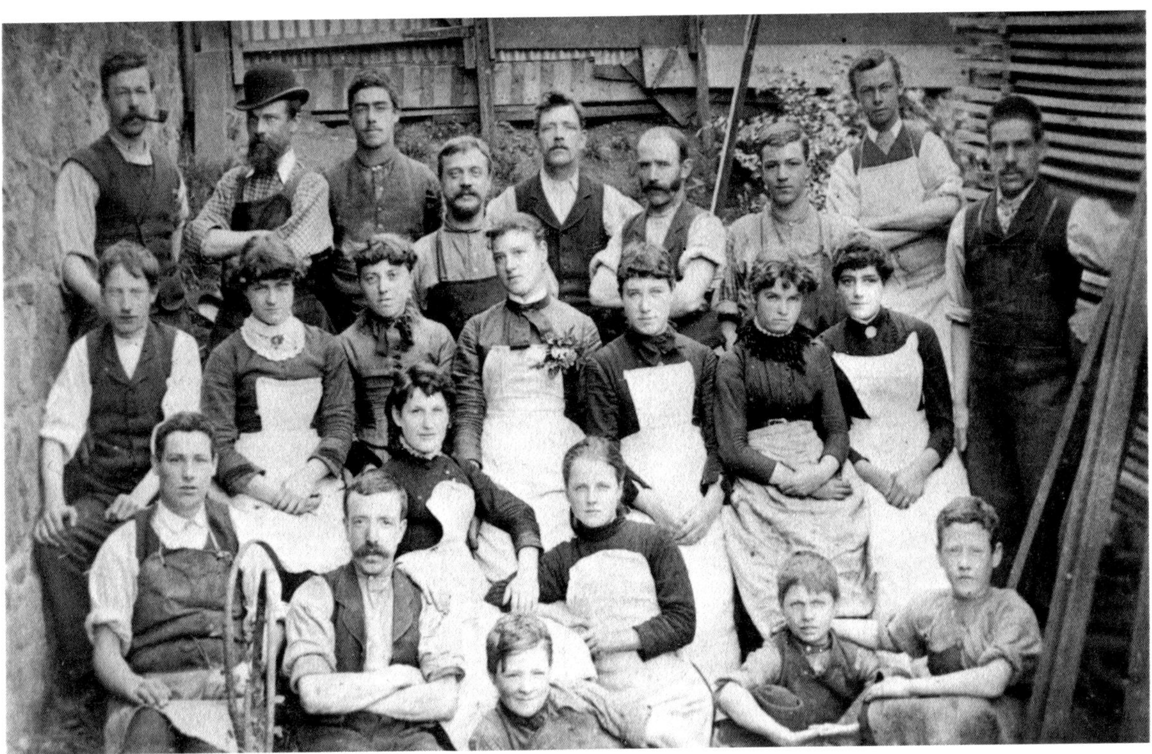

photographs as postcards over the next 30 years, while also taking many new originals. It was also the era of the box-brownie camera used by amateur photographers, although if any locals owned one, their work might now be found only in second-hand shops. People often had postcards made using photographs they had taken. Even moving images had been shown in Dunedin by 1900. We do not know if the first global smash-hit, 'The Kiss', was ever shown in the city, but by 1908 moving images had captured the working classes and begun to beguile the children of those who considered themselves their social betters.

ABOVE **Robert McKinlay's loyal workers,** probably in the early 1890s, after he re-opened his shoemaking business at the request of his old employees. McKinlay ran his businesss on terms satisfactory to his workers, and they proved loyal and industrious. There are 11 men, eight women, and three boys, presumably apprentices. McKinlay stands alongside. Limiting the ratio of boys to skilled men, especially finishers, had been an issue in this industry throughout the 1870s and 1880s, culminating in a rash of strikes in 1886–90. Employers wanted to reap the savings made possible by new technologies by bringing less-skilled workers, even boys, on to work once done by skilled men. When the skilled men resisted, trouble began. Turbulent industrial relations lasted until the 'Fighting' bootmakers crashed to defeat in 1892. Thanks to the newly elected Liberal–Labour government, a swag of new laws, most famously the Industrial Conciliation and Arbitration Act, encouraged the revival of trades unionism, first among the skilled, then the unskilled.

Courtesy McKinlays Footwear Ltd.

II

When Wendy Harrex and I decided at the last minute to exclude from *An Accidental Utopia?* the photographs I had spent months collecting and researching, the idea for this book was born. Like all ideas, it has taken on a life of its own. Although the larger transformations of the landscape and the polity remain of interest, the primary focus is on workplaces, workers, and work.

I have been attracted to the use of photographs in historical works ever since I returned home after undertaking my PhD in the United States, over 40 years ago. Ray Knox had launched *New Zealand's Heritage* and pioneered the dense use of illustrative material to help convey a mood or feeling of the ways in which the past had been another country and the ways in which it remained familiar. Some images even provided evidence that no other source provided. Keith Sinclair and Wendy Harrex took the idea further in *Looking Back: A Photographic History of New* Zealand (1978). I became a convert to the approach, a form of 'thick description', long before Clifford Geertz made the concept fashionable among scholars. A small number of my publishers, notably Brian Turner and Barbara Larson at John McIndoe *(Relics of the Goldfields* and *A History of Otago*), Bridget Williams (*The People and the Land*), and Elizabeth Caffin at the Auckland University Press (*Building the New World*), encouraged me in this enthusiasm. I have long used paintings, cartoons, advertisements and photographs to contextualise and illustrate the fundamental processes of change that transformed New Zealand.

One of the pleasures I have had in researching and selecting images for this book is that I have slowly come to know and appreciate the work of the painters, cartoonists and photographers who made it possible. In the late 1860s and early 1870s, Joseph Perry and Joseph W. Allen, both flawless technicians, captured not only the transformation of the landscape into a new sort of city, centred on single-unit homes, but the dramatic story of industrial growth. In the late 1870s and '80s, Frank Coxhead and Alfred Burton, also consummate technicians, were interested in the same story. And then, around 1900, one photographer after another began to document the dramatic expansion of industry.

The short-lived partnership of Armstrong & Greer, who (fortunately for me) persuaded several important firms to commission an album in 1899, was especially skilful at organising their subjects and capturing the streetscapes. At Donaghy's, perhaps because there were women working alongside men, then most unusual if not unacceptable, they produced some wonderful shots. A few years later, the Dunedin Drainage and Sewerage Board commissioned W.H. Ombler to document the construction of the Musselburgh Pumping Station and the new outfall pipes for sewage and stormwater. He also took some memorable images of the navvies at work digging vast trenches for the newly made Monier pipes that carried sewage and stormwater from the Flat to the pumping station and from there to the sea. Thanks to the growing use of photographic illustrations in papers such as the *Otago Witness* and the *Auckland Weekly News*, a host of other photographers, notably S. Collins and Guy Morris, also documented the dramatic story of progress.

In the nineteenth century, the railway best symbolised progress. While there are many photographs of lines being laid and opened, and plenty of locomotives, little has survived that shows the work of the railway workshops. In the 1920s, a railwayman with a passion for photography further enriched the archive by taking some 30 shots of the Hillside workshops. Alfred Percy Godber, a brass turner transferred to Hillside from Petone on promotion to foreman, documented 'then' and 'now' when the workshops were demolished in 1926–27 and replaced by a new and modern complex.

OPPOSITE **The constant flow of new technologies transformed the world of work unevenly.** John Wardell, a tea importer and family grocer, had his shop on Dunedin's main street, George Street. He prospered in the 1900s and went into partnership with his brothers. Eventually the firm expanded into Wellington and became the country's first delicatessen, selling imported cheeses, coffee, and top-end products from England, such as golden syrup and Marmite. Wardell Brothers & Co still relied on horse-drawn delivery vehicles in the mid-1930s, when this photograph was taken, although by then they were unusual. The driver remains unidentified, the photographer unknown, but the curators are confident that the street is in Wellington.

Alexander Turnbull Library, F 269981/2, 026998-F.

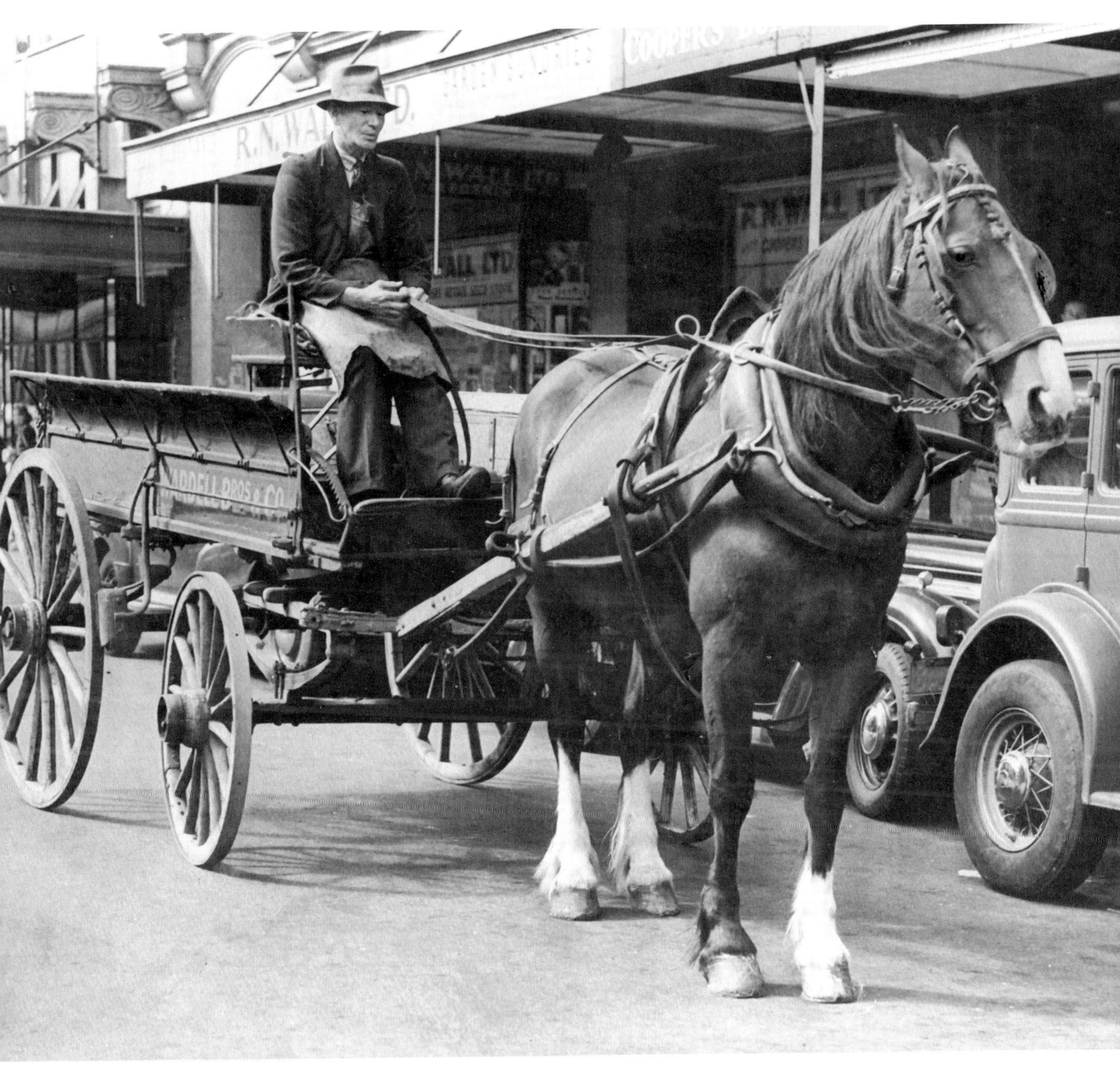

Both major political parties of that era shared Godber's vision of a modern, efficient nation. Their achievements are not the focus of this book. But the state, like the borough councils, actively encouraged the transformation of wilderness into city and farm, constructed the railway network, and in the 1890s took several innovative steps to ensure that the extremes of wealth and poverty that characterised the old country did not put down roots here. That same government also actively intervened in the country's labour markets to protect the weak and vulnerable, including working men.

CHAPTER 1

From 'the Swamp' to 'the Flat'

THREE STREAMS crossed the salty swamp that separated Otago Harbour from the Pacific Ocean and at the same time linked the Otago Peninsula to the mainland. A fourth came down the Glen. Before European purchase, Māori travelled from the settlement at Otakou near the mouth of the harbour to catch eels, crayfish, pūkeko, ducks and weka. The main route south for Māori was a path that ran along the inside of the extensive sandhills bordering the ocean. This track was used well into the period of European settlement, and was still well formed in 1870. While you could bypass this swampy area to the south or north, the rest was hard going at best and impenetrable at worst. As one Australian wit remarked, 'in the summer it is under water, and in the winter it is not to be found'.

Purchased in 1844 as part of the Otago block, settled by Europeans between 1849 and 1852 as part of suburban Dunedin, thanks to W.H. Valpy, a wealthy Englishman who migrated for the sake of his health, the slopes facing northeast soon produced grain, potatoes and other vegetables. Gardeners and farm labourers found employment and many began to buy their own small plots overlooking the swamp. The first European track, a wet path, ran along the harbour foreshore, linking the town to Andersons Bay. The first formed road edged around the Mornington–Maryhill Rise, although before it climbed Caversham Valley a fork ran diagonally across to Valpy's farm at Forbury, which in turn connected to a path Valpy's people built to the beach. Caversham, close to this 'Swamp Road' and with a slightly higher elevation than the rest of the Flat, produced the beginnings of urban settlement. A limeworks exploited the local sandstone, although it proved too soft to be useful for building, and the first few dwellings and businesses were established.

I

It did not take immigrants from rural Britain long to realise that the streams would have created alluvial fans with rich potential. And so it proved, although removing fallen trunks and branches proved very hard work. Within 50 years the immigrants' knowledge of drainage, reclamation and the science of horticulture saw the swamps transformed into productive land for agriculture and horticulture, almost 12,000 years of Eurasian history compressed into roughly 30 years. As the population began to grow, some of the early farmers and nurserymen began subdividing their holdings. In the last quarter of the century – but in St Kilda not until the 1920s – urbanisation proceeded rapidly. Whereas Dunedin was an early Victorian city, the northern part of the Flat reflected the late Victorian era, when much of Caversham, South Dunedin and Kensington were settled. The southern area of the Flat, including St Clair and most of St Kilda, began urbanising during the Edwardian era, although only in the 1930s did they reach their maximum populations. At that point, St Kilda was the most densely populated borough in New Zealand. By international standards, of course, it was not densely populated at all.

Construction of the Main Trunk Railway (1873–79) saw industry and people move to the northern parts of the Flat. Cheap land and proximity to the railway, plus two streams, proved irresistible. In the 1880s, gas pipes were laid along the main streets for the purposes of lighting. Some gas was even reticulated to private homes. The introduction of a horse-drawn tram service in 1879 from the city, via Hillside Road and David Street, to the township provided yet another modern amenity that began the process of transforming this independent township into a suburb. Extensive stables on David Street replaced the earlier nightsoil depot, no longer acceptable in proximity to a residential area. In 1881, horse-tram services to St Clair and St Kilda were inaugurated. Paradoxically, urbanisation here retained strong rural characteristics: hens, cows and horses were everywhere, the cows often free to wander and a source of income for the owner. Twice a week, large mobs of sheep and cattle from the farms to the north of the city and the Otago Peninsula were driven along the Main South Road to the abbatoirs at Burnside.

During the 1880s and 1890s, the population of both Caversham township and South Dunedin grew. New subdivisions were opened, with housing built usually by speculative builders. The retail strip along the Main South Road in Caversham Valley and South Dunedin's shops in King Edward Street consolidated between the 1880s and 1905 to become the two main retail and business centres serving the Flat, with South Dunedin's becoming quite substantial. The concentration of large industries attracted workers and their families, while 'spec' builders increasingly emphasised the health-giving qualities of the area, ocean breezes and sunlight. Whereas the subdivisions of the 1870s and 1880s had narrow streets and small sections, often one-sixteenth of an acre, those opened up in the 1890s and 1900s tended to have wider streets and more substantial sections. The streets were

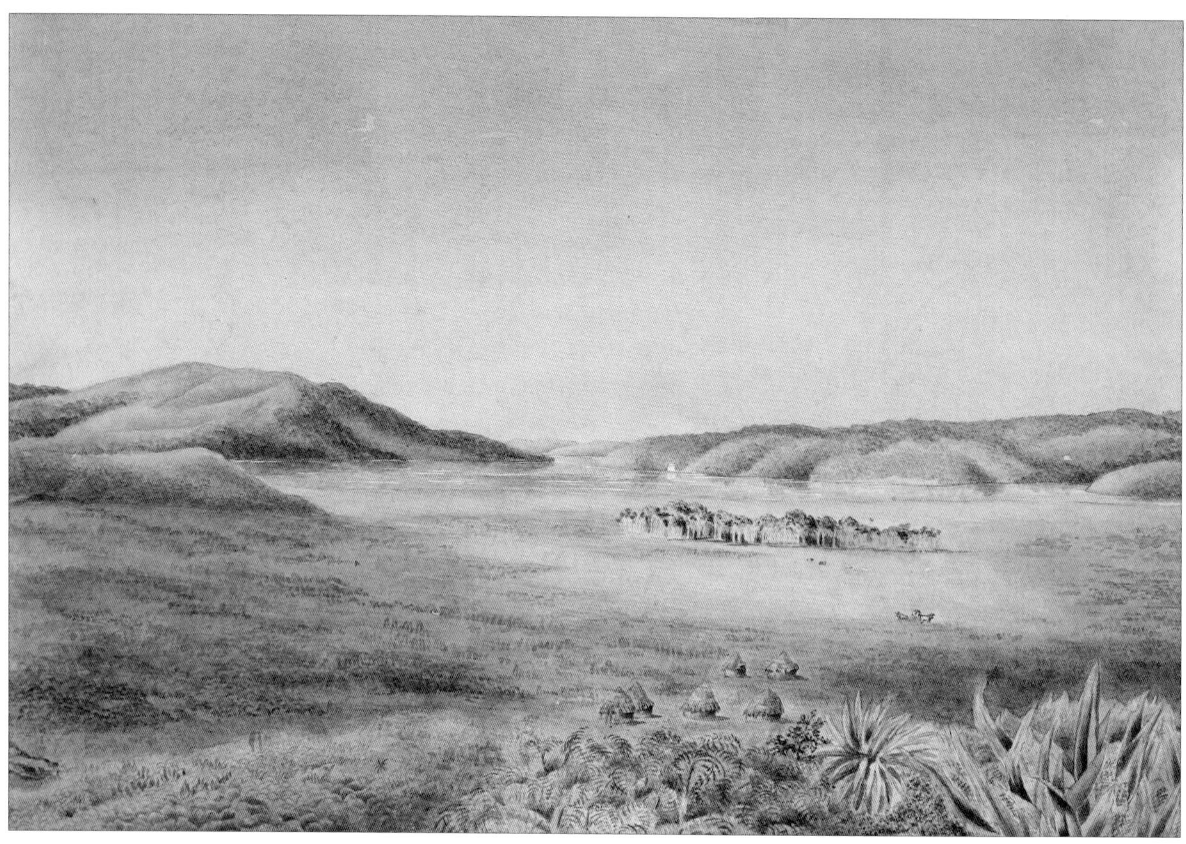

ABOVE **The 'Swamp Road'. Miss Ellen Penelope Valpy's watercolour, *Dunedin Harbour and Andersons Bay*,** shows a bush remnant in the middle of the Flat, her father's wheat stooks at the foot of Caversham Valley, and a horse and cart travelling along the path to her home. Her father, William H. Valpy and his wife, Caroline (née Jeffreys), settled in 1849 with five of their six children on a farm that he developed on his suburban lot. Valpy, descended from a long line of scholars and until 1836 a judge with the East India Company, decided to emigrate for the sake of his health. He brought a large staff of servants, a sawmill and a flourmill, loads of farming equipment, and took up several pastoral runs as well as his farm on the Flat. He named Forbury after his own birthplace, although this name was used for various places across the Flat, and Caversham after his mother's birthplace. As the little colony's most eminent and wealthy Anglican, he helped make the Flat the preferred destination for settlers who disliked the Free Church and for those who wanted an eight-hour working day. After Judge William's death in 1852, much of his fortune was lost and some of his children, including Ellen, fell on hard times. She married a cousin, who deserted her with a small son when pregnant with a second. Painting was one of the few ways a woman of gentility could earn money.

Toitū Otago Settlers Museum.

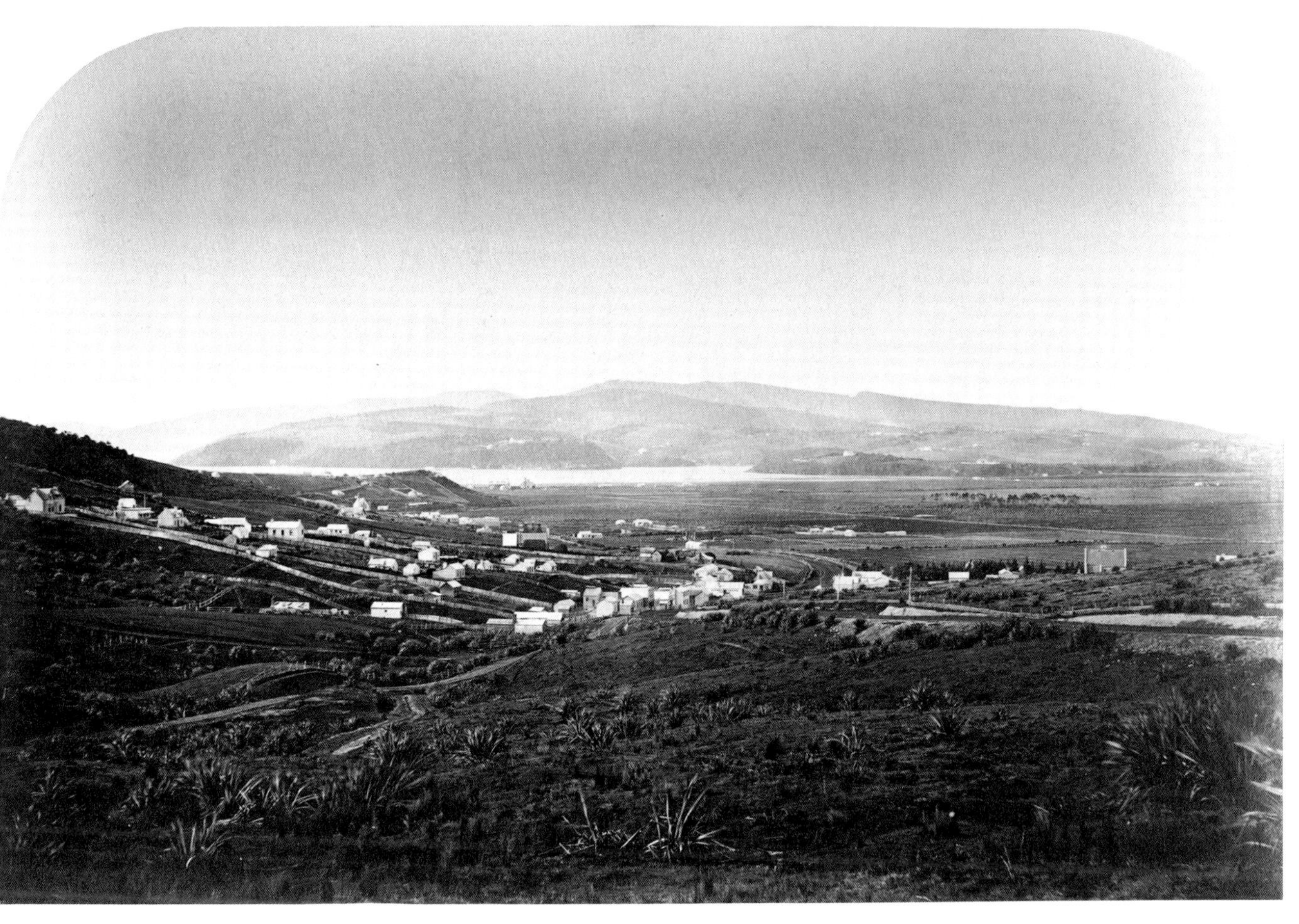

ABOVE **Joseph Perry's photograph of the 'Township of Caversham',** looking towards the Otago Harbour, 1865, captures the way in which the first settlers scattered around the small nucleus that constituted the centre of the township, the centre comprising an hotel, post office, blacksmith's smithy, general store and school. Caversham began as a staging post on the road south from Dunedin to the Taieri and Tokomairiro Plains, South Otago and Southland and inland to Central Otago. Miss Valpy's remnant bush might possibly be visible mid-right.

Hocken Collections, Joseph Perry Photograph Album No. 12; image S08-247f.

still unpaved and passing traffic, as well as the mobs of cattle and sheep driven along the Main South Road towards the abbatoirs at Burnside, raised storms of dust in dry periods and churned the road to mud in wet periods.

In Caversham Borough, settlement now spread back from the township towards Kensington and up the hill towards Lookout Point. In South Dunedin, by contrast, it spread west from King Edward Street towards Forbury Road. At Forbury Corner, where Hillside and Forbury Roads met, Searles' two-storeyed Waterloo Hotel faced Kew's Primitive Methodist Church. Both were handsome brick buildings. By World War I, substantial villas on larger sections dominated the newer areas of Caversham and South Dunedin.

With the exception of the seaside settlement at St Clair, the southern half of the Flat retained a rural character. Although by the 1880s nobody tried to grow wheat or oats on the Flat, Chinese market gardeners, mostly ex-gold miners, worked about 50 acres straddling Caversham and South Dunedin. Some lived in little huts on the gardens, mainly to protect them from thieves and larrikins, but in the 1890s most lived in a large wooden house in Caversham township. It seems that the Chinese provided most of the European settlers with their vegetables. There were also several substantial gardens, nurseries and orchards centred on St Clair flat. In the early 1900s, when the Borough of St Kilda had a population of around 700, John Fleming's farm still dominated much of the area. Fleming, a produce and provision merchant who had taken over both the business and the farm from his father, bred racehorses, including the locally famous 'Gypsey family'. It is easy to forget that racing, gardening and horticulture were industries, but nobody thought them worth photographing, or if they did nobody considered the photographs worth keeping.

It is also easy to forget that it was the *flatness* of the Flat that rendered the area distinctive and gave it its name. The stretch of land between harbour and ocean was by far the largest area of flat land close to the city. The creeks running down the Glen, Caversham Valley and Playfair Street gully, and Allandale gully, attracted industries. The higher land at the northern end of the Flat, where the Main South Road and the Main Trunk Railway ran, was quickly identified as some of the city's best land for industry. It was not only close to the railway and the road, but the port was less than a mile away. The Flat's proximity to the sea and its health-giving ozone also appealed to residents of Dunedin's squalid central lanes and alleys who sought a healthier environment. It provided a second chance for those unhappy with the outcome of their first chance. It also provided a more congenial environment for anyone who disliked the dominance of the Free Church in the city centre. No form of Presbyterianism was ever dominant on the Flat.

By the early 1900s, however, it had become clear that there was a stiff price attached to living in an independent borough. Neither Caversham nor South Dunedin could afford the sort of amenities, whether electric trams or modern drainage systems, that Dunedin's ratepayers could afford. In

OPPOSITE **Caversham from Rockyside, 1875.** A lot has happened since 1865. This image looks towards the ocean and nicely shows the train steaming past the Immigration Barracks, where from 1872 all new immigrants were housed on arrival, and the two-storeyed Benevolent Home, established by the Province in 1869 for indigent old men and women (with provision made later for unwed mothers to deliver their babies). Extensive areas of farmland, the St Kilda swamp and the towering sandhills are also visible. Larrikins throughout the country amused themselves by setting fire to the indigenous sand cover, pīngao; one of the main tasks of the Ocean Beach Domain Board was to identify and establish another grass that would flourish in the sand but not burn.

Hocken Collections, *New Zealand Herald*, 9 April 1875, Pt. I.

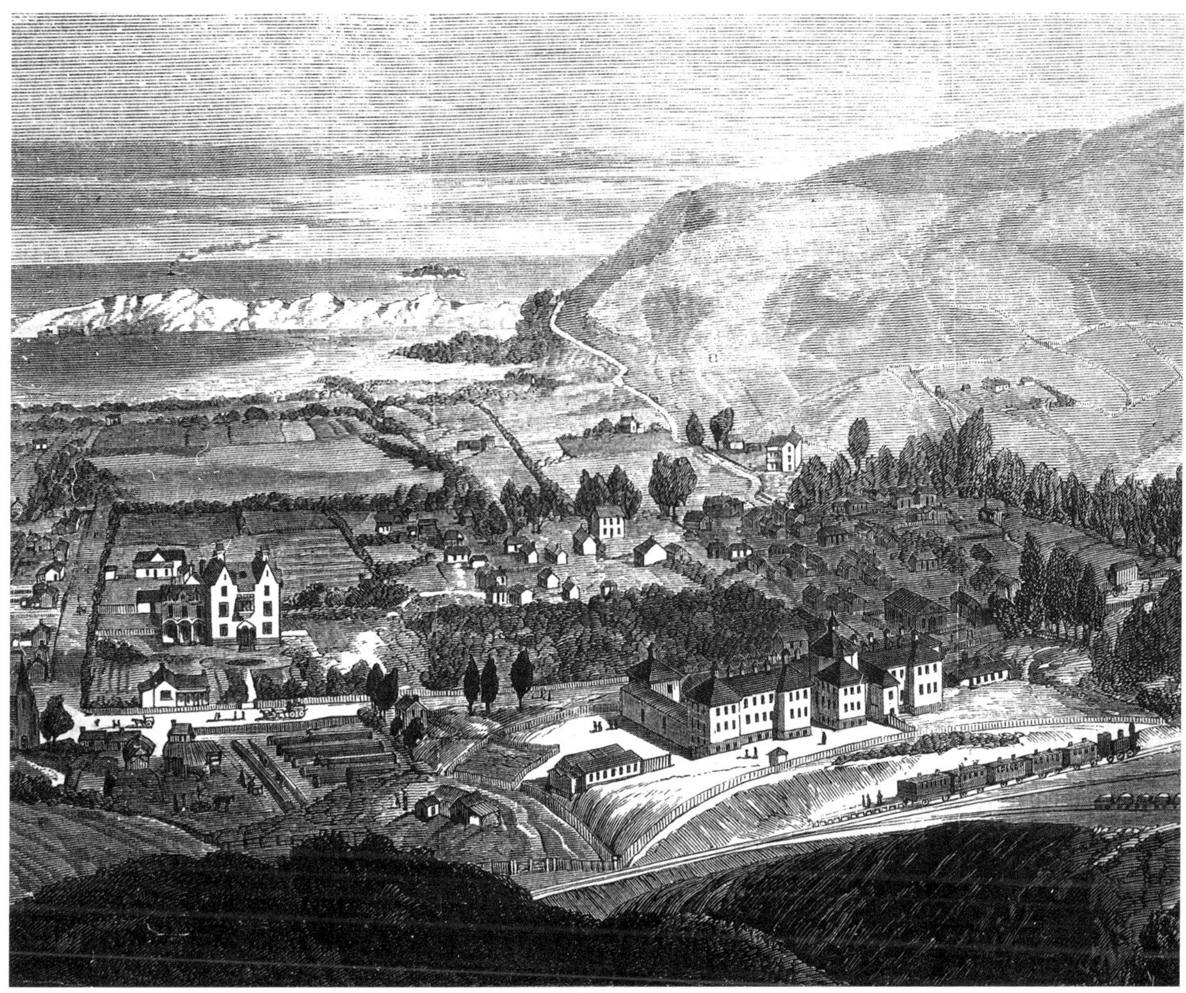

1904, the ratepayers of Caversham voted – with only the township's booth opposing – to amalgamate with Dunedin in order to obtain an electric tram service. In 1905, South Dunedin's ratepayers followed suit. Over the next few years the City provided a range of modern amenities, including an electric tram service, modern systems for removing sewage and stormwater, and reticulating clean water and even electricity (which the City generated at Waipori, the country's first extensive hydroelectricity scheme). Not only were the streets dug up to install the various pipes required, but the principal thoroughfares were widened to accommodate the double tram tracks. In the process, the City experimented with various methods for sealing the surfaces of major thoroughfares. By the Musselburgh quarry near St Kilda's eastern boundary, a gigantic pumping station was also installed and a small army of navvies dug their way through rocky Lawyer's Head and also built a gravitational tunnel through the sandhills at Tahuna to provide outfalls. The Ocean Beach Domain Board, a veritable law unto itself, also tried to protect the Flat by stabilising the sandhills that stood sentinel between the turbulent Pacific and what had once been salt swamps.

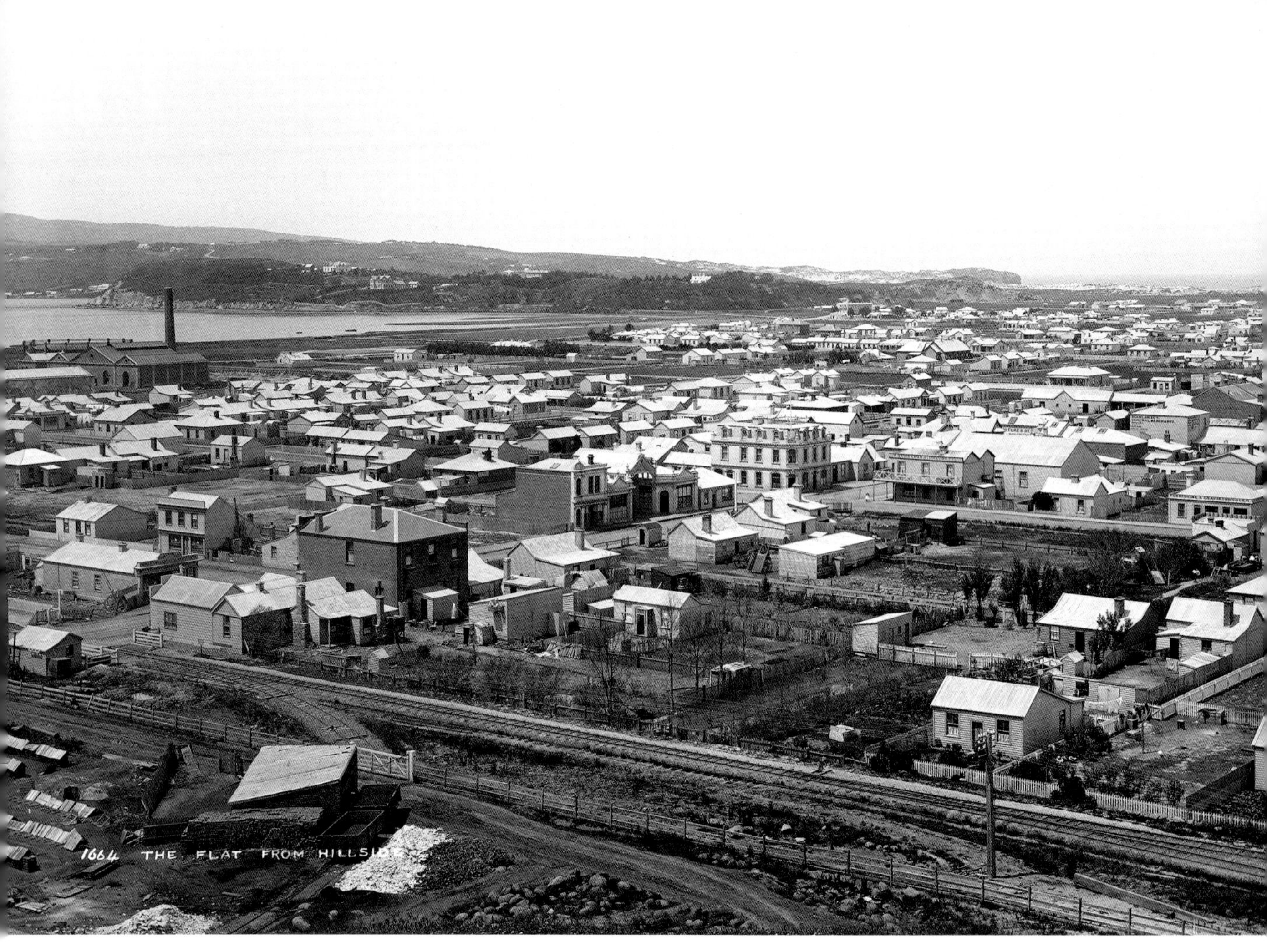

ABOVE **'The Flat from Hillside', c. 1886.** The single-track Main Trunk Railway line is in the foreground, the municipal gasworks to the left, and Ogg's Corner (later Cargill's) is flanked by the most substantial building in view (Naumann's Hall being almost opposite). Ogg's Corner was named for a famous local publican and sportsman, and became the major commercial centre on the Flat.

The two-roomed cottage plus lean-to is already ubiquitous, although most enjoy a sizeable section and a backyard shed, which would have accommodated the earth closet if nothing else. Although many of these cottages housed families with between seven and 10 children, most children left home when they left school at age 12 (only strictly enforced from 1895). The houses represented a considerable improvement over the tenements and row houses of Britain, let alone the slums of Manchester, Glasgow or Dublin, where each room often housed one family. Alfred Burton took this photograph, based on one by J.W. Allen.

Te Papa Museum of New Zealand, C012452.

II

The northern and southern halves of the Flat developed quite differently; by the time the area was fully urbanised (around 1930), the traces of these differences marked the character of each. The northern, naturally drier, part was built up earlier and from the beginning industrial workshops and commercial enterprises were located close by and among residential streets, especially in the earlier built-up areas. People who lived here often also worked here. St Clair, St Kilda and Kew, by contrast, were developed exclusively as residential, with the pattern of streets of houses being broken only by the considerable number of parks and recreational facilities. Most residents in paid work travelled out of these suburbs to work six days a week, increasingly on electric trams travelling on paved roads above a spaghetti of pipes carrying clean water, stormwater, sewage and gas. Even the two major retail areas were in Caversham and South Dunedin, with little beyond local corner shops opening in St Kilda and St Clair. In these newer suburbs, home and family were truly enclaves, separate from the working world. When the lack of reserves in Caversham and South Dunedin was recognised in the 1910s, land had to be bought back to create playgrounds, unless given.

The predominance of industries in close proximity to a residential area, and particularly the influence of their largely male workforces, defined the character of the older areas of the Flat in both class and gender terms. Masculine bonds tied the Flat together as a whole. From these very local workplaces, large or small, men formed various kinds of union: craft and trade unions, but also lodges, churches, affiliations centred on the several pubs, and sports clubs. There were also informal networks based on pubs, a passion for the popular gambling game known as 'two-up', and other forms of gambling. The national drive to make rugby an amateur game met several sizeable obstacles on the Flat. These networks, we might argue, were important in what made the area distinctive rather than typical. (The concentration of these workplaces and their corresponding affinities in Caversham and South Dunedin may explain the fact that the term 'the Flat', while technically covering all of the flat land, is most strongly associated with these areas.)

Women, in a fashion peculiar to this area, formed networks both further afield and closer to home. Some young, unmarried women worked on the Flat, but many more – especially as domestic work fell more and more from favour – travelled into the city to work, meeting at the tram stops or on the trams, or as they walked into town, and certainly as they worked together in clothing or biscuit factories, offices, or department stores. Reversing the more common pattern, therefore, southern Dunedin retained its working men within the locality, and sent its young women off to the city or farther afield to work in either the Roslyn or Mosgiel Woollen mills. Married women, and others whose work kept them in the household, had different networks, much more characteristic of women elsewhere: their links were likely to be neighbourhood-based, much more tightly local than those of either men or women in the paid workforce. It was they who held together

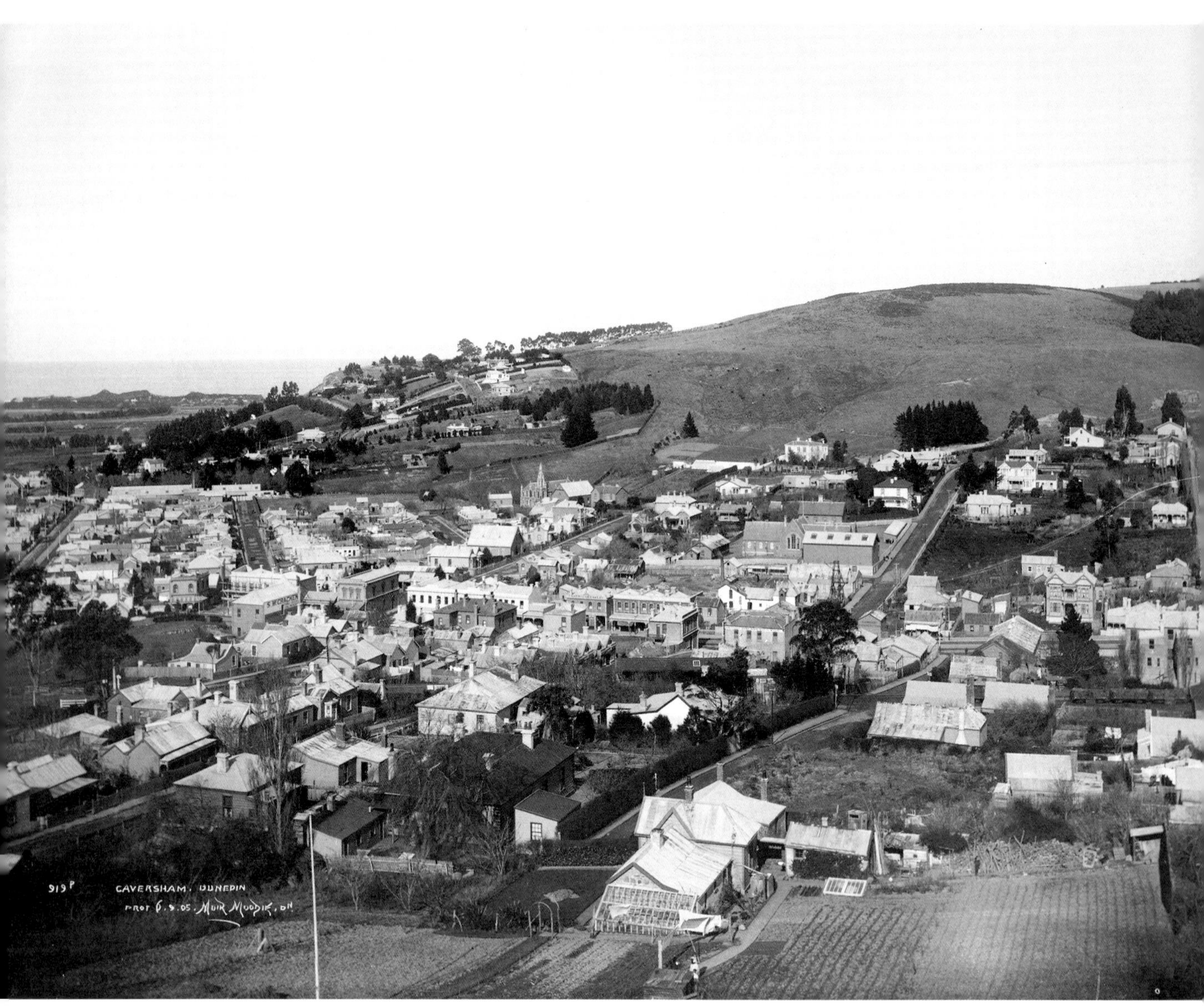

Alfred Burton's two shots of Caversham in the 1880s from the north, looking south (this page), and the east, looking west.

ABOVE **Burton took this photograph of Caversham township from Caversham Valley**, probably in about 1888–90 (Muir & Moodie reissued it later). The two-storeyed shops which flank the Main South Road today are clearly visible in the centre, as is the original Caversham School (long since known as College Street School). When the photograph was taken, Caversham township and tram terminus had become the heartland of the small-scale manufacturers, dealers and self-employed craftsmen. The bare hill that dominates the background, then John Sidey's farm, later became Kew suburb; to the left, just beneath the trees, was Forbury Corner. Sir William Barron's house is just visible. After Sir William's death in 1916, his daughters lived there until they gave the home away to become the Prince Edward Convalescent Home for Children. In the distance, across the Flat, lies St Kilda, Forbury Park and the market gardens being just visible. Beyond that lies the Pacific Ocean.

Te Papa Museum of New Zealand, BB2356.

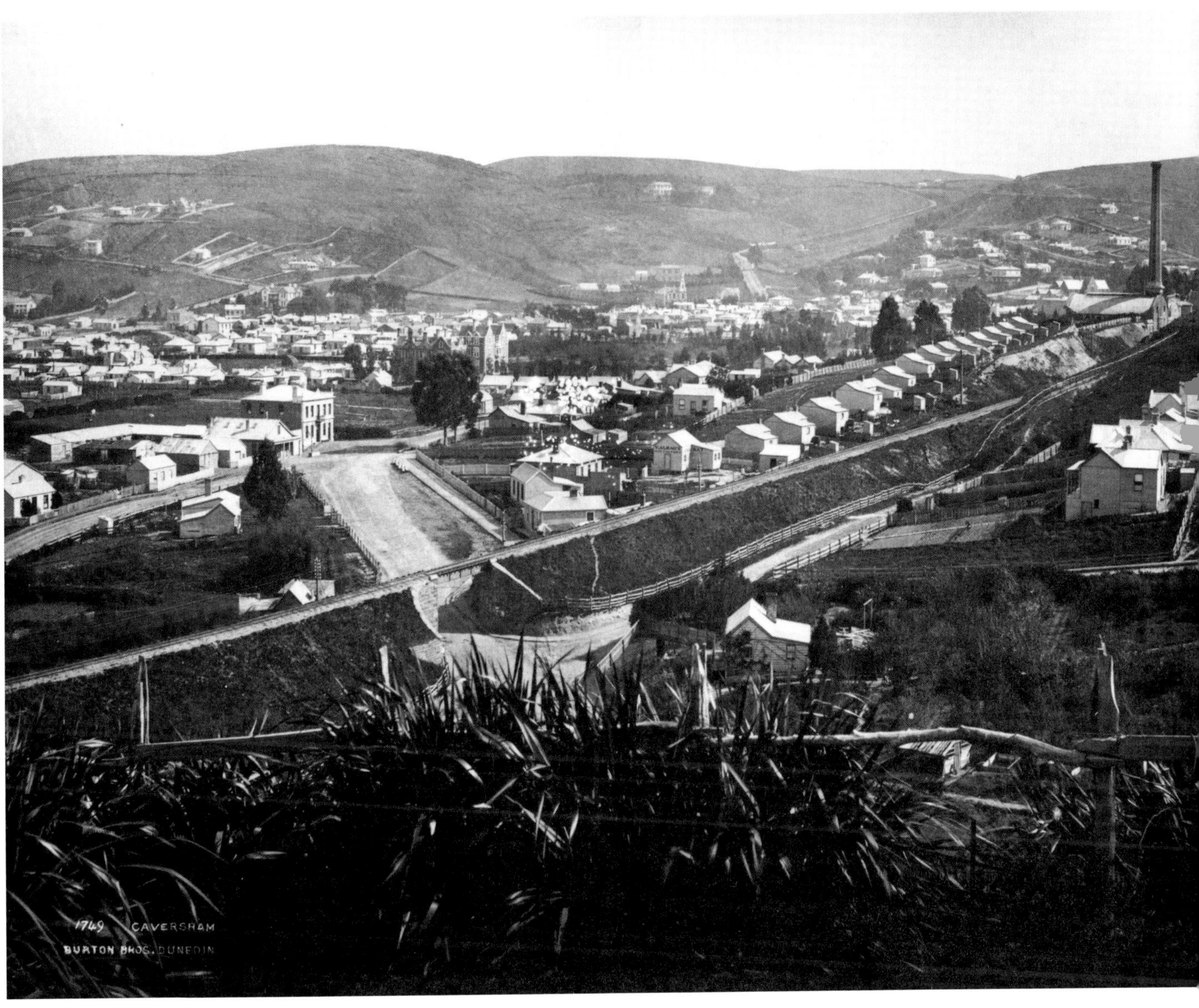

ABOVE **Alfred Burton took this shot just up the hill from McKinlay's boot factory on the southwestern slope of the Eglinton spur** (which runs down from Mornington). The Main South Road is visible in the foreground, with the single-track Main Trunk Railway, and the chimney of Hutchison's Caversham gasworks visible to the right (the row of cottages and privies parallel to the railway may have been built for labourers at the gasworks). At the nearer end of that row stands the hatting factory of Vincenzo Almao. In the centre, one can see the Parkside Hotel – Parkside being the name of the subdivision – and the two-storeyed brick Benevolent Home is clearly visible further along the Main South Road. The extensive open areas were paddocks for grazing horses and the gardens and orchard of the Benevolent Home.

Te Papa Museum of New Zealand, BB2307.

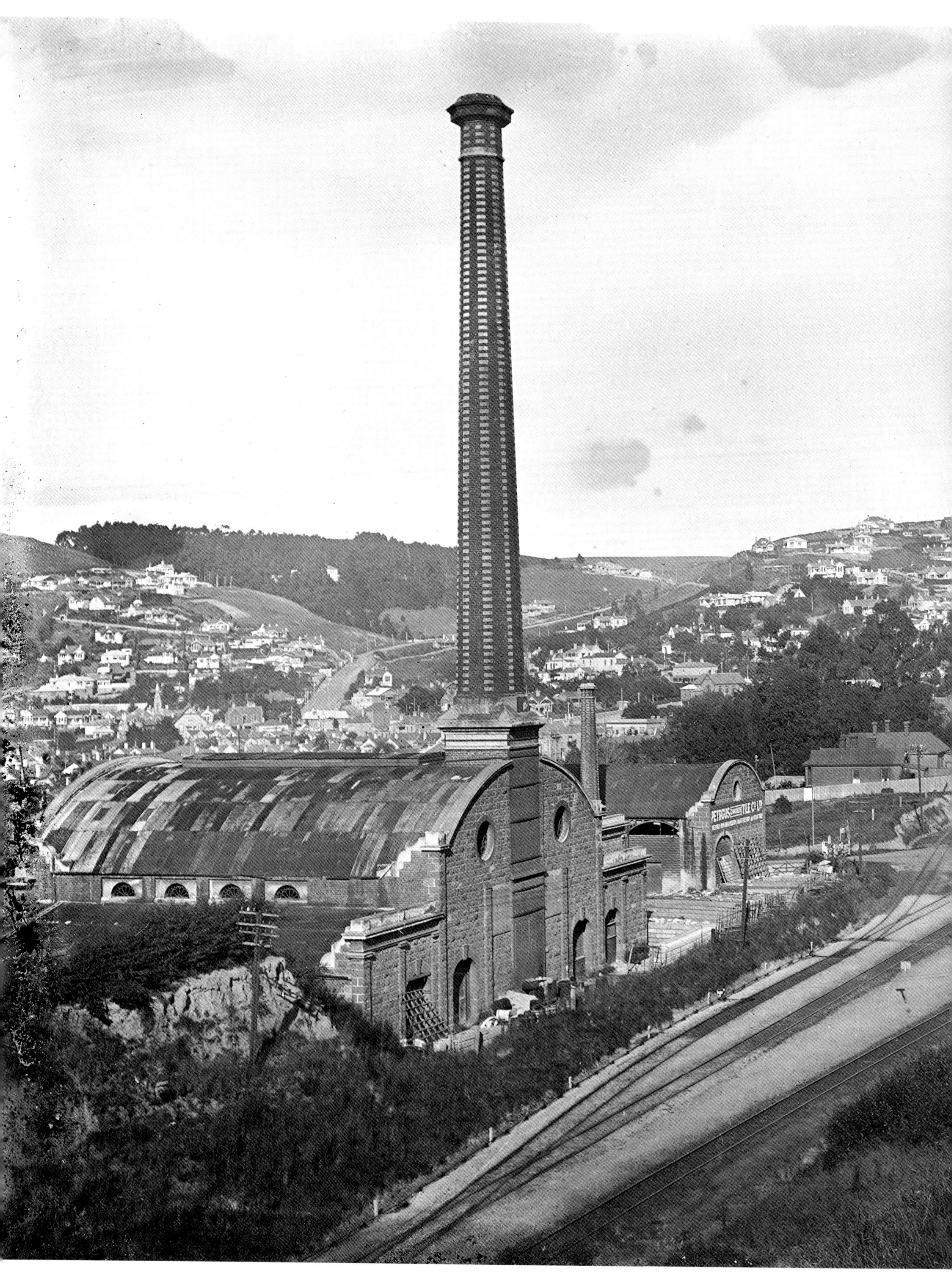
TILE Co LD

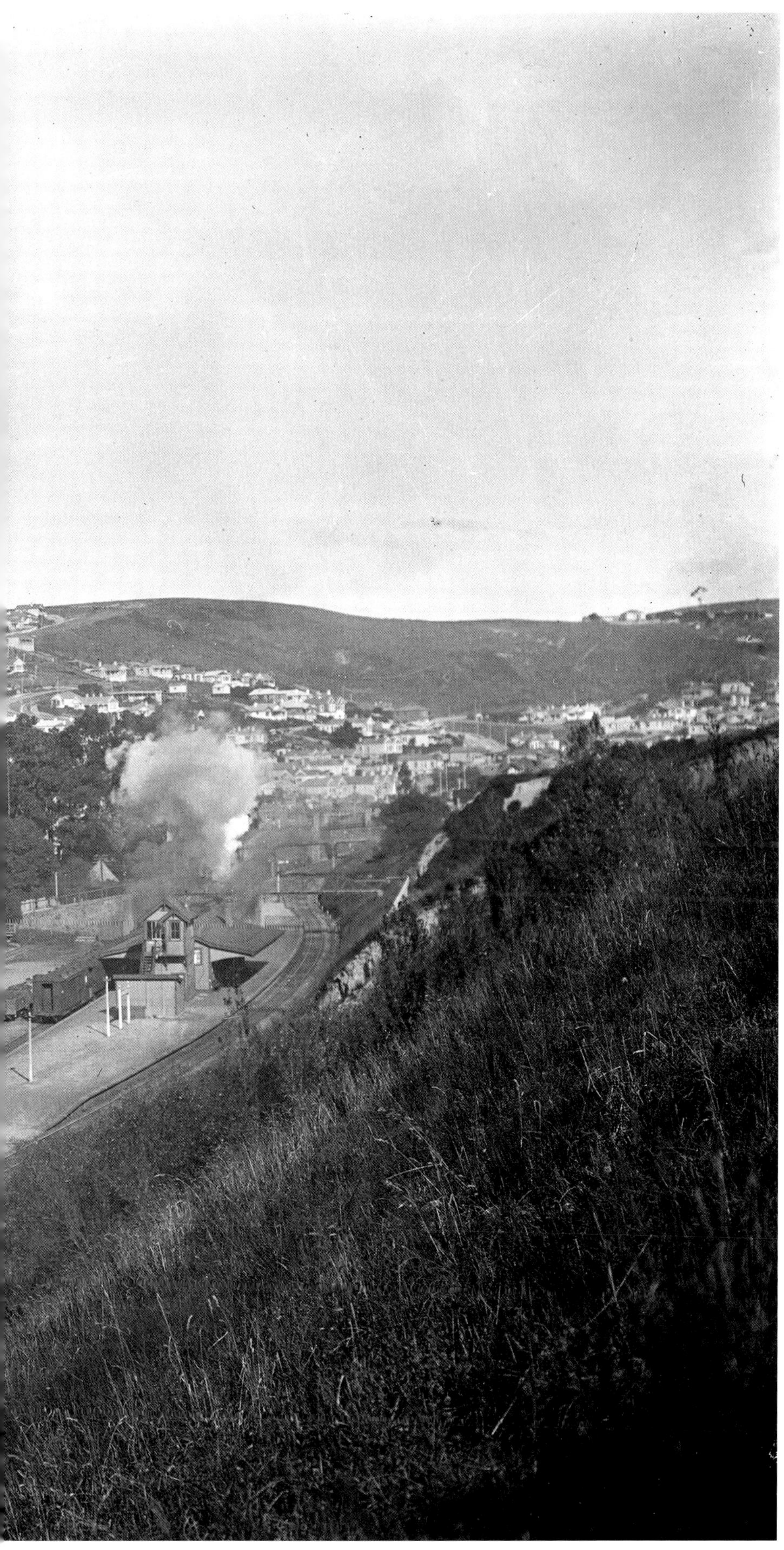

LEFT **The towering chimney of Stephen Hutchison's Caversham gasworks** dominated the skyline from 1881 until 1953. Hutchison supplied Caversham, Mornington, Roslyn and later St Kilda with gas for street lighting. In 1907, the City bought and closed the works. Except for the winter of 1909, thereafter the gasholder was used for storage only and most of the buildings were let for various industrial purposes. In the heyday of the gasworks, two shifts of men operated 12-hour shifts seven days a week. Besides employing stokers, trimmers and yardmen, the gasworks company also needed a tester, an engineer, a furnaceman, a purifier and a scrubber. Gasfitters (usually tinsmiths), blacksmiths, bricklayers, meter readers and repair staff were also on the payroll. When Albert Percy Godber took this photograph, in 1925 or 1926, the chimney still dominated the skyline and emphasised the industrial character of the area. The new Caversham Railway Station, built in 1910, is visible to the right.

Alexander Turnbull Library, A.P. Godber Collection, Part 3 of a three-part panorama, G 1326 1/2.

more immediate communities, through daily conversation (or conflict), by exchanges of children's clothes or food across back fences, by mutual child-minding, by providing support and help in times of need, or by fostering the illegitimate child of a neighbour's daughter. In this period no local photographer, to the best of my knowledge, found housewives or households interesting enough to photograph.

III

Seen from Maori Hill or Opoho, the Flat seemed homogeneously working class, blue collar. But that was only ever an outsider's view. The Flat was always much more complex than that. Kensington and eastern South Dunedin conformed most closely to the outsider's view, being dominated by unskilled men and their families. But their sons and daughters who became skilled or white-collar workers, or even entered one of the fast-growing semi-professions such as nursing or teaching, often preferred to stay on the Flat, even if they moved to Caversham or Kew, St Kilda or St Clair, or further afield to Musselburgh, Tainui or Andersons Bay. Away from Kensington and eastern South Dunedin, one always found unskilled workers and their families, but they no longer dominated. The skilled men were most numerous elsewhere, excepting St Clair where white-collar men and their families were most numerous. No neighbourhood or street, other than the very shortest ones, was homogeneous in terms of occupation, class, religious denomination or even nationality in this period, when over half of all adults had been born in the northern hemisphere.

Cumulatively, however, if you added up the Flat's largest employers and the small employers, the great majority were skilled men made good. If you add to them the skilled employees, you have over half the male workforce in every suburb but St Clair.

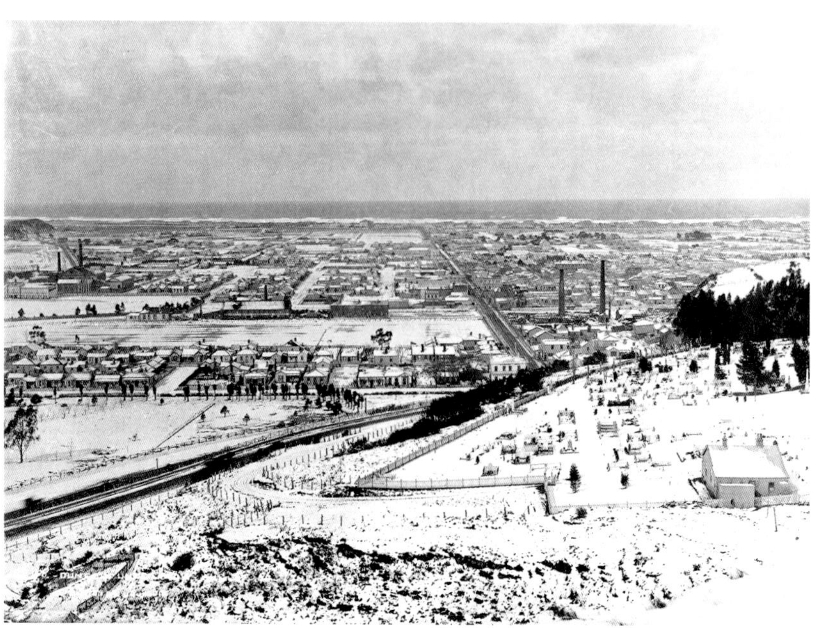

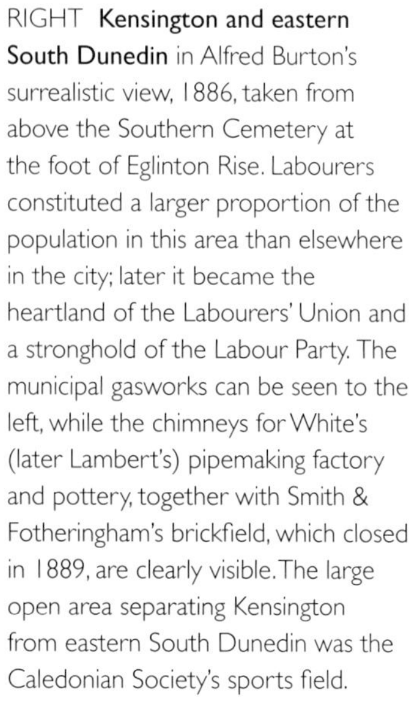

RIGHT **Kensington and eastern South Dunedin** in Alfred Burton's surrealistic view, 1886, taken from above the Southern Cemetery at the foot of Eglinton Rise. Labourers constituted a larger proportion of the population in this area than elsewhere in the city; later it became the heartland of the Labourers' Union and a stronghold of the Labour Party. The municipal gasworks can be seen to the left, while the chimneys for White's (later Lambert's) pipemaking factory and pottery, together with Smith & Fotheringham's brickfield, which closed in 1889, are clearly visible. The large open area separating Kensington from eastern South Dunedin was the Caledonian Society's sports field.

Te Papa, Museum of New Zealand, C.011717

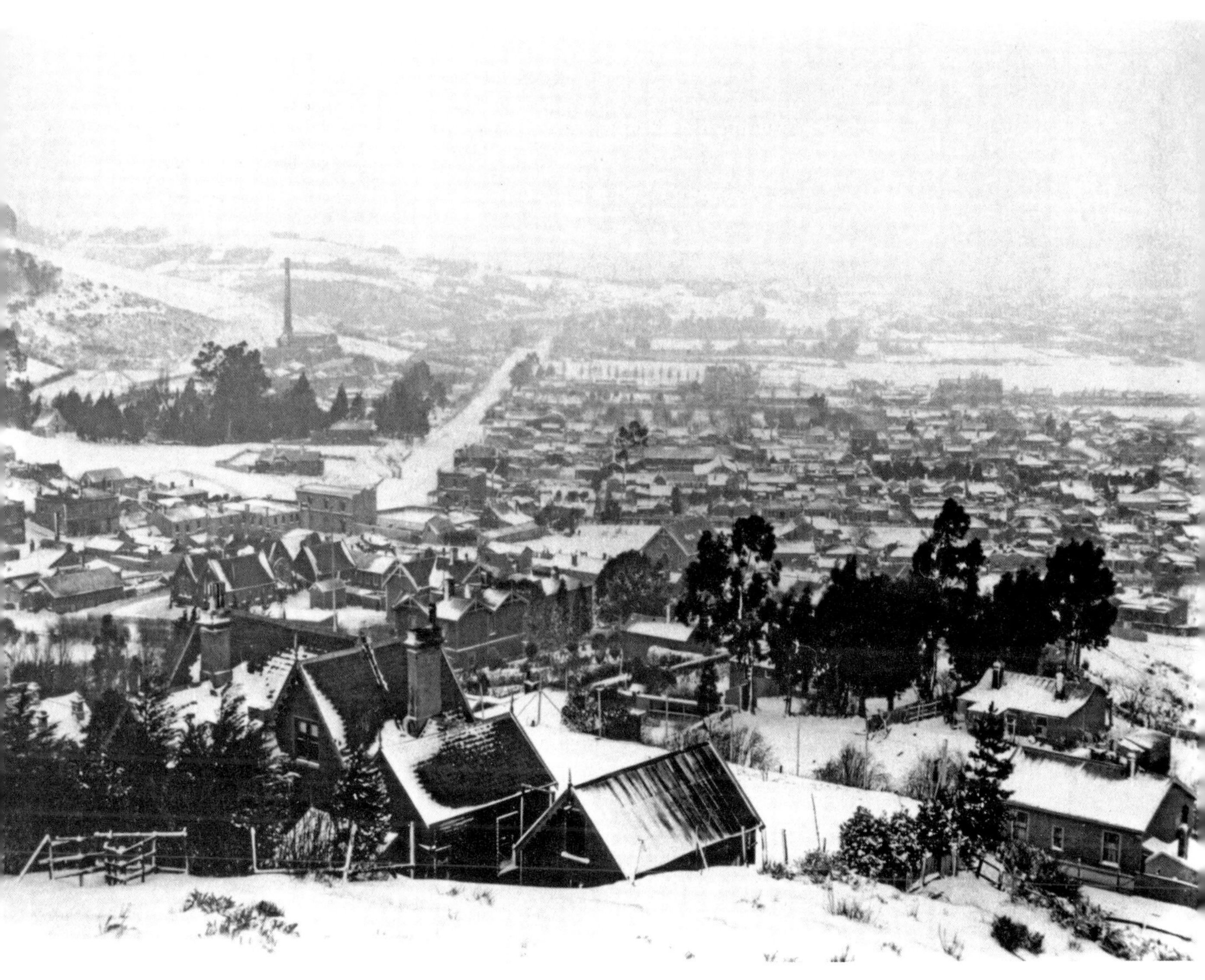

ABOVE **The substantial villas on larger sections in the newer part of Caversham** dominate the foreground in the picture above. The Caversham gasworks are beyond, mid left. This and the photograph on the next page remind us that outdoor workers, especially the men of the building trades, often lost quite a lot of time in winter, even if their apprentices had lots of fun when it snowed. Lost time equated to lost earnings, although the high proportion of families who owned, rented, or simply used land to grow vegetables and fruit, and perhaps run chooks, were able to compensate. The great snow storm of 1901, captured here and on the next page by Frederick James Lake, a tinsmith whose hobby was photography, briefly brought greater Dunedin to a stop. OVERLEAF we see a snowball fight outside Wootton's Hairdressing Saloon in Caversham.

Hocken Collections, A.N.L. Clark Collection, c/n E5329/38A & c/n E 5329/37A.

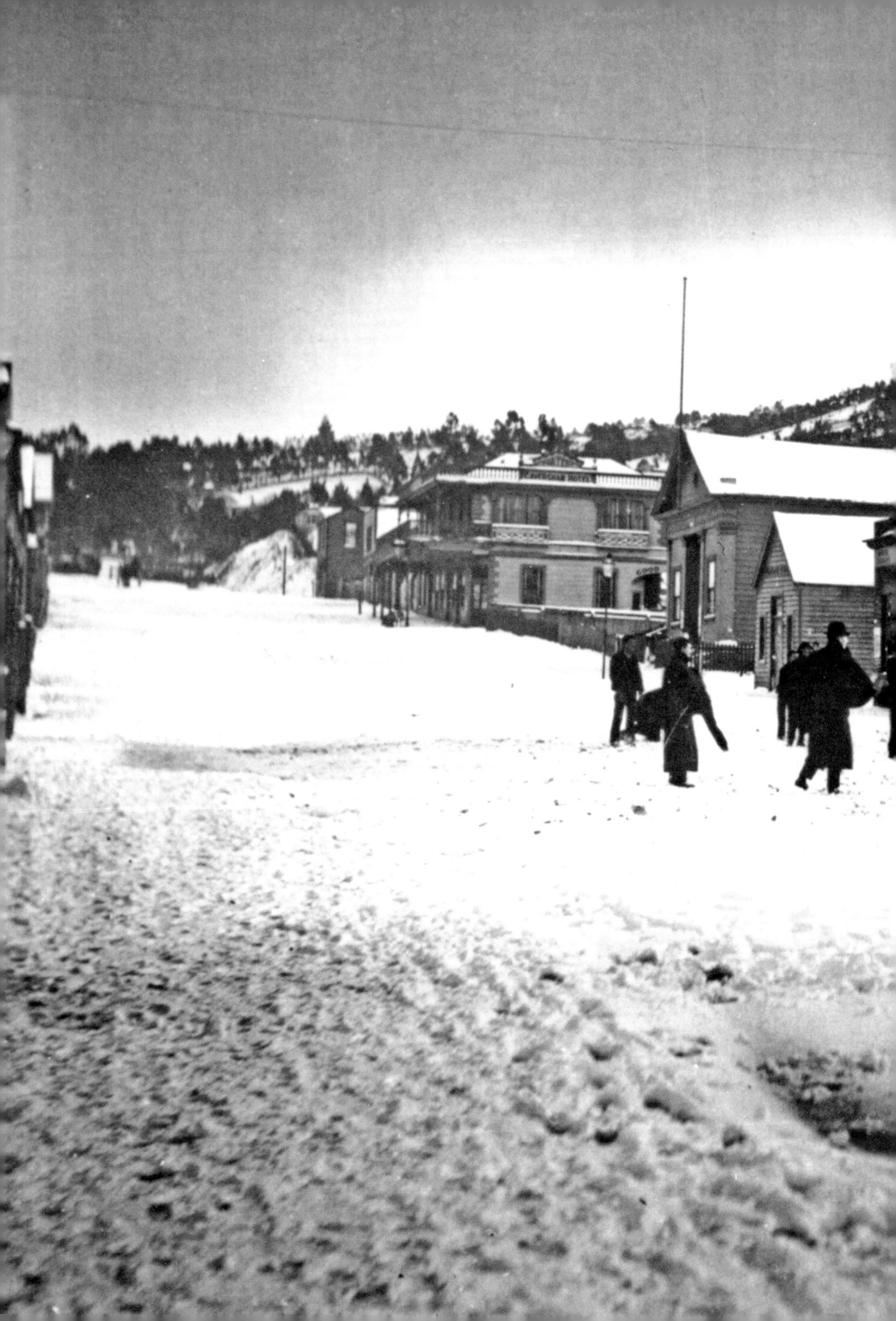

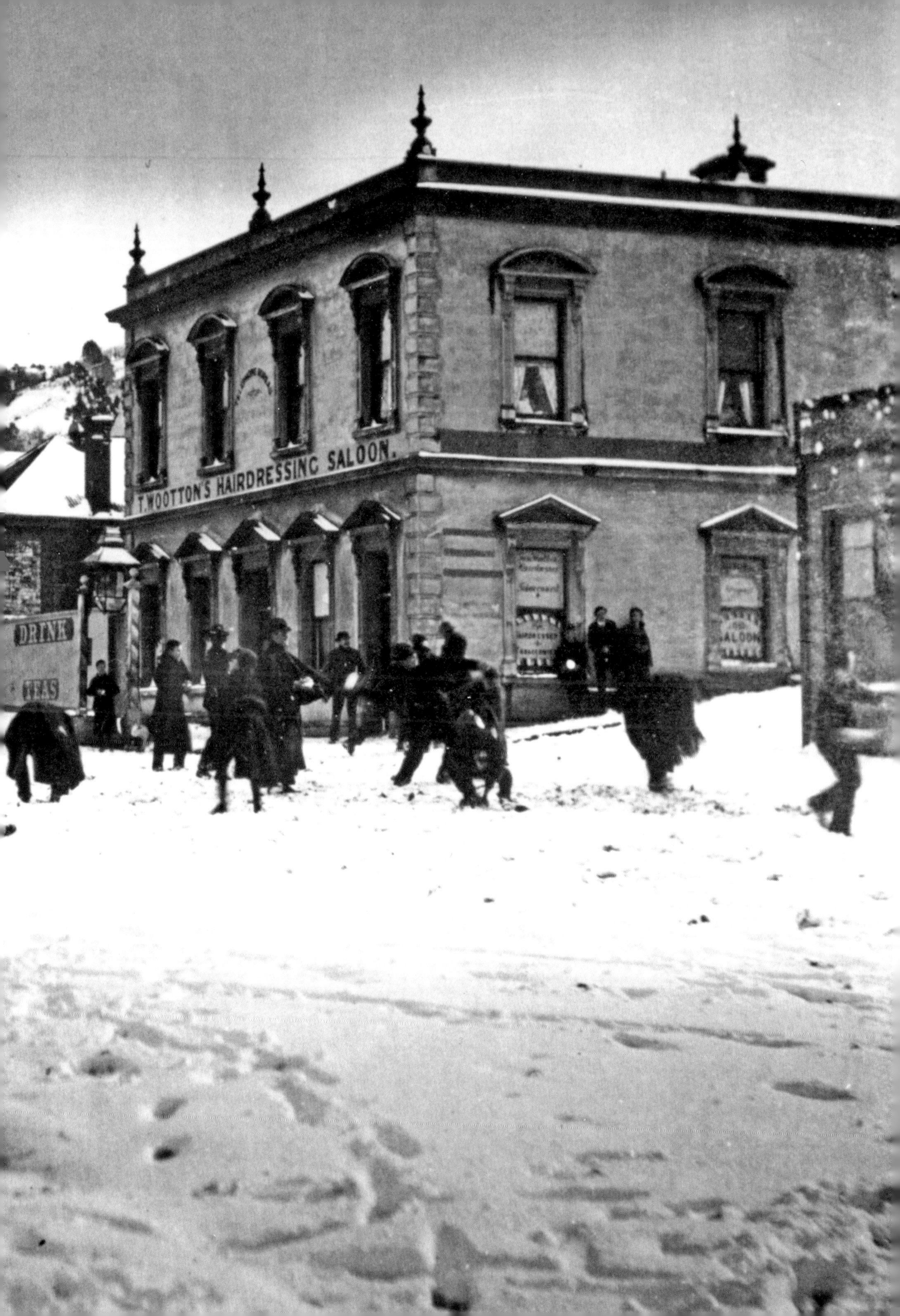
T. WOOTTON'S HAIRDRESSING SALOON.
DRINK
TEAS
SALOON

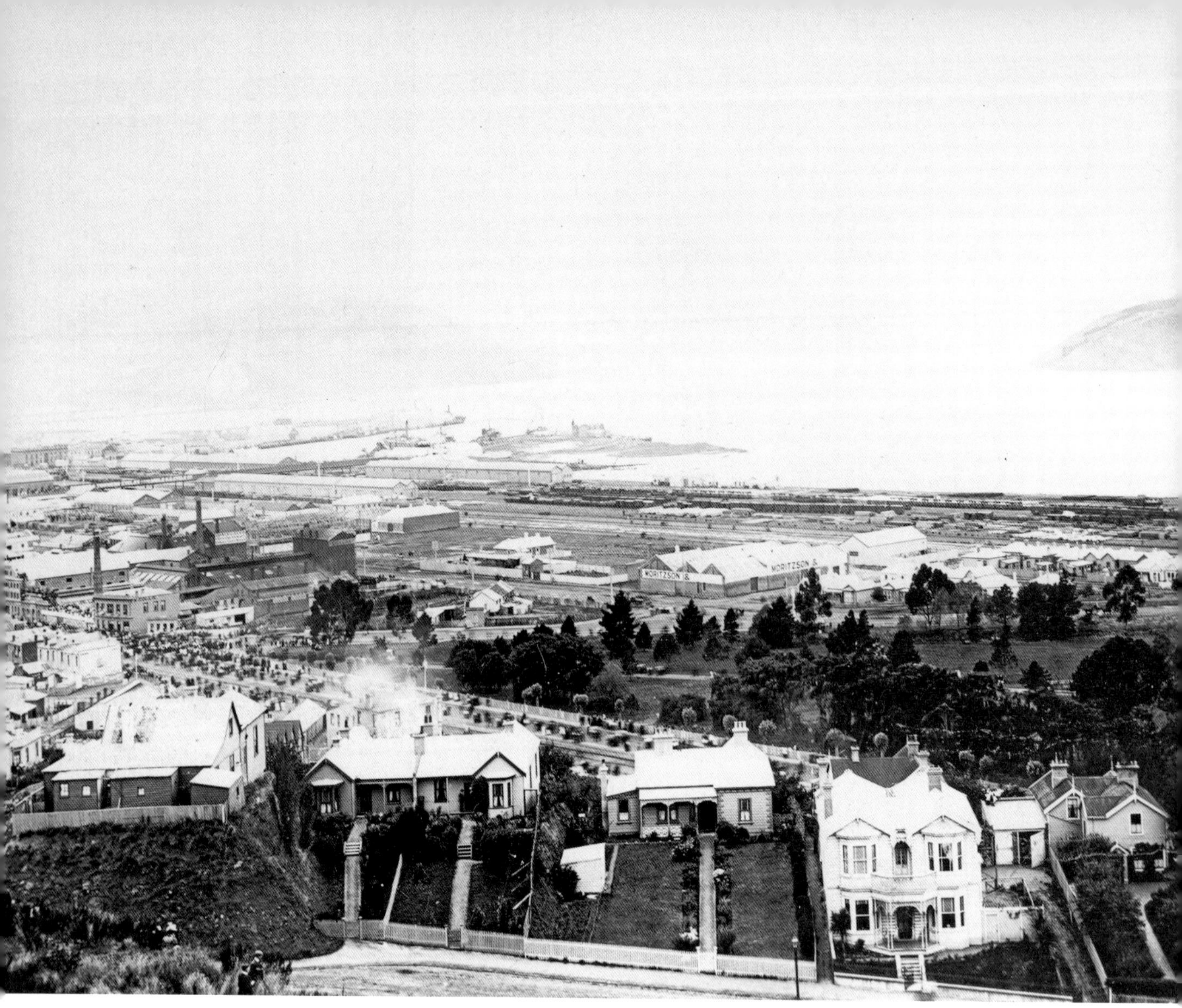

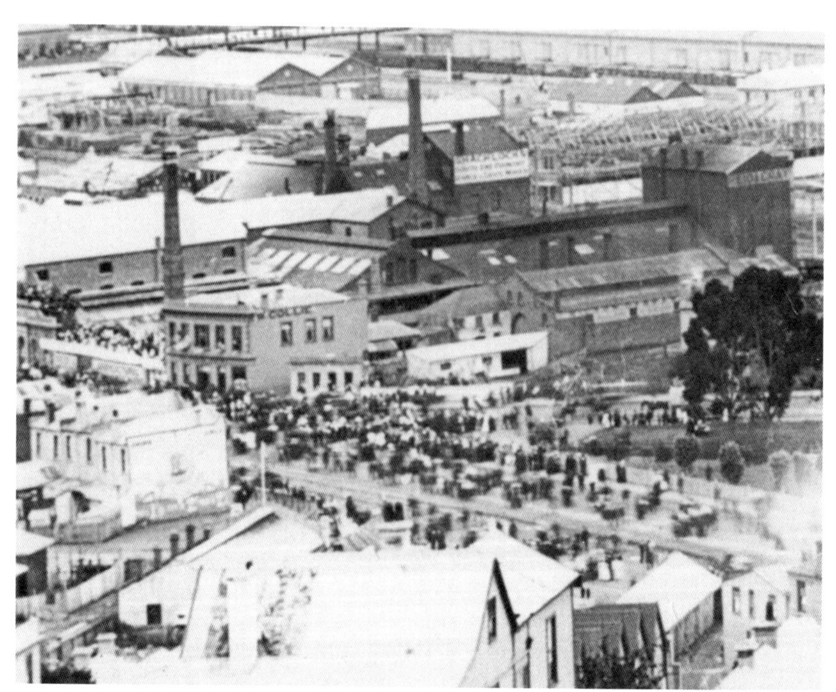

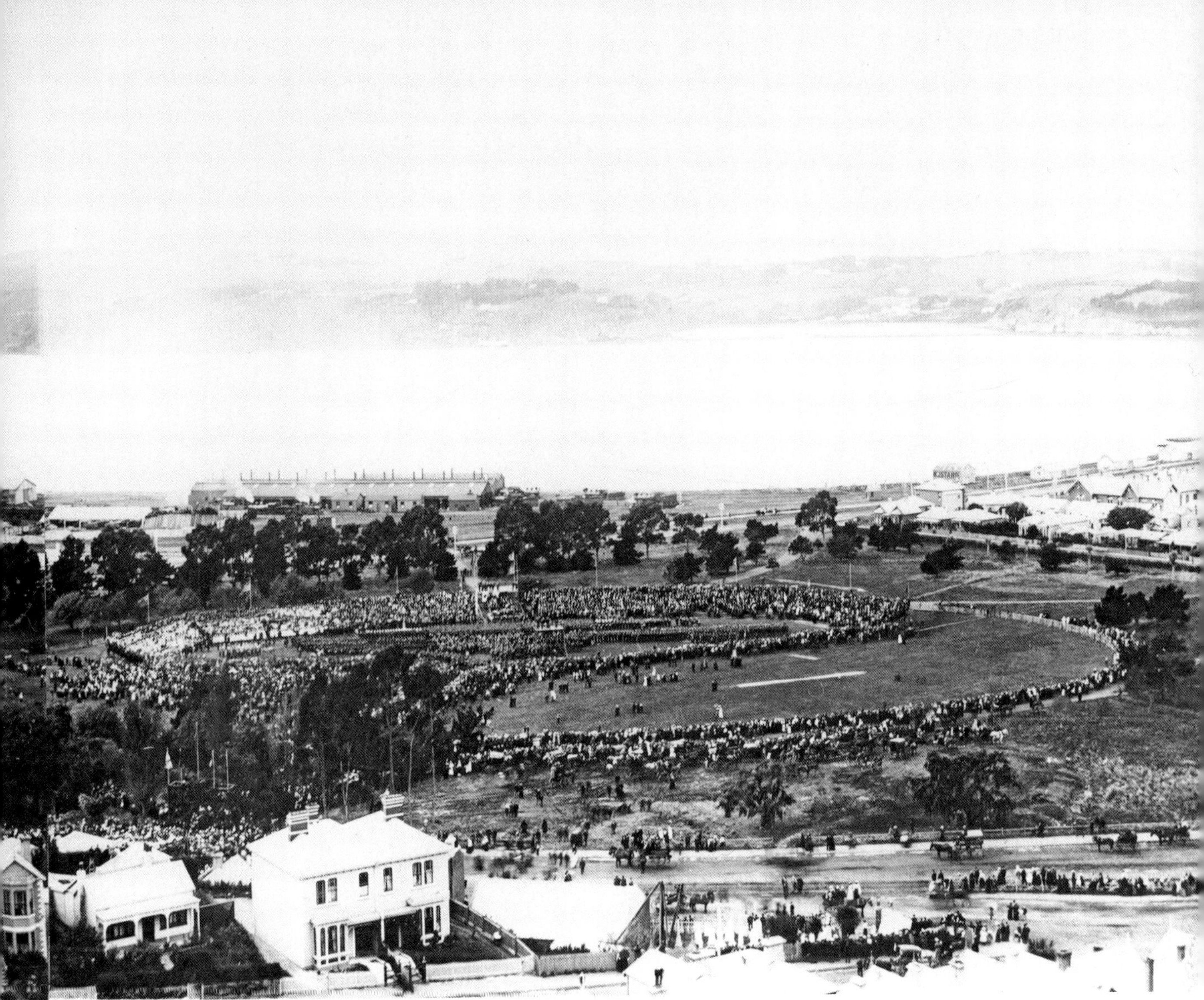

Another image taken in 1901: these two photographs from a three-part panorama – photographer unknown – pushed the technology to its limits. Perched somewhere on the eastern flank of the Eglinton spur, Part 1 is one of the best shots of the industrial area on the reclaimed harbour foreshore, and Crawford and Cumberland streets west of the city centre, where warehouses and factories clustered (far left). One can see Shacklock's, Reid & Gray, and Cossens & Black, and Moritzon's to the right. Part 2 shows the Oval itself, where a large crowd watched the Imperial troops parade, and beyond the Oval one can see the railway running sheds, where the locomotives were housed and maintained. Although we do not know whether those parading were regular troops from the British Army, or one of the Otago Contingents of Volunteers for the Boer War, the war between Imperial forces and the Boers in South Africa (1899–1902) aroused a wave of pro-Imperial jingoism and New Zealand nationalism.

Hocken Collections, c/n F165/12, 85.353.

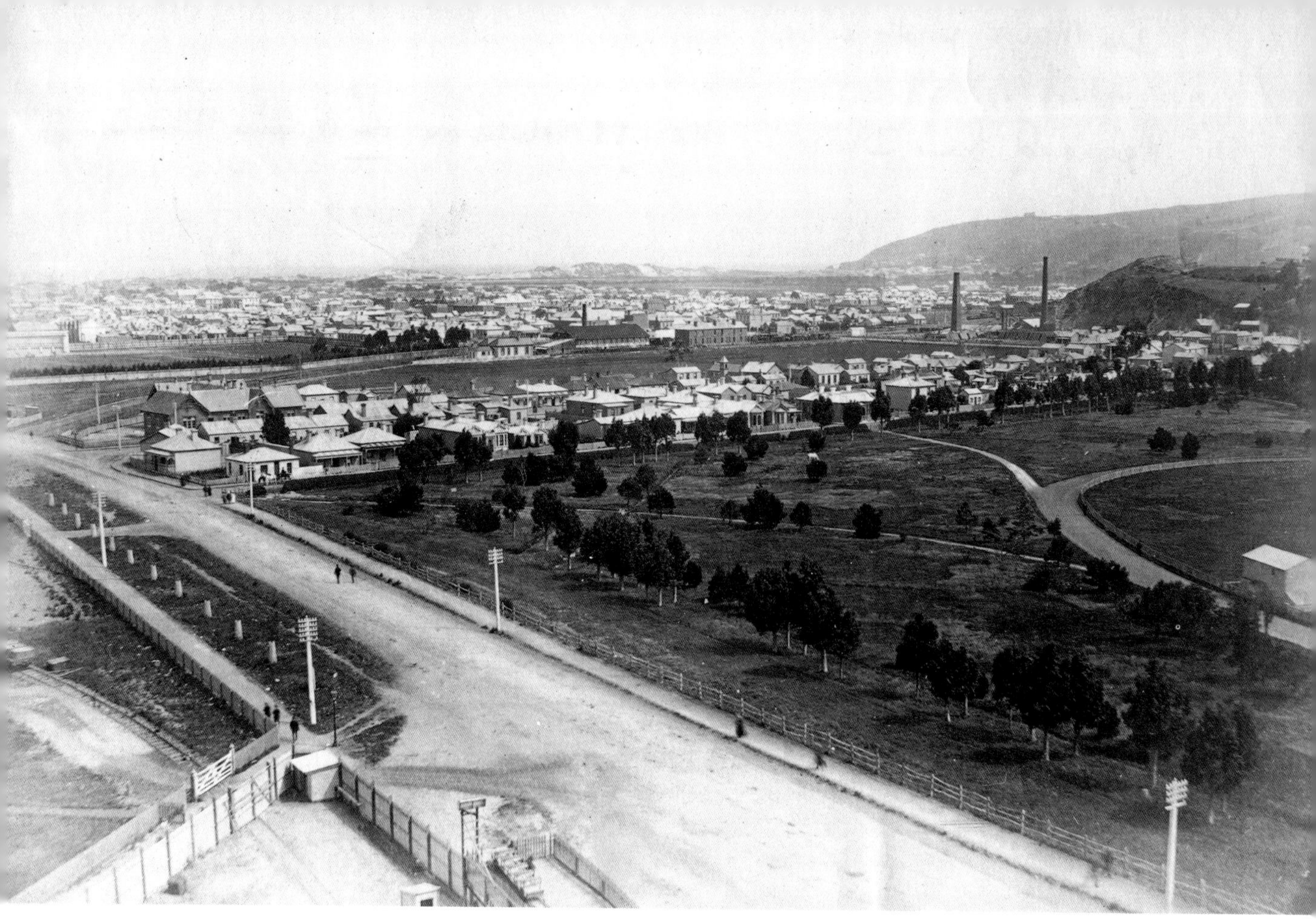

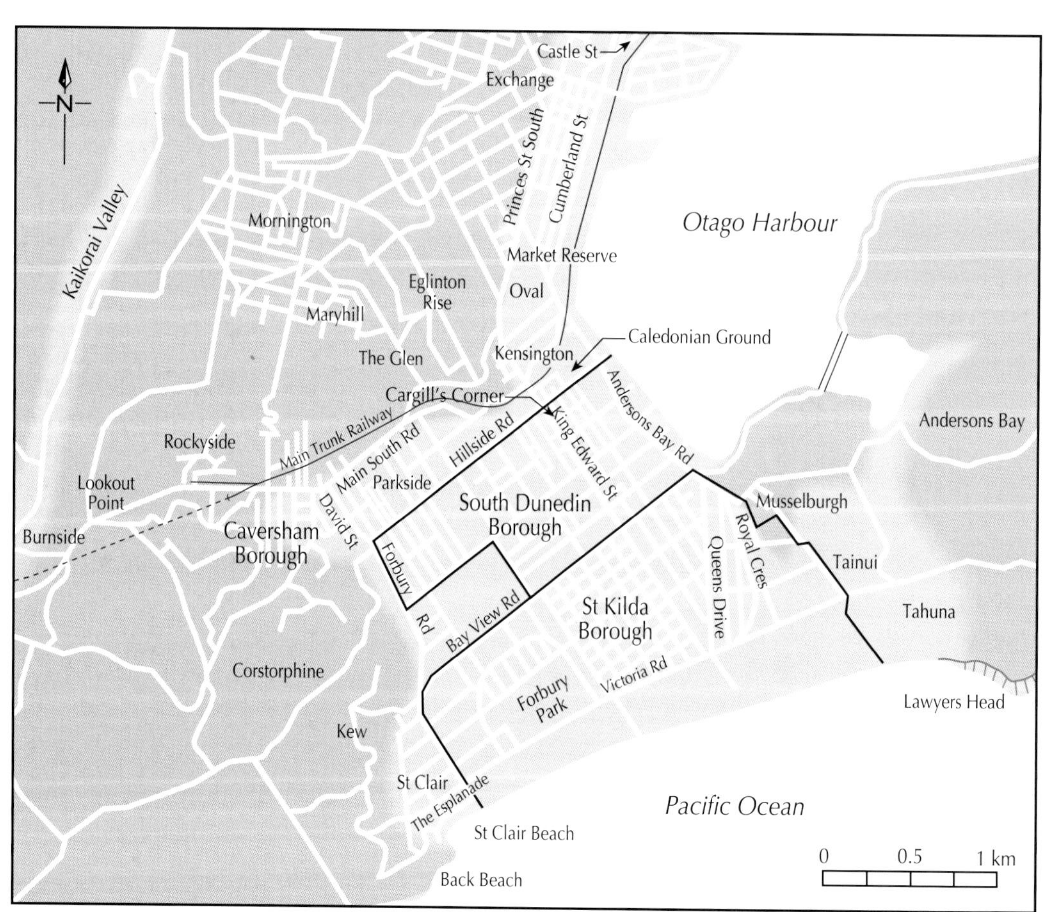

N
Castle St
Exchange
Princes St South
Cumberland St
Otago Harbour
Kaikorai Valley
Mornington
Market Reserve
Oval
Eglinton Rise
Maryhill
Caledonian Ground
Kensington
The Glen
Cargill's Corner
Andersons Bay Rd
King Edward St
Main Trunk Railway
Hillside Rd
Andersons Bay
Rockyside
Main South Rd
Parkside
Musselburgh
Lookout Point
South Dunedin Borough
Royal Cres
David St
Burnside
Caversham Borough
Tainui
Forbury Rd
Queens Drive
Bay View Rd
St Kilda Borough
Tahuna
Corstorphine
Forbury Park
Victoria Rd
Lawyers Head
Kew
St Clair
The Esplanade
Pacific Ocean
St Clair Beach
0
0.5
1 km
Back Beach

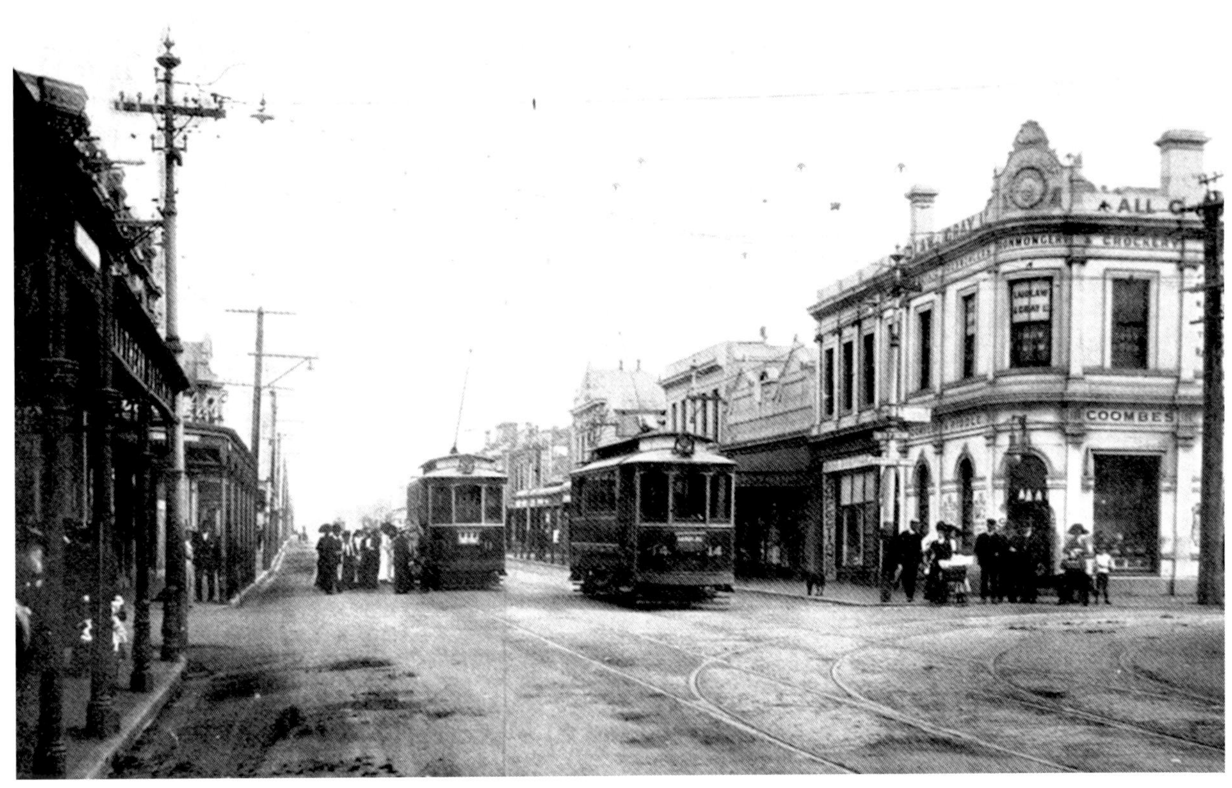

ABOVE **Cargill's Corner** (this name replaced Ogg's shortly after South Dunedin Borough voted to amalgamate with Dunedin City in 1905). The young women with their full-length skirts are about to climb onto the St Kilda tram to travel home at the end of the day. They had either been shopping or, more likely, working in one of the local stores and shops, Cargill's Corner having become the commercial hub of southern Dunedin. In the morning, at lunchtime and in the evening, there was a tram between Cargill's Corner and the Exchange every $2\frac{1}{2}$ minutes.

Otago Witness, 20 March 1912, Allied Press.

OPPOSITE, TOP:
David de Maus's unusual photograph (c. 1888) cleverly captures the Oval, Kensington and the Caledonian Society's grounds ('Cally'). Kensington School is just visible at the left edge of Kensington, on Andersons Bay Road (foreground), and Lambert's Pipemaking and Sanitary Ware factory and chimney are visible on the far side of the Cally. If one zooms in using digital technology, the firm's name is clearly legible. The tall chimneys to the right belonged to Smith & Fotheringham's brickfield. In the far distance is the Pacific Ocean.

Hocken Collections, New Zealand and South Seas Exhibition Album (1890), S12-005.

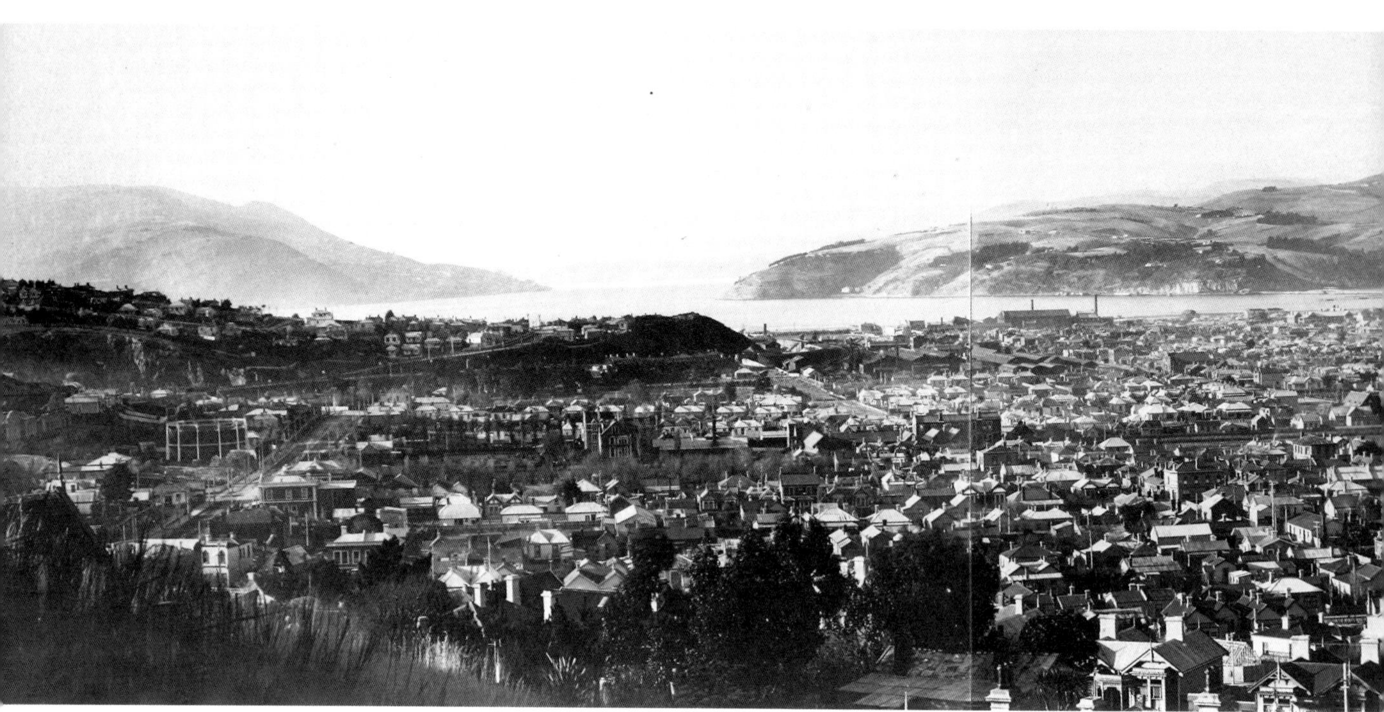

ABOVE **The industrial suburb realised.** This unusual photograph of southern Dunedin has been taken from either Morrison Street or Peter Street in Caversham township. The Main South Road runs past the old Caversham gas retort to the left, flanked by the railway, and the Hillside Railway Workshops and the Kensington industrial district are just to the right of the Eglinton spur. The central portion of the picture has

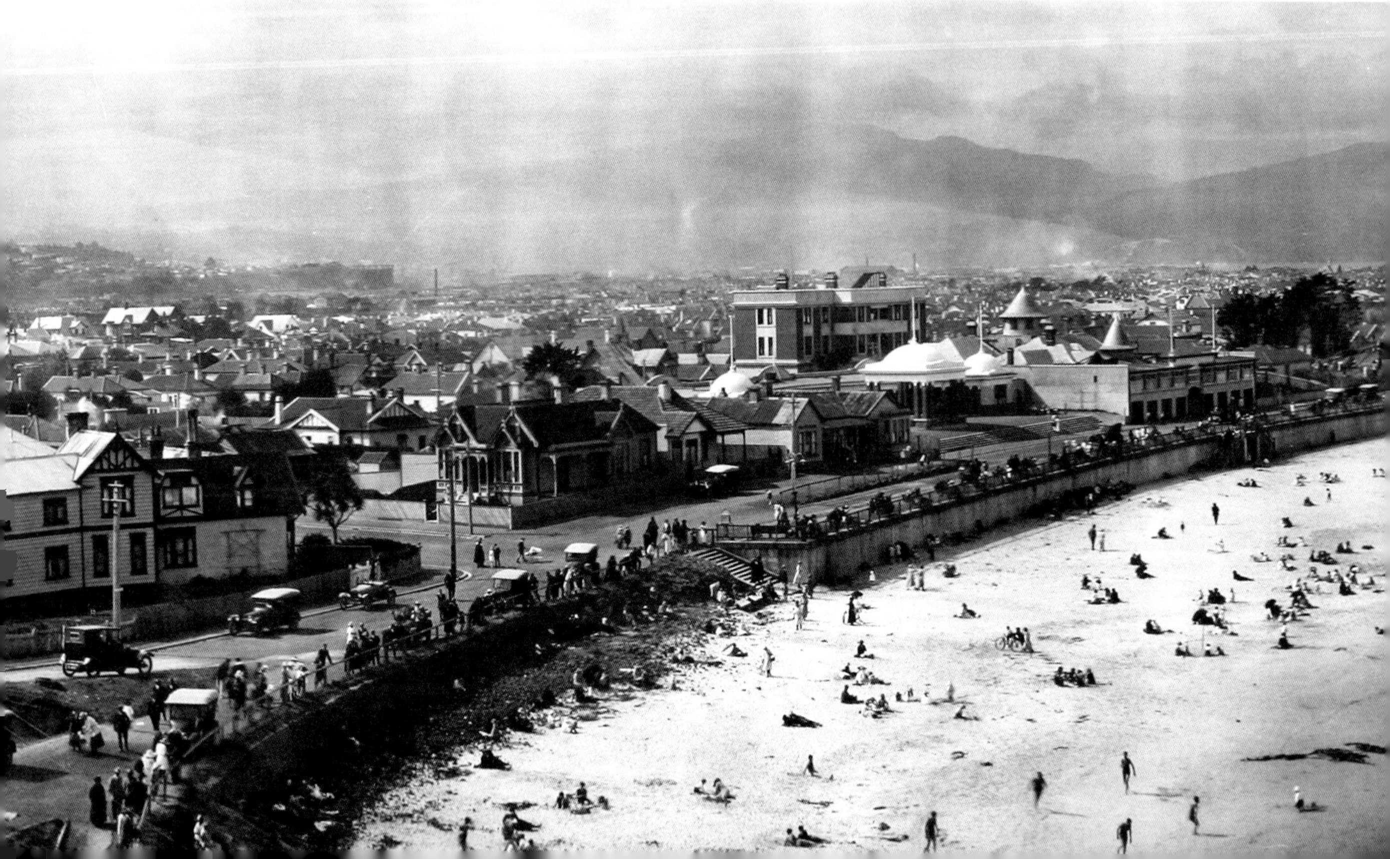

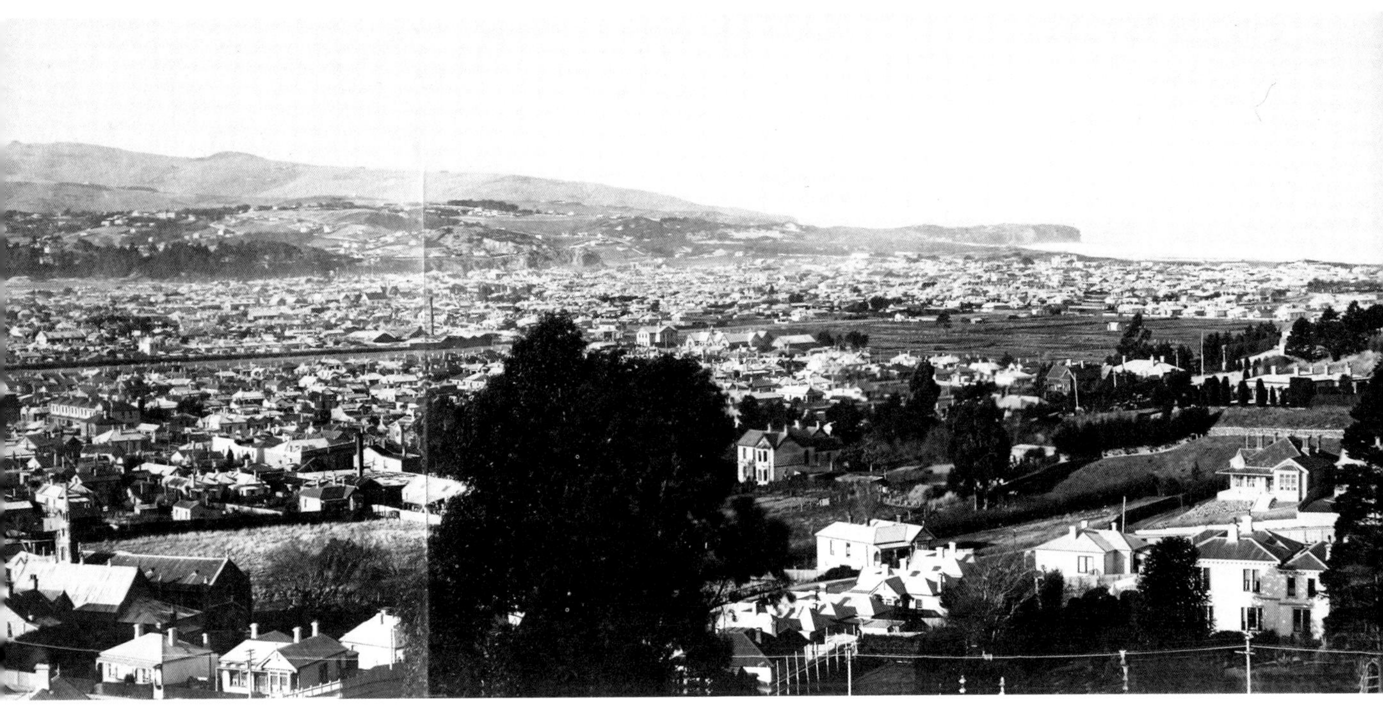

Vauxhall and Waverley in the distance and pans across Caversham township and South Dunedin, Bathgate Park and Donaghy's rope walk being visible as a dark straight line about halfway across the Flat. To the right, one sees lower Kew, the last Chinese market gardens dominate the mid-foreground, and then St Kilda on the right and Andersons Bay on the left lie in the distance.

Courtesy Dunedin Public Library and the late Reg Graham (for copying).

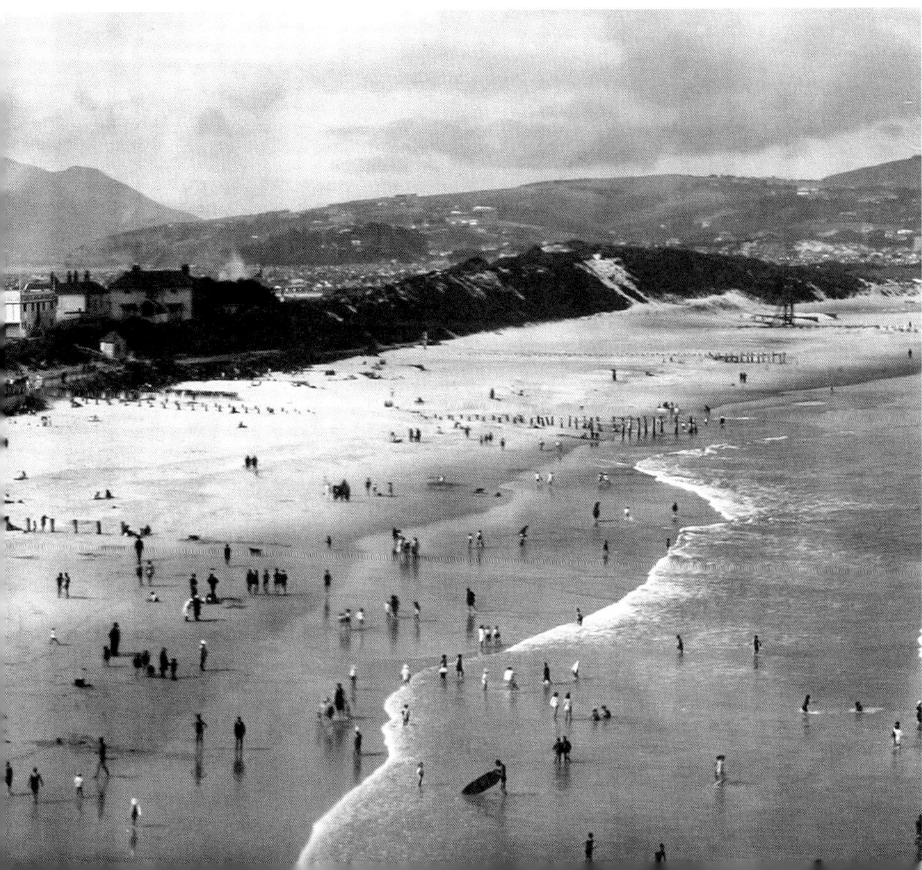

LEFT **At play.** The panorama, taken from somewhere above the salt-water swimming pool, captures the extent of St Clair's development by 1912, and particularly the concentration of amenities to cater to the weekend crowds who came to enjoy the beach. Note the unusual high-rise accommodation house, Majestic Mansions, which was mainly patronised by young male clerks and office workers. The pavilion, built in 1912, burnt down a year later and was never replaced. St Clair beach remained a very popular weekend destination, however, and the Army band that manned the artillery battery on St Clair rise often performed at Second Beach.

Toitū Otago Settlers Museum.

CHAPTER 2

Factories, Shops and Offices

A map showing the locations of businesses appears on page 65.

OPPOSITE **The founding father of McKinlay & Sons' bootmaking factory at Hillside,** who died in 1907, with sons (left to right) Walter, 1867–1903; William, 1874–1941; and Robert, 1861–1914. There is no photograph of the original factory. The firm is now run by the fifth generation of McKinlays.

Robert McKinlay, a Paisley blockmaker by trade, arrived in Dunedin in 1873 and used his Masonic connections to find work with Alexander Inglis, an old friend who owned and ran a department store, cutting out soles with a handpress. He began learning about leather. In 1879 he went out on his own account, operating from his scullery, and sold soles to bootmakers around town. When asked to attach some soles to uppers – known as bottoming uppers – he hired a clicker. In 1881, Sargood's hired him to manage their bootmaking factory and he moved in with his own machines and men. In 1885, when Sargood's and Inglis lowered wages and the men struck both shops, around 20 disgruntled employees persuaded

DURING THE BUOYANT 1870s, industries and their workers moved to Kensington and Caversham townships, both of which had railway stations. As the navvies and engineers kept the Main Trunk Railway moving south, the advantages of moving into the area grew. Dunedin City's decision to locate its gasworks immediately to the south of Kensington, and the colonial government's decision to site the region's major railway workshops at Hillside, to the west of Kensington, created an industrial hub.

Both gasworks and railways were symbols of the Industrial Revolution. Proximity to railway and port also saw private businesses – McKinlay's shoe factory, the 'Blue Bell' oatmill, W.M. ('Bricky') White's pipe factory, Smith & Fotheringham's brick yards and several smaller workshops, including Sam Lister's printery and J.H. Hancock's extensive yards (he was a wood and coal merchant) – situate themselves nearby. At the Glen, the tannery had long since closed, but Hugh Fox's quarry had exposed the whitish sandstone that still frames the view from the west.

To the east and west, two other concentrations of industry were established at the same time. Less than a mile to the west, Caversham township, long a convenient stop on the Main South Road, became a small industrial centre. The inevitable quarry and two brickfields, a brewery, and several small workshops – bakeries, Todd & Brown tailors and a dressmaker, milliner, confectioner, two fruiterers, Wootton's hairdresser, Wilkinson's pharmacy, three substantial groceries (one of them also a significant wood and coal merchant) – and the tram terminus itself made the township a bustling centre. It also boasted Porter's famous Caversham Hotel, Caversham School, a Methodist church, and the Caversham Town

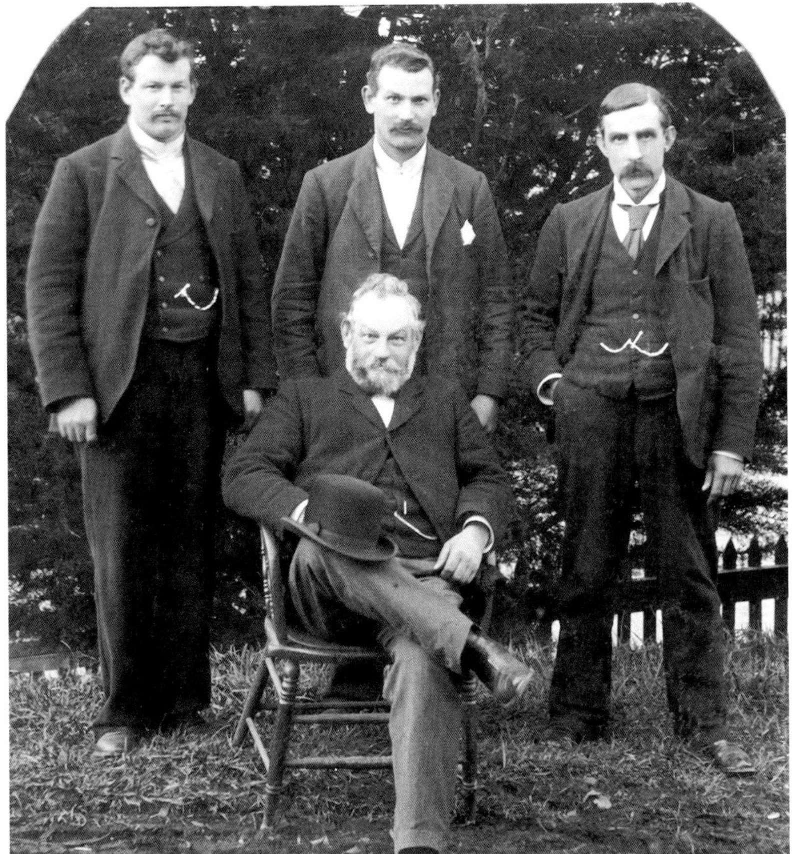

McKinlay to leave and set up on his own account. He returned to his old premises at Hillside.

Genial and generous, yet 'a man of sterling worth', he dealt fairly with his workers and they rewarded him with their loyalty. Armed with an 'Arrow' brand to signify good quality work manufactured under good conditions, the business flourished. When the 1889 New Zealand and South Seas Exhibition in Dunedin ended, he bought the dining room and established a full bootmaking factory with its own clicking, machinery, 'rough stuff' and store departments. The rooms were commodious, well lit and well ventilated, and gas stoves were installed to keep the factory warm. Even after a sharp downturn in 1892–93, caused by increased local competition and cheap imports, McKinlay had 100 men and women working for him. It was at about this time that he took on his oldest son as a partner.

Otago Daily Times, 30 June 1894, obituary in the *Australasian Leather Trades Review*, 2 March 1908, and interview with Bill McKinlay, his great-grandson, June 2012.

Hall. In Rockyside, which slopes steeply to the north, stood the Briggs family's brewery and George Methven's engineering shop. Further east, on the northern side of the Main South Road, was the railway station and Stephen Hutchison's Caversham Gasworks; on the southern side stood the two-storeyed wooden Immigration Barracks (now rented out to the New Zealand Wax Vesta Company), a handsome Baptist church, an extensive area of market garden and, a little farther along, the two-storeyed brick Benevolent Home. A few years later, Alfred Morris joined these businesses when he moved his dubbin manufacturing business from Green Island to Caversham (dubbin was used to waterproof, soften, preserve and renovate anything made of leather, including footwear).

Several workshops and the stables for the tram horses were located along David Street, which connected the township to Forbury Corner (a five-minute walk south). Apart from the Caversham Bowling Club's handsome grandstand and greens, one found a sizeable bakery, a confectioner and an asphalter. Charles Thorn's workshop may have been the biggest. A joiner by trade, Thorn specialised in making coffins and embalming corpses (his recipes are held by the Hocken). He was also quite a substantial 'spec' builder. At the foot of the Playfair Street gully there were at least three substantial suburban estates, Peter Calder's quarry and a sizeable nursery. At Forbury Corner, where David Street met Hillside and Forbury Roads, the Waterloo Hotel, the Primitive Methodist Church and the Manchester Unity Order of Oddfellows hall dominated. On the other side of David Street was the New Zealand Wax Vesta Company's factory and Robert Murray the bootmaker, whose elaborate mechanical shoemaker, imported from

OVERLEAF

The Railway Workshops at Hillside opened in May 1875. Until 1897, they specialised in undertaking repairs of locomotives and wagons. In 1899, the government considerably expanded and modernised the workshops; over the next five years electric power, a Bessemer furnace for making steel, two new scrap furnaces, and new shops for the blacksmiths and the boilermakers were introduced. The spring-makers and the painters' shops were also extended. Hillside now made not only locomotives and wagons but also all wheels for the entire country. With a workforce of 400 men in some 12 trades, Hillside was the second-biggest engineering workshop in the country, and substantially larger than the next largest in Dunedin, A. & T. Burt. Photographed by F.A. Coxhead.

Hocken Collections, S07-182a.

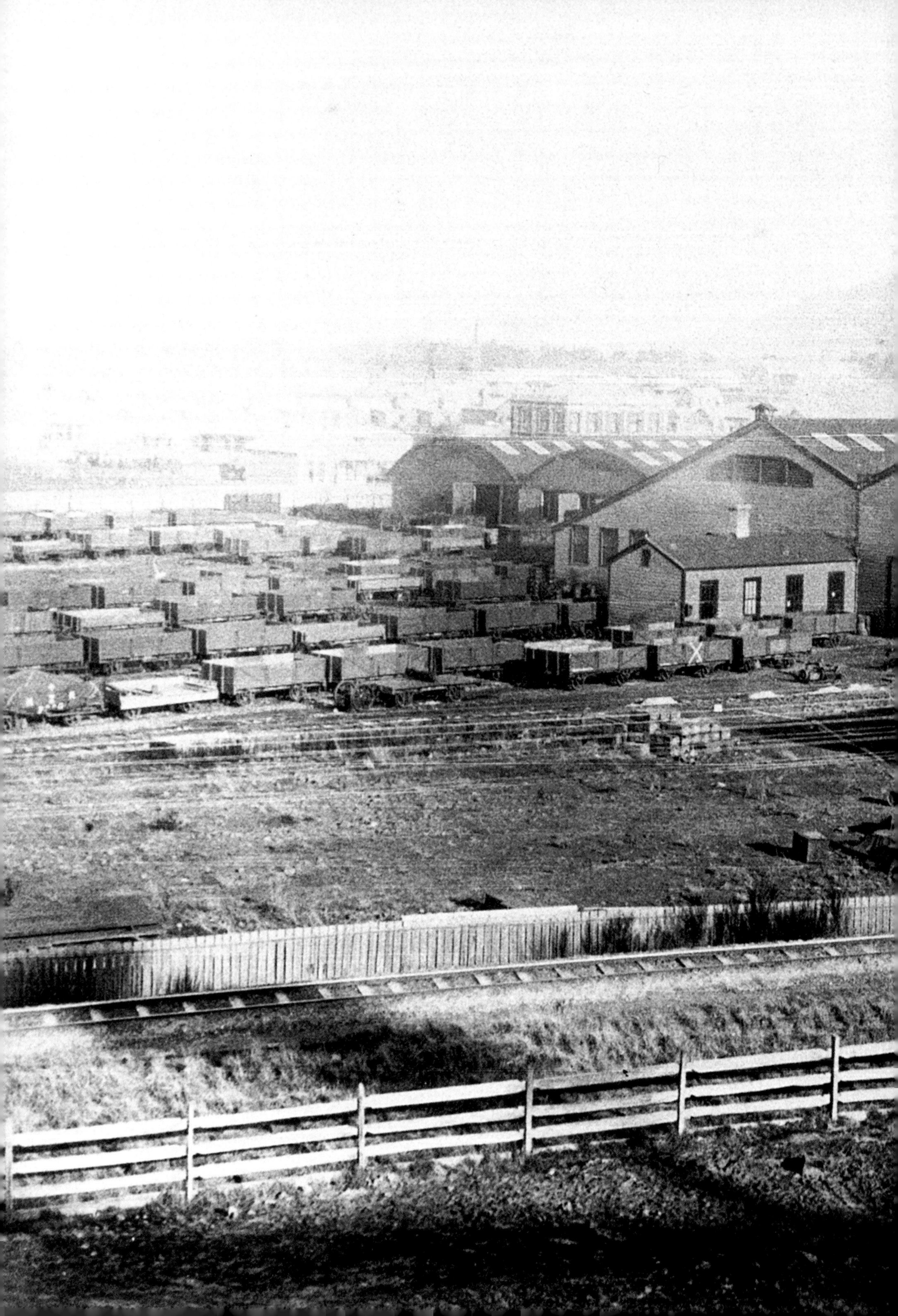

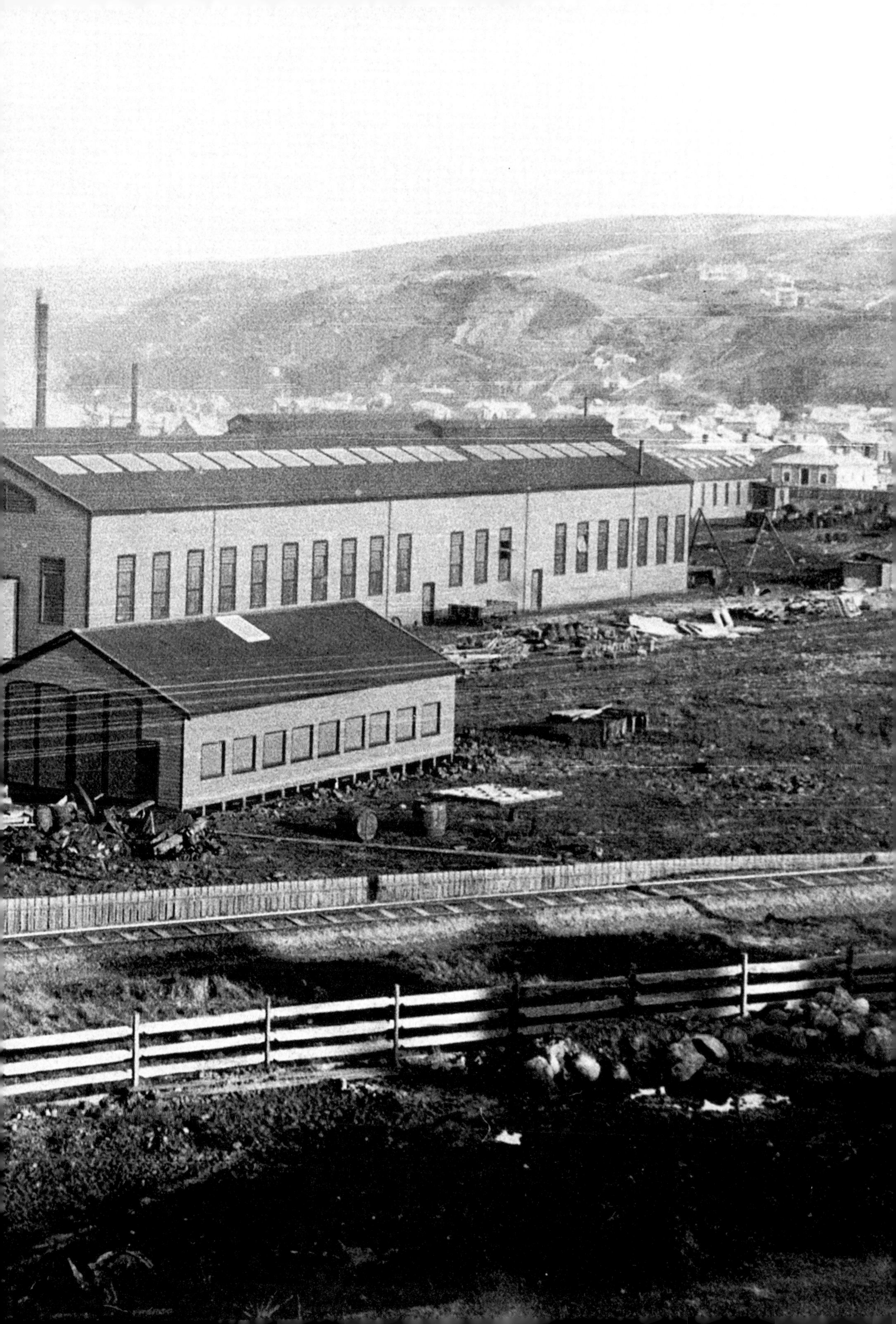

RIGHT **McSkimming & Son Ltd, c. 1929.** Although Frank Petre – 'Lord Concrete' – pioneered the use of concrete in the 1870s and 1880s, brickmakers took some time to recognise its potential. McSkimming's of Benhar opened a concrete-making factory across Otago Harbour at Pelichet Bay around 1900 and manufactured Monier Concrete Pipes for the city's new sewerage system. (Joseph Monier, a Frenchman, first patented reinforced concrete in 1867.) Later Gerald Shiel at Forbury followed suit, although he had to go out on his own because his brothers could not grasp the potential of concrete. McSkimming's, like Lambert's before them, gave high priority to maintaining a stable workforce. (Incidentally, the Romans knew how to make concrete and used it extensively, but all knowledge of the product and the craft disappeared from Europe until the nineteenth century.)

Hocken Collections, McSkimming's Ms-AG-246-A, 85-1837, c/n E 3002/79.

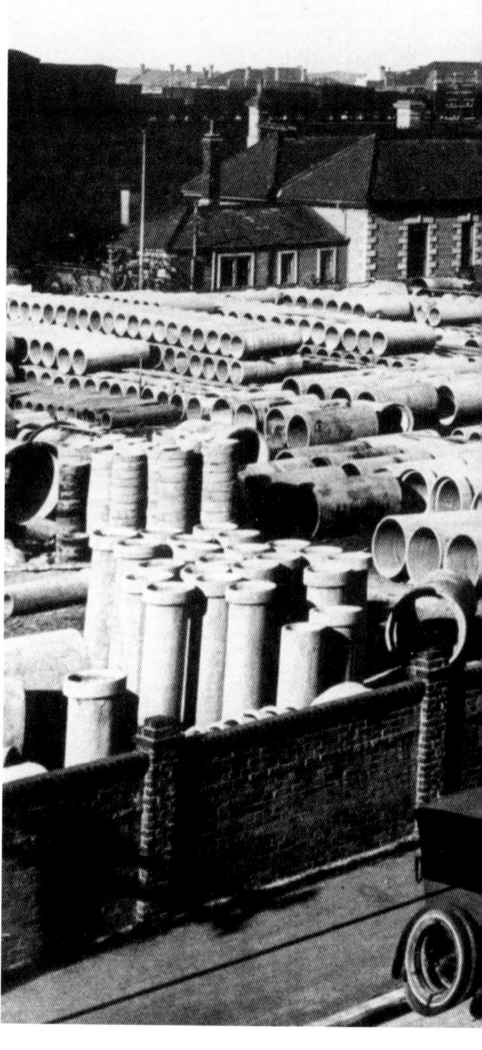

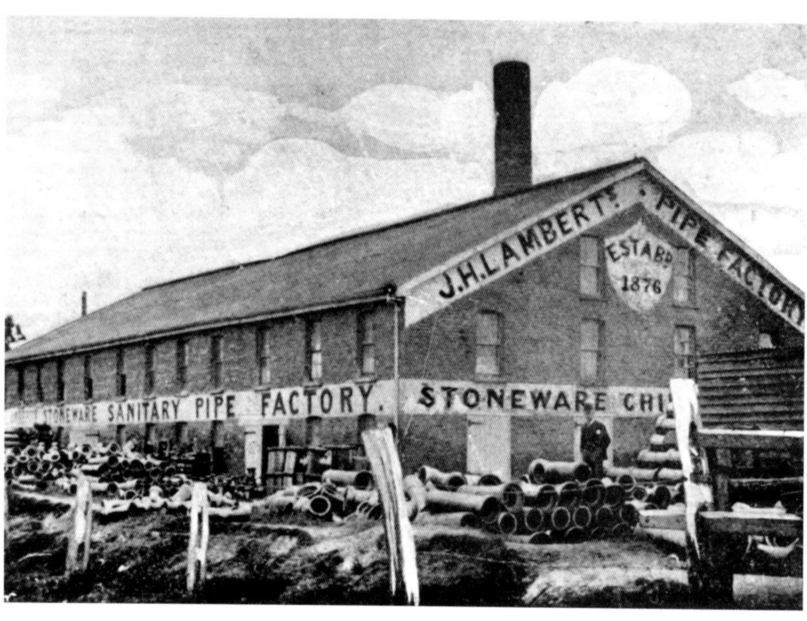

ABOVE **J.H. Lambert's Sanitary Pipe Factory, c. 1904.** John Halls Lambert, pipe and pottery manufacturer, was the third generation to enter this line of work, and established his first factory in North East Valley in 1862. In 1888, he bought the Kensington pipe factory, which had been established and run by W.W. White since 1876 on land rented from the Caledonian Society. Lambert and his family also moved to Kensington. At the Kensington factory there were two large permanent kilns, each with 10 fire holes, and an ordinary circular down-draught kiln. A 10-horsepower horizontal steam engine drove three pottery wheels, the pipe-making machine and the clay grinder, which prepared fine clay for bottle making. Demijohns, acid bottles (for the NZ Drug Company), glazed pipes, butter crocks, limestone filters, chimney pots, garden vases, border tiles and various other items were produced. Lambert had four daughters and three sons, all of whom entered the trade and took over the business.

Cyclopedia of Otago and Southland, p. 355.

RIGHT **Fletcher Bros Head Office, Cameron Street, Kensington,** with the Kensington train station visible behind, c. 1910. James Fletcher, who arrived from Scotland in 1908, quickly saw the opportunity to challenge Thomson, Bridger & Co, whose joinery factory in Bond Street enjoyed something of a monopoly. Fletcher and Morris, at first a local house-building firm, established their own joinery factory in 1909 and quickly went from success to success. In 1919, by when it was tendering for and securing projects throughout New Zealand, it became Fletcher Construction Company and moved from Kensington to the city. Although the firm moved its headquarters to Wellington in 1923–24, it built the Dunedin and South Seas Exhibition complex.

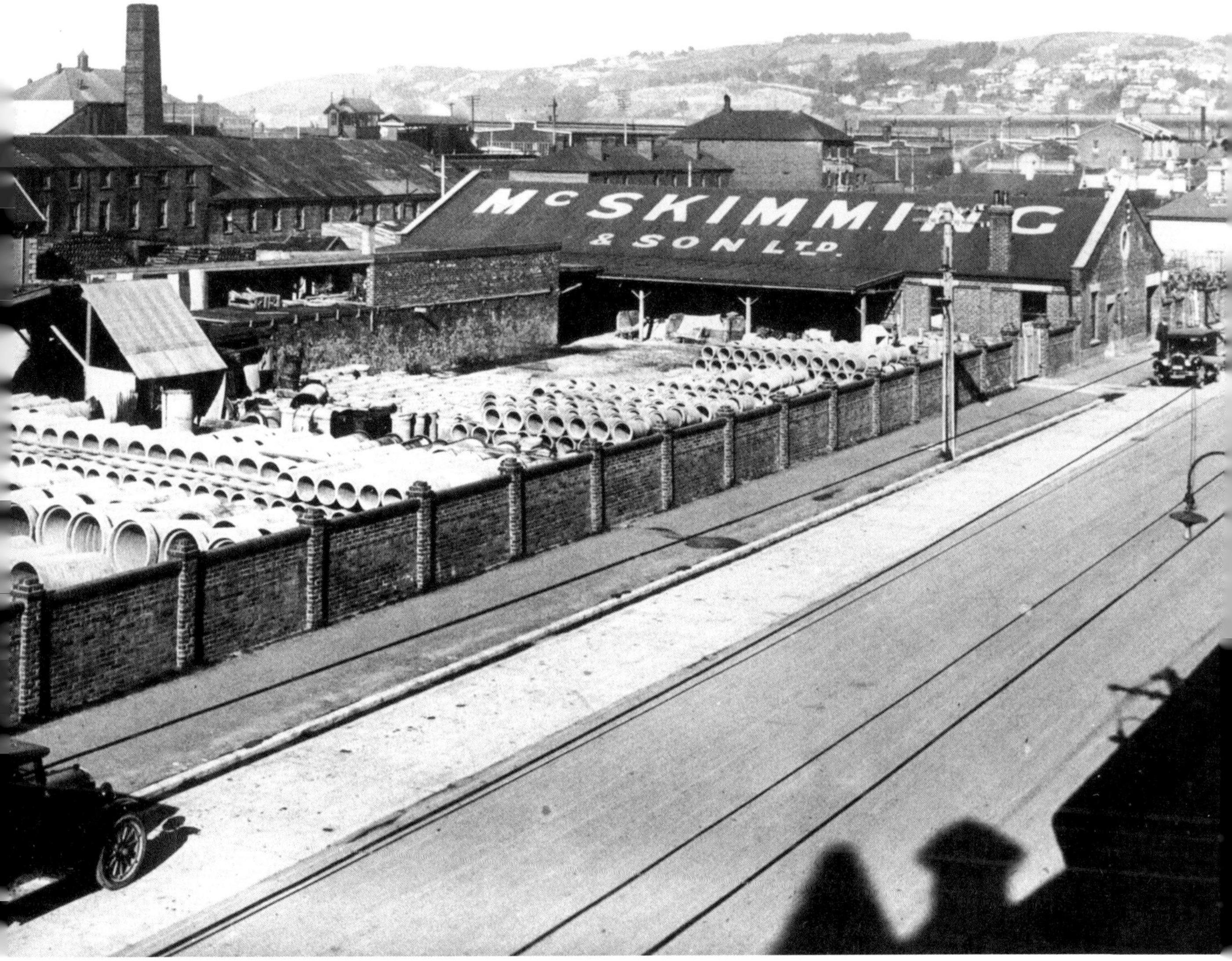

Courtesy Fletcher Challenge Archives.

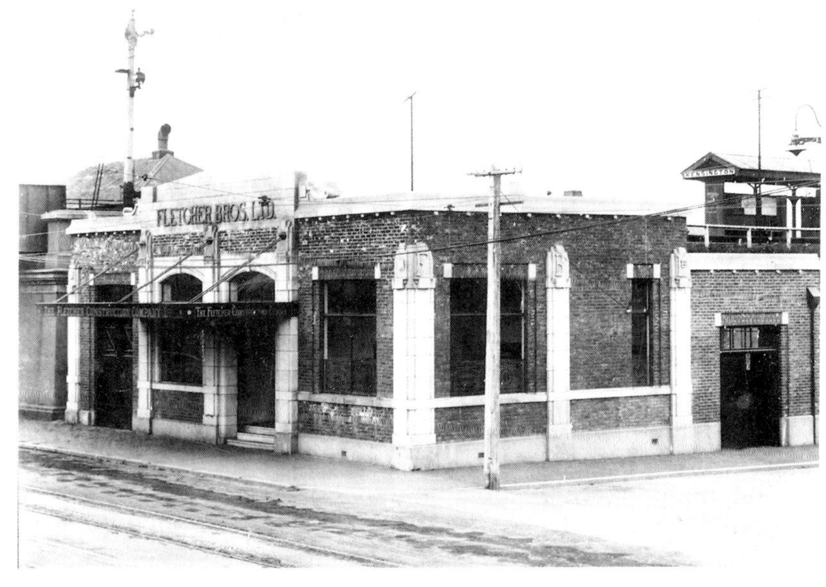

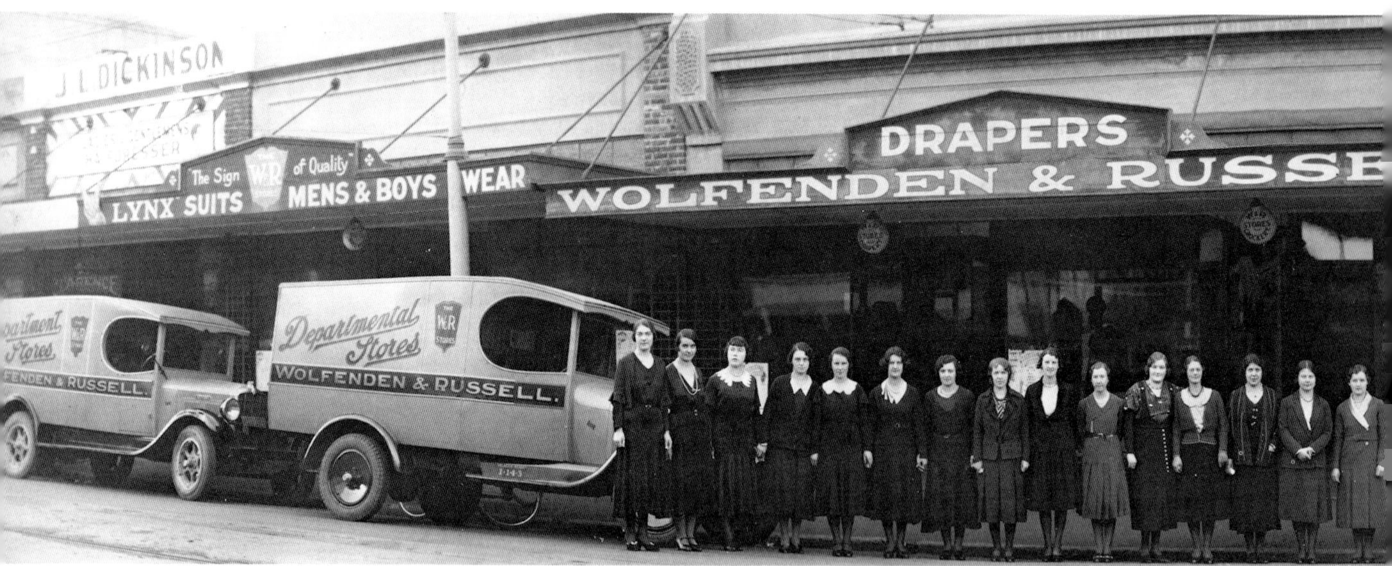

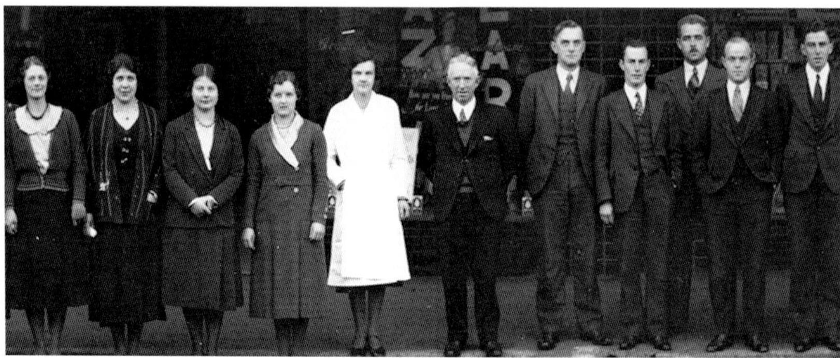

ABOVE **Wolfenden & Russell, department store,** King Edward Street, South Dunedin. When still in their twenties, Harry Russell and Edward Wolfenden, who had served their apprenticeships as a grocer and a draper respectively, went into partnership in 1911 and bought a grocery store. Both men belonged to the Caversham Brethren Assembly; many of their employees were also Brethren. Over the next decade, they expanded until they ran a department store with specialist departments for crockery and glassware, fine china, and eventually drapery, menswear and ladies' wear. Like all stores, they delivered, running several horse-drawn carts, most of which were replaced by vans in the 1920s. This photograph of some of the staff in front of the storefront dates from 1934.

Courtesy Philip Gilchrist, Ted Wolfenden's grandson.

Germany, fascinated children across the generations, when he remembered to wind it.

There were relatively few large industries in South Dunedin other than the city's gasworks and Donaghy's rope and twine factory. From 1895 onwards, the towering brick chimney of Donaghy's rope works – an important local industry to this day – marked the centre of the Flat on Bradshaw Street, its rope walk running alongside Bathgate Park. These industries dominated the new urban townscape on account of their physical extent, high chimneys and larger labour force.

If you took the train or climbed Lookout Point to Burnside, as many Caversham residents did each day (the walk took about 20 minutes), you would find several large industries clustered within about 500 metres of each other, including the New Zealand Refrigerating Company's abattoir and freezing works, Kempthorne Prosser's chemical works, the Otago Iron Rolling Mills, Harraway's oat mill and a sizeable tannery. Not far away again was the Abbotsford Tileries, an innovative firm that was one of the few to manufacture roofing tiles. Near Abbotsford, at Fairfield–Green Island, several small coalmines operated, Walton Park being the largest. These collieries provided most of the industries on the Flat and Princes–Crawford streets with their coal.

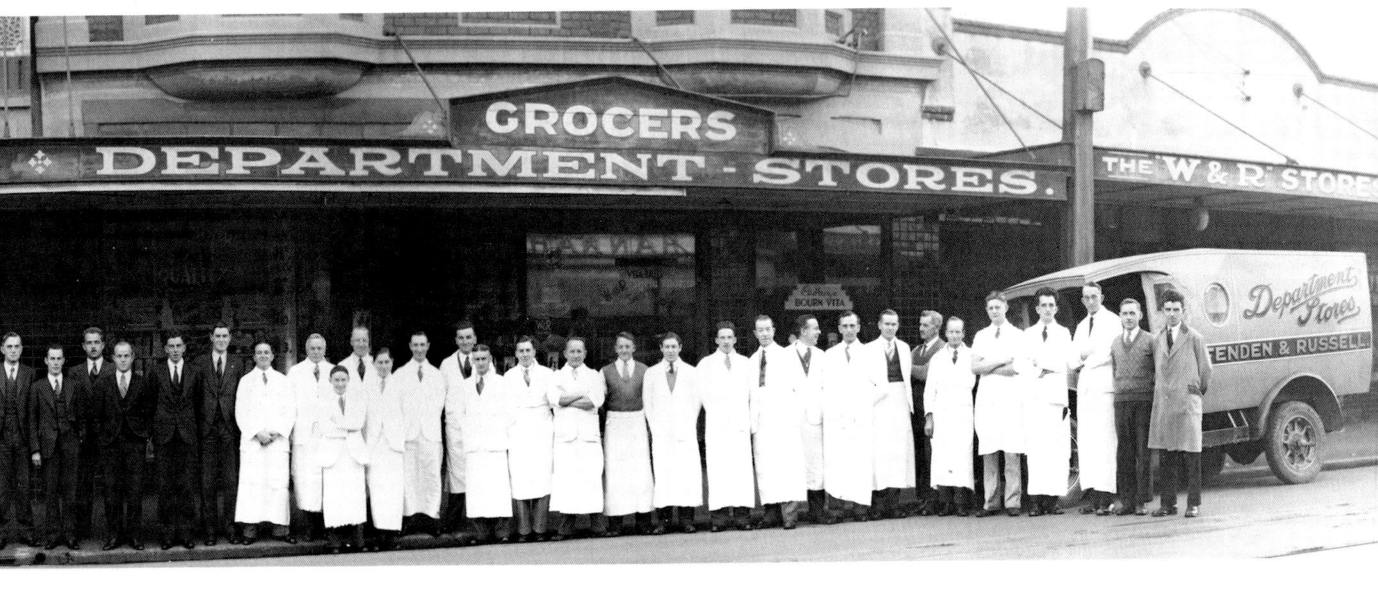

OVERLEAF

Dunedin City Corporation Gasworks, c. 1898 (the bottom photograph runs on from the right-hand end of the top photograph). Gas lighting had been introduced in London in 1812. In 1862, in order to light the streets at night, the Dunedin Gas Light and Coke Company was formed, with Stephen Hutchison as engineer, building and operating the works. Built on a two-acre site on Andersons Bay Road, the business comprised a retort house with a cast iron roof, condensers, a purifying house, a governor and the gasholder (a hollow cylinder 60 feet in diameter and 20 feet deep, closed at one end, inverted in a tank of water which formed a seal). After the City bought the gasworks in 1879, a new retort house and gasholder were built. (The cast-iron legs still survive, as does the brick chimney, which may also date from the early 1880s.) In the late 1890s, the City installed a carburette water-gas plant and built new buildings to house the generator and the purifiers. As well as a new two-storeyed brick house for the manager, a new and larger gasholder was built, although the unstable land proved unable to cope and in 1915 it was moved to Wilkie Road.

Manufactured gas for lighting and heating was one of the first and most significant products of the Industrial Revolution that brought radical change to the lives of countless millions around the world. By burning coal in long horizontal tubes – a dirty, dangerous and toxic occupation – one could manufacture both gas, which could be reticulated to households and used to light streets, and coke, a by-product widely used for domestic heating. When Hutchison sold the works to the City in 1879, something went wrong and in fury he built the Caversham gasworks and competed successfully for almost 30 years: the decision whether to buy gas from Hutchison or the City reduced South Dunedin almost to civil war. In the late 1890s a considerable effort was made to expand the use of gas, with the trading department opening its own retail store and introducing the sale of ovens by hire purchase. Despite an explosion and fire in March 1903, the City modernised and extended the works and bought out Hutchison's Caversham company in 1909. I am grateful to Peter Petchey for his help in interpreting this picture and explaining the process.

DCC Archives.

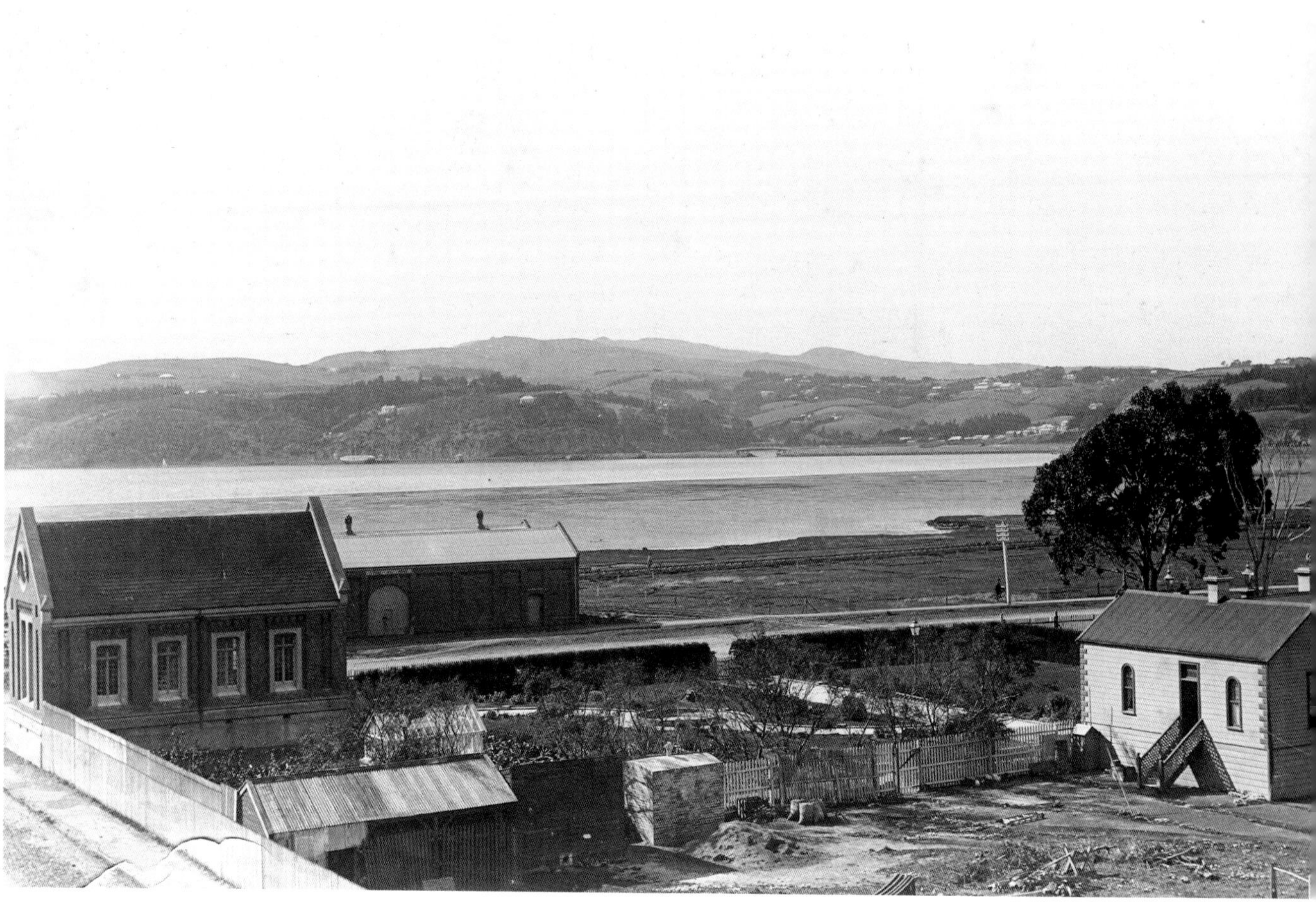

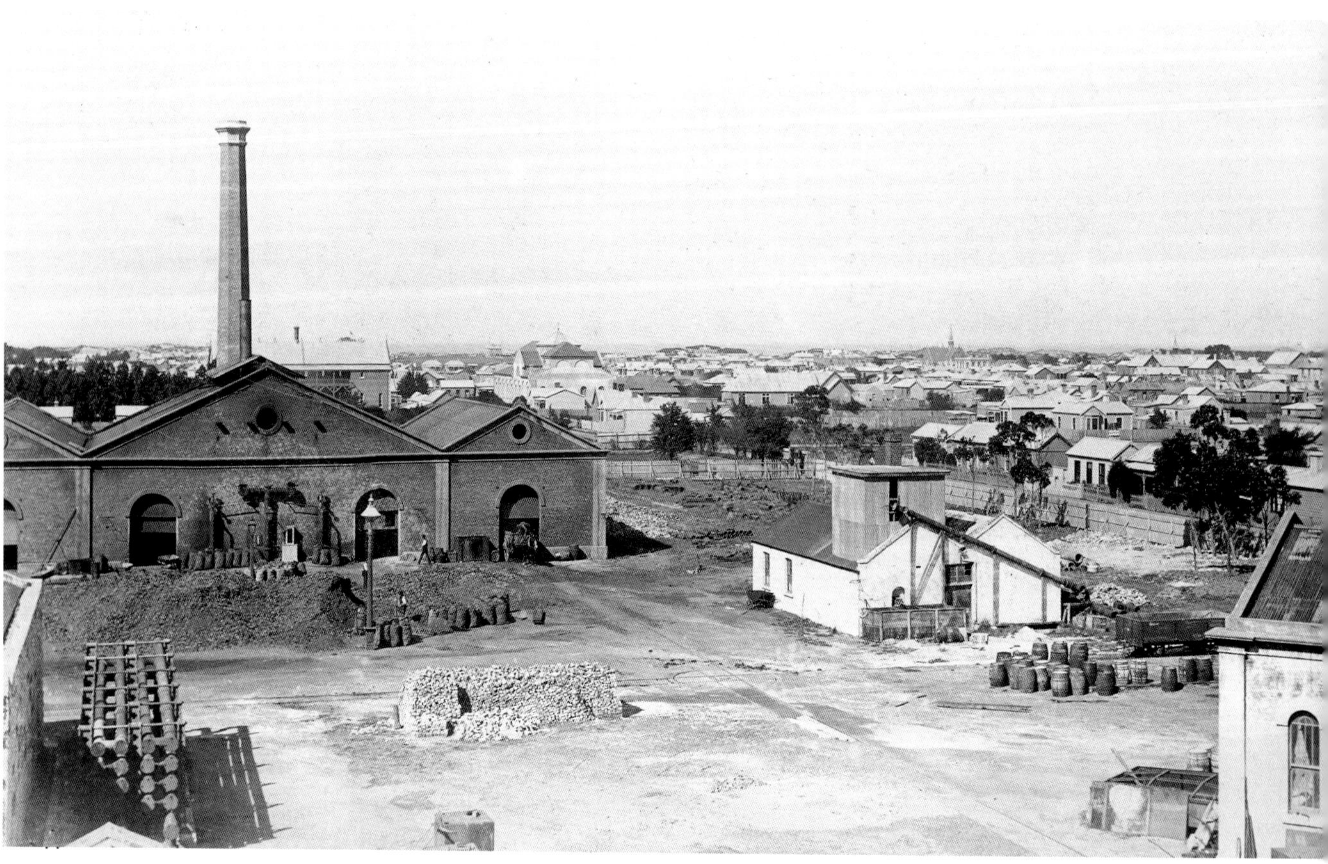

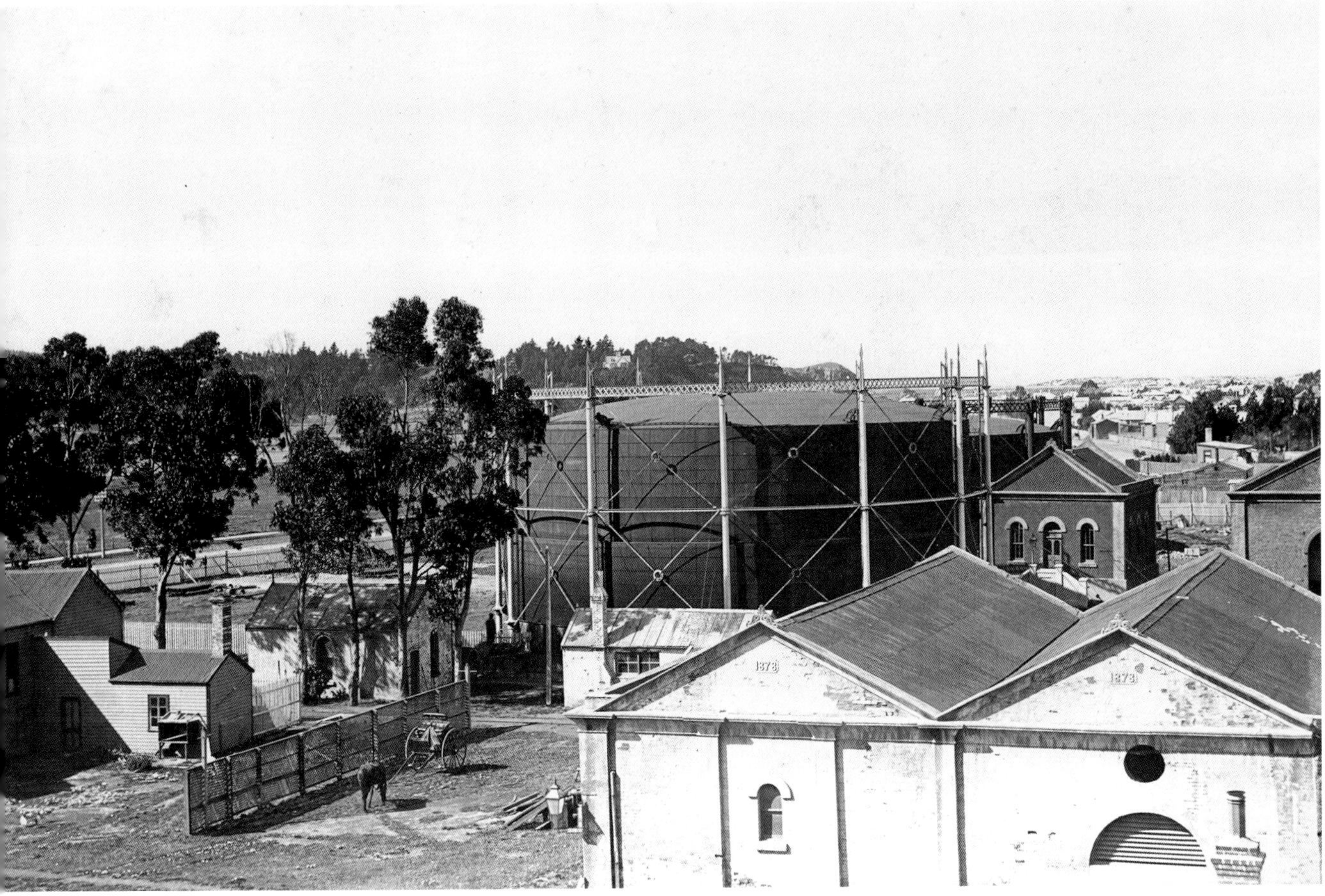

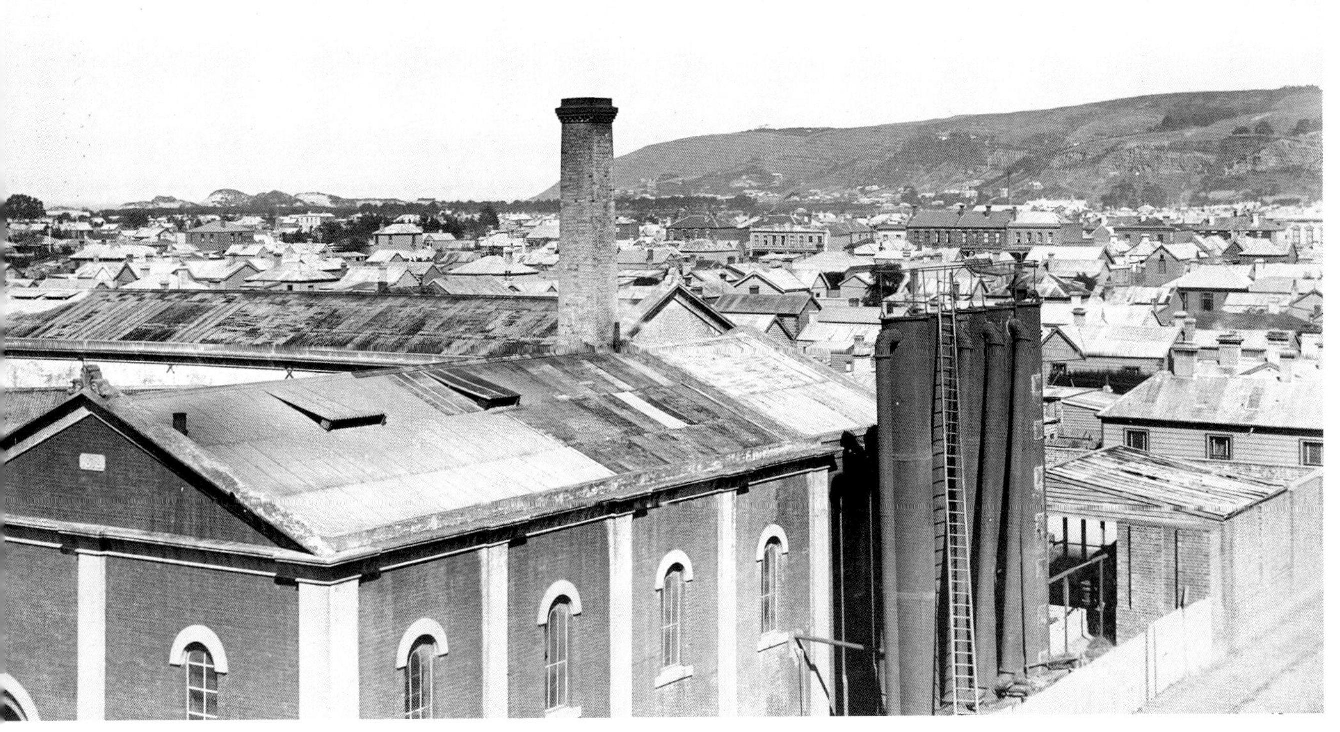

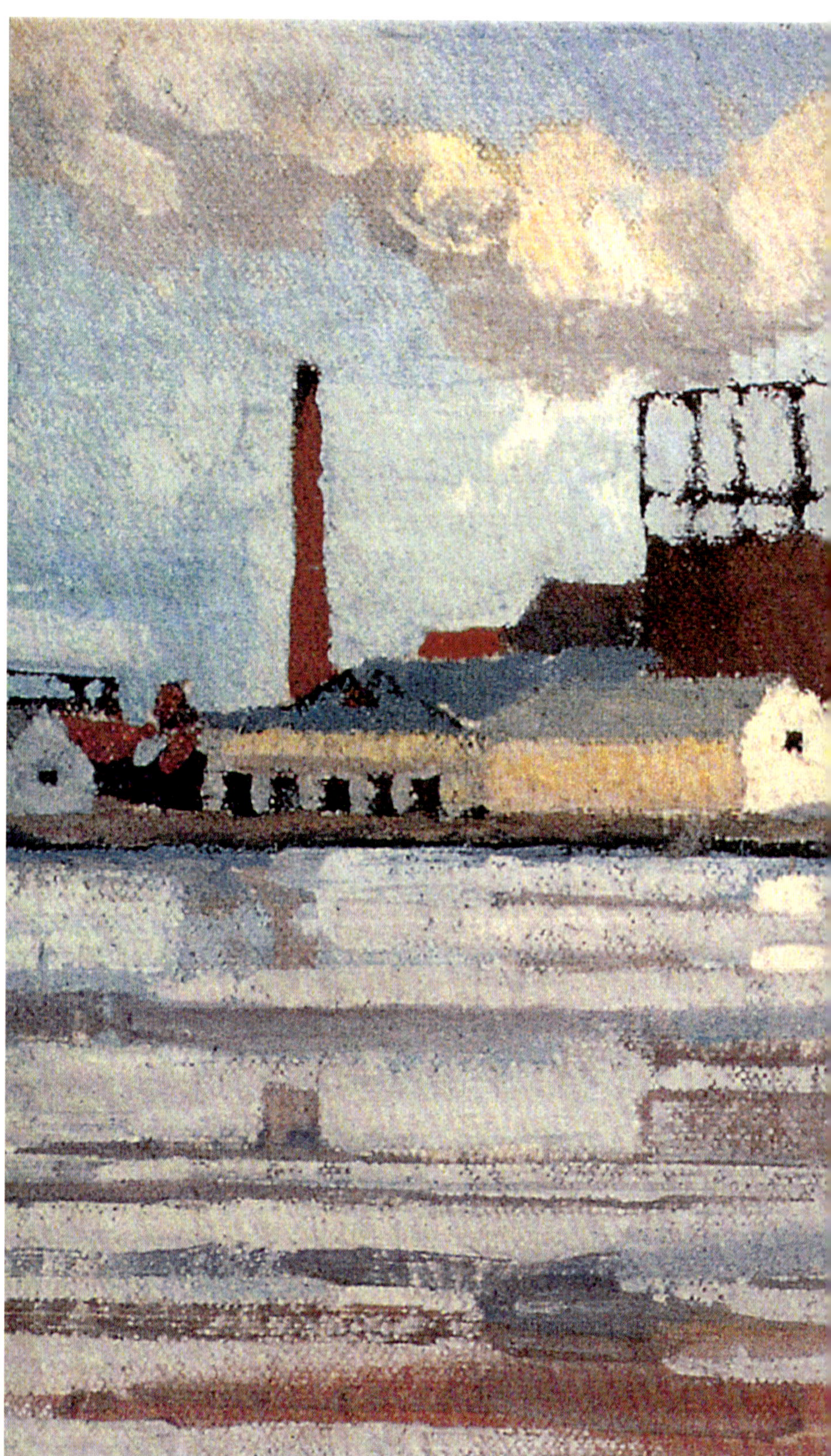

The gasworks seen from the harbour's edge. Doris Lusk, *Gasworks and foreshore*, Dunedin, 1936, oil on canvas, 290 x 430 mm, donated to the Hocken Library by the artist, 1979.

Until 1967, when she moved to Christchurch, Lusk lived on Kew Rise, overlooking the Flat. She became well known for her paintings of man-made, functional structures, such as this, set in landscape. The genre of en-visioning machines and industrial structures in gardens and landscapes, so fundamental to the colonisation of the New World in the nineteenth–twentieth centuries, has been largely ignored by art historians in this country.

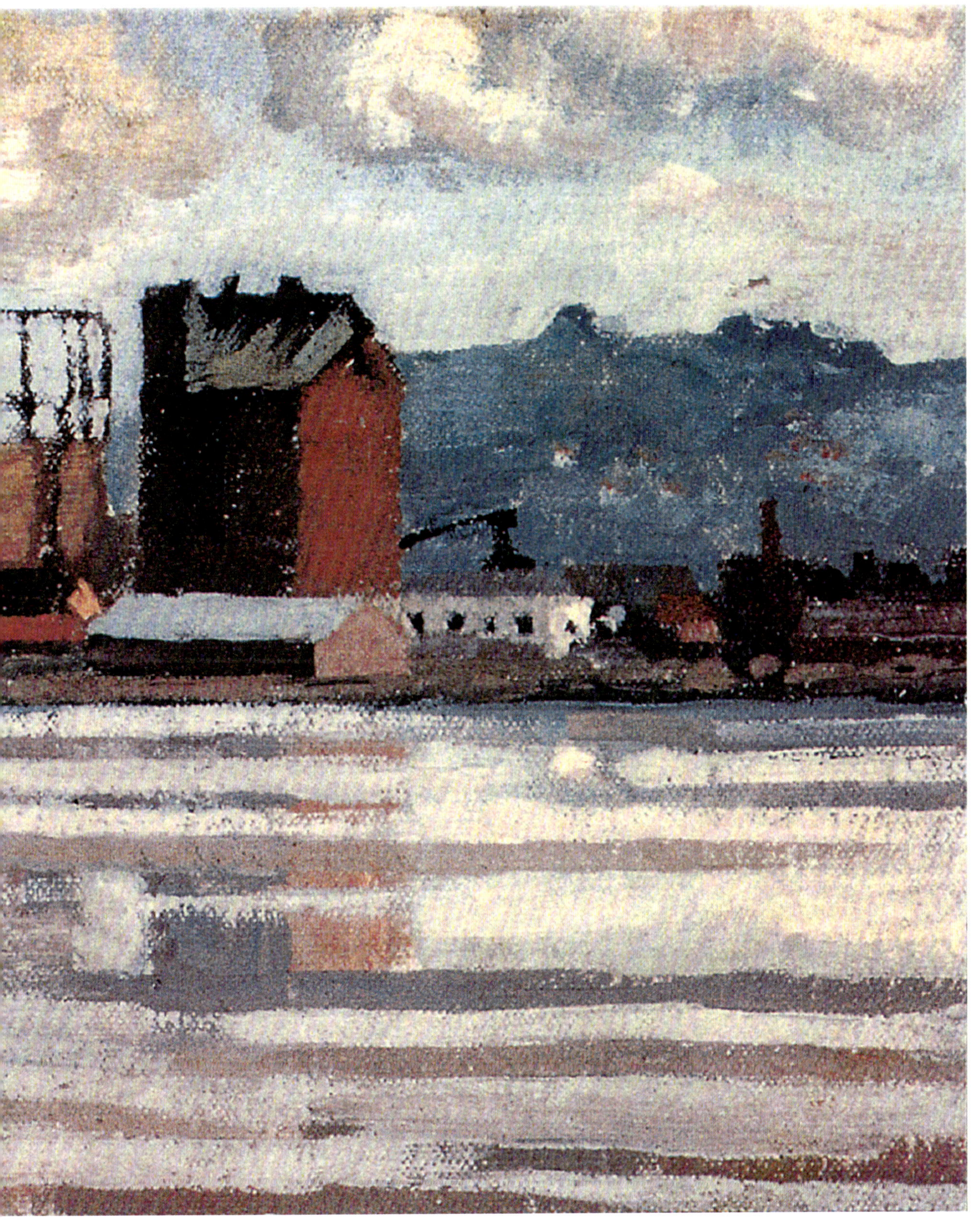

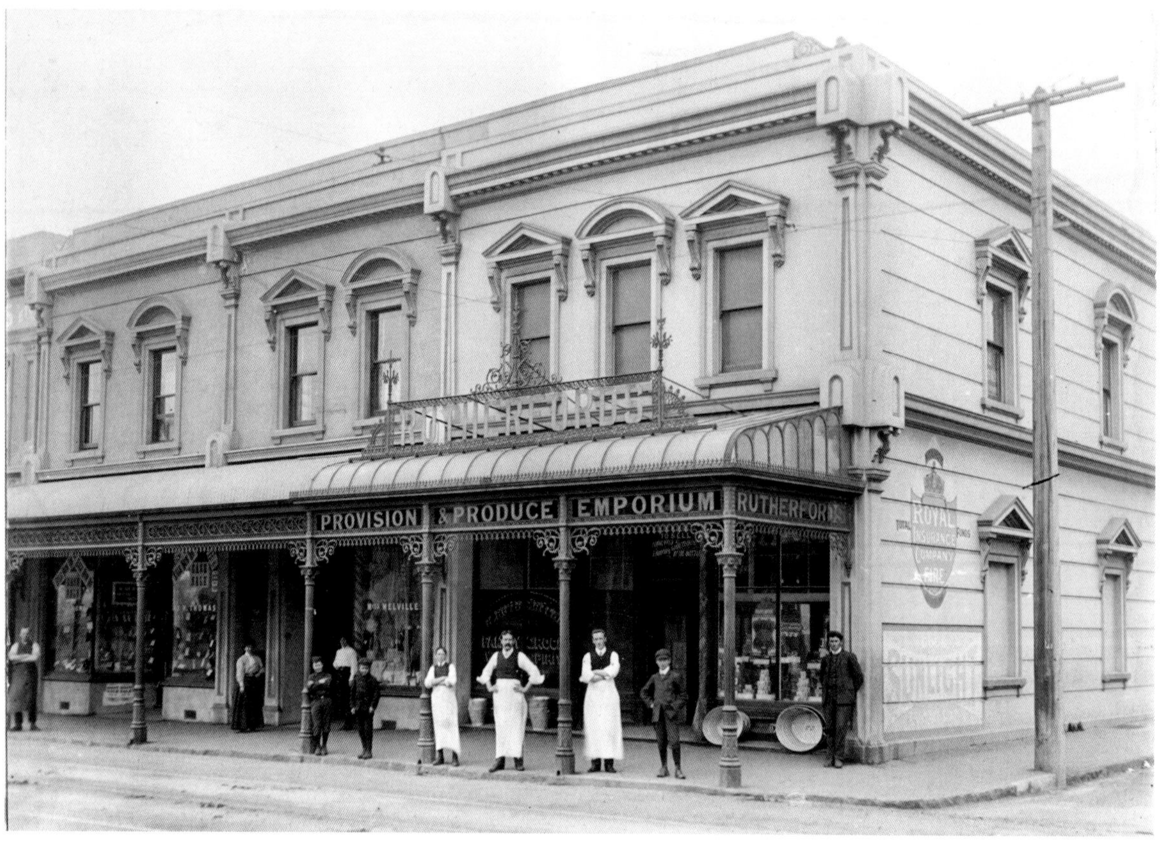

ABOVE **In Caversham village, Rutherford's and McCracken's General Stores** faced off across the Main South Road. Robert Rutherford extended his two-storeyed brick store (above) in 1905 to occupy the entire block (his brother Peter had opened the first grocery there in 1876, but Robert later bought him out). Sam McCracken had traded from his site since c. 1882, and built his two-storeyed brick shop (opposite) c. 1900. It did not take much capital to set up as a grocer, and many women did so when they lacked a male breadwinner and had to provide for themselves, but most of these businesses collapsed within a short time. By contrast, successful grocers such as Rutherford and McCracken established local dynasties.

The successful grocer considered himself a master craftsman and had served an apprenticeship. Both Robert Rutherford and Sam McCracken trained their own sons. In this age, the grocer knew every customer by name, asked about their family, and could remind them of what they needed. They also provided a delivery service, often conducted by one of their sons, a system of credit, and exchanged gifts at Christmas. Special treats for small children were expected by customers, and Rutherford was famed for his generosity.

McCracken had started as a grain, coal and wood merchant, later branching into groceries, whereas Rutherford started as a grocer and later added a coal and wood business. It seems likely that they catered to distinct constituencies. Although both were Presbyterians, Rutherford was from Scotland while McCracken came from Ulster. They differed in their attitude to alcohol: Rutherford had a bottle licence, McCracken, a teetotaller, did not. Rutherford was also keen on the horses, played a prominent role in the Gun Club, and strongly supported the Liberal Party. McCracken was active in the church, disliked the Liberals and had close links with the Brethren, also strong in Ulster. According to the family, when Ted Wolfenden married Annie McCracken, Sam's daughter, she and her mother became Brethren.

In the photograph of Rutherford's (at the corner of Playfair Street and the Main South Road) are from far left: Joe Maloney, bootmaker; a couple of strays who grabbed the chance to be photographed; Miss Maxwell, a small-scale draper who specialised in needles and threads; W. Tyrell, Rutherford's apprentice; Robert Rutherford himself; and William (Bill) Hudson, the assistant, one of a large local family that eventually boasted two members of Parliament, both Labour. Bill Hudson was later taken on as a partner. Robert's three children also entered the grocery trade. Robert jnr later became manager, John the traveller, and Jessie kept the books and looked after her father after her mother's death.

I am grateful to R.J. (Jack) Rutherford for lending this photograph and identifying the people in it.

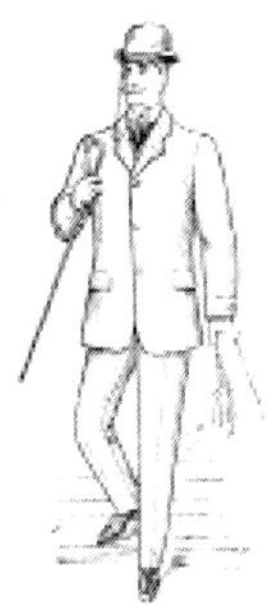

It will Pay You to Call and See our New Shipment of
SUMMER and AUTUMN TWEEDS

TODD & BROWN

The Up-to-date TAILORS

145 Rattray Street, DUNEDIN.
Main Road, CAVERSHAM.

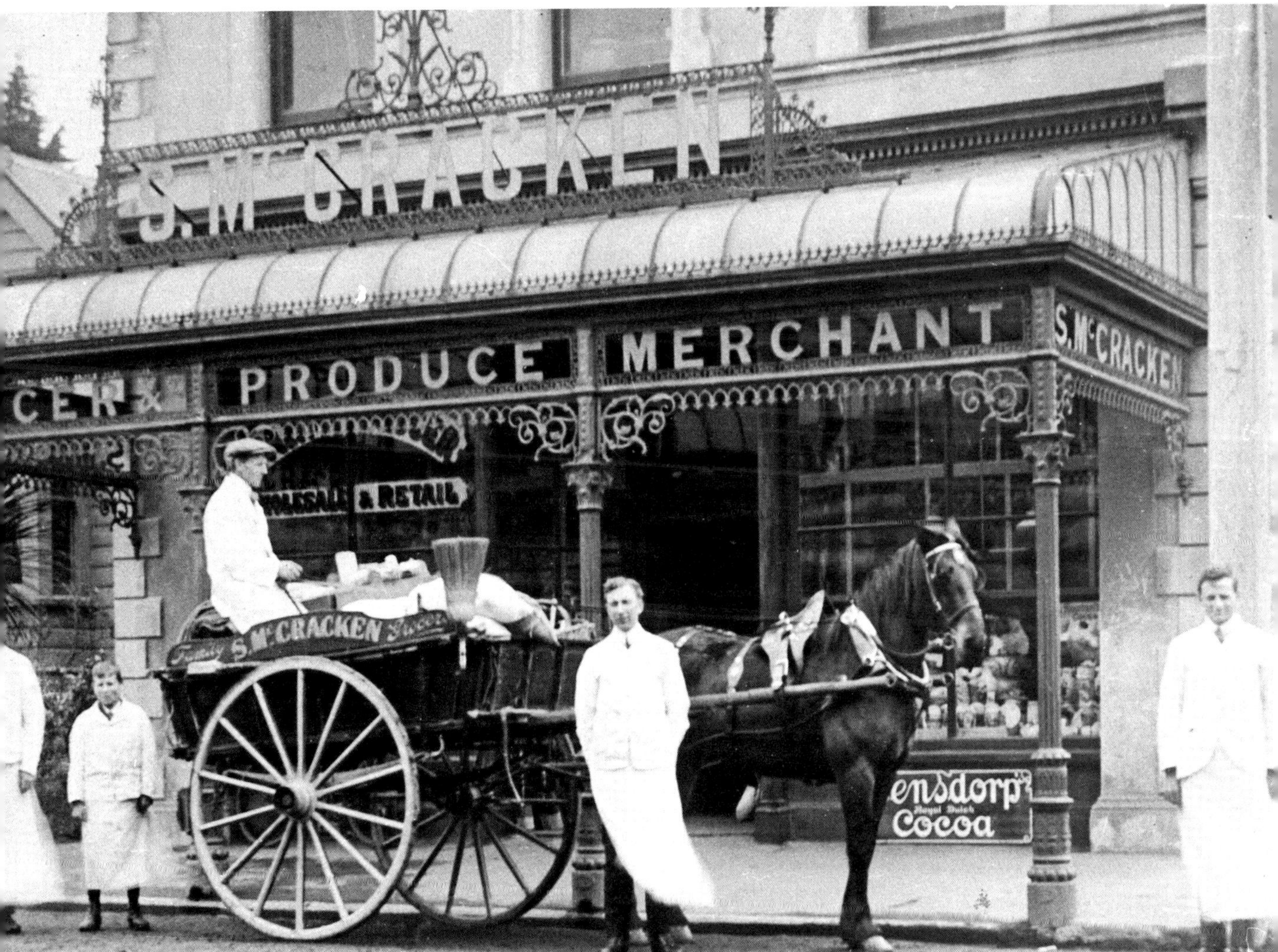

ABOVE **In front of McCracken's Store** are from left: Tom McCracken, two unnamed assistants, Billy Glidden (killed in World War I) and Sam McCracken.

Hocken Collections S13-350a.

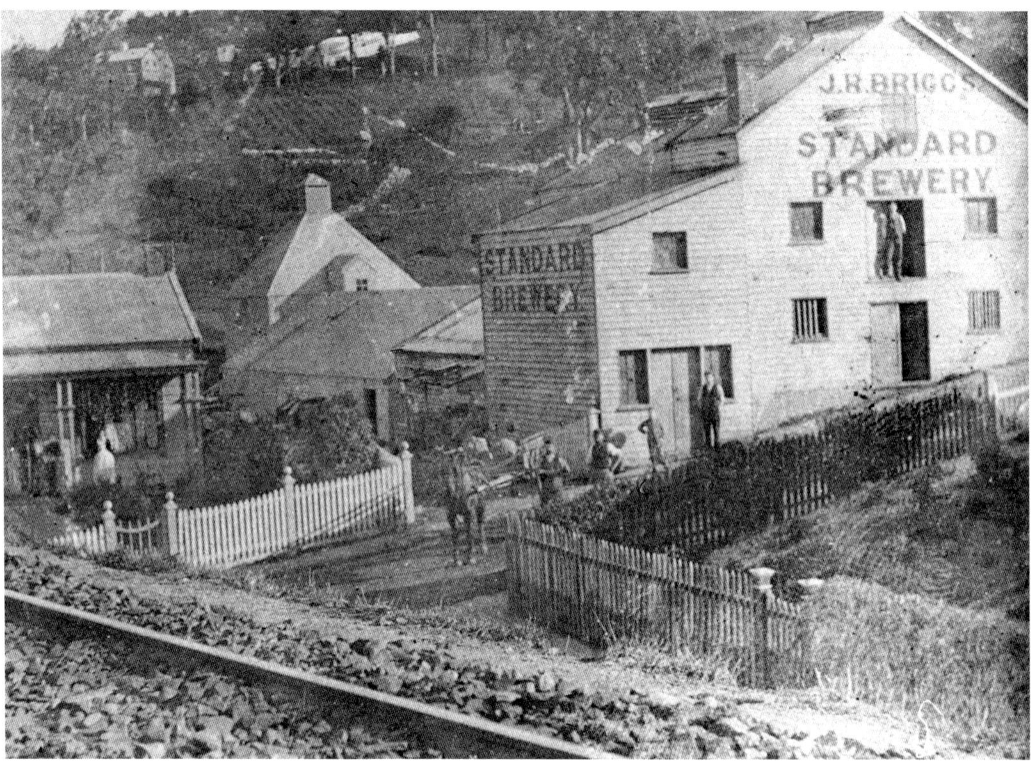

ABOVE **J.R. Briggs' Standard Brewery, Sydney Street, Caversham, c. 1888,** with members of the Briggs family on the verandah of their cottage, Sydney Street. Briggs' brother-in-law, Edward Cochrane, had the Caversham Brewery nearby, between Playfair and College streets, but he sold it to Alexander Cowie in 1890 and left for Queensland. Both breweries also boasted a malthouse and kiln. They were typical of the small but substantial businesses so common at this time and place. Every brewery needed a maltster, a brewer, someone to operate the kiln and several labourers. If they made their own barrels, they also needed carpenters and coopers. The two breweries, one owned by an Englishman, the other by a Scot, shared water rights from a spring owned by the Railways Department. A third brewery had closed in 1894, but these two companies survived until the 1920s.

Hocken Collections, 76.1141.

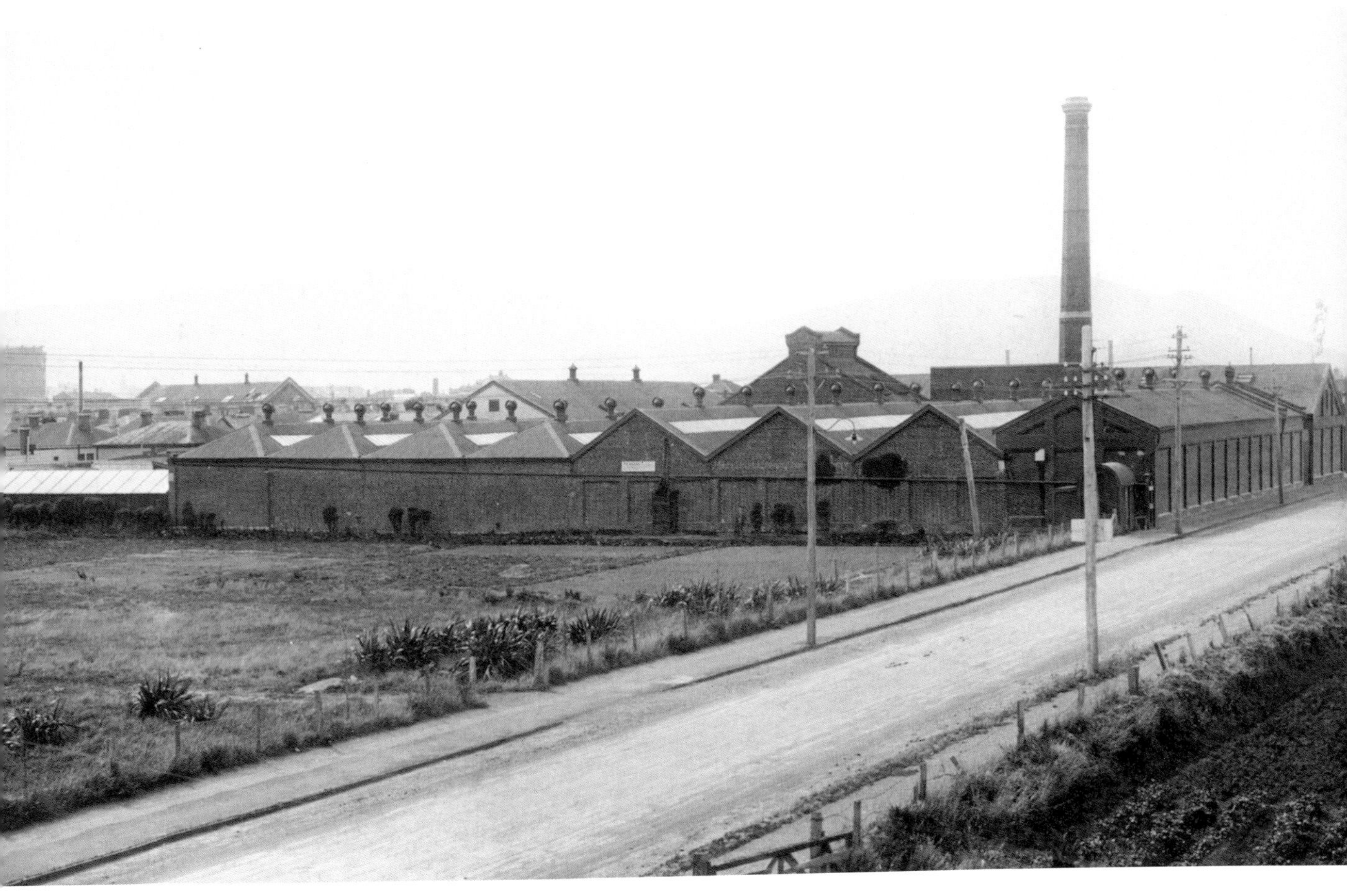

ABOVE **Donaghy's Rope and Twine Company** was founded by John Donaghy, of Geelong (Australia), in 1876. The 1870s was a buoyant decade in New Zealand, and the factory moved to its present site two years later. Before long, Donaghy sold to three Dunedin businessmen, one of whom, Alfred Lee Smith, was a prominent Catholic from Yorkshire and later a prominent Liberal (who served in the Legislative Council from 1898 until 1905 and was a close friend of Joseph Ward, one of the Catholics in the Lib–Lab cabinets). After importing new machinery and moving into purpose-built brick premises in 1888, Donaghy's became a public company in 1889. In the 1890s the firm boomed, its single-ply binder twine becoming enormously popular with farmers. In the mid-1890s Smith, the company chairman, re-built the ropewalk in brick and added new buildings for seaming and roping twines, as well as a new boiler house with a towering chimney. Its whistle pierced the air four times daily, once stampeding Mary Irvine's cows into the factory! Donaghy's chimney dominated the Flat until 1921, when the works were electrified. The company also operated factories in Invercargill and Auckland.

Work in a rope factory was extremely dusty and, in this period, dirty. Coal-fired boilers drove all of the machinery and the raw rope fibre had to be lubricated in whale oil, the smell of which permeated clothing, hair and the pores of the skin. The whale-oil tanks had to be cleaned every two months, a filthy job: the worker stripped to his underpants and climbed into the tank. The factory also had to be cleaned annually, the men being organised into two teams, 'whites' and 'blacks'; the blacks cleaned the flues by crawling through them, while the whites whitewashed the walls.

Hocken Collections, Donaghy's Photograph Album, Ms-AG-202/425.

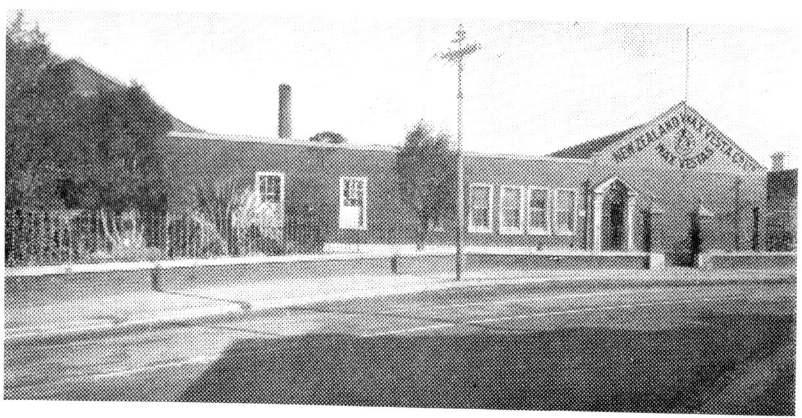

ABOVE **The New Zealand Wax Vesta Company's factory, David Street,** near Forbury Corner, c. 1927 (the earliest close-up photograph known). Established in 1895, as the first match factory in the country and the only one in the South Island, it began manufacturing in the old Immigration Barracks but, due to rapid expansion, shifted to this purpose-built modern factory in 1902. The brick building, with a fire-proof asphalt floor, was airy, well ventilated and well lit. It included a boiler room, two large workrooms (one for the waxing machine and the other for packout), various specialist rooms, cloakrooms for men and women, and toilets.

The first purpose-built factory burnt down in 1913 but the directors found temporary premises while the factory was rebuilt. The business was founded in 1895 by Robert W. Rutherford, who managed the factory until 1921, John Watson, the engineer, and Rutherford's father, also named Robert. It became a limited liability company in 1898. During the 1880s, Robert W. had been Caversham's photographer. He was a cousin of Rutherford the grocer.

Hocken Collections 76.518.

OPPOSITE **The Shiel brothers' quarry and brickfield at Forbury, 1900,** with the market gardens visible in the background. These are the only known photographs of either the quarry or the brickfield, and both are by an unknown photographer named Stuart. Little is known about this extensive business but Charles Shiel, one of a large Irish-Catholic family, learned the trade working for Smith & Fotheringham and established a brickfield at Caversham after they closed for want of clay. In 1899–1900 he went into partnership with one of his brothers, William, and moved to Forbury, where they had identified an excellent source of clay and stone. By 1910 the brickfield occupied 10 acres. An overhead conveyor system transported clay and bluestone across Forbury Road to the brickfield itself. Apart from clay, the quarry yielded mainly aggregate for road construction, excavated by pick and shovel, as can be seen in the photograph, with the occasional use of dynamite. The Shiels also had a quarry in North East Valley and a share in the Blackhead quarry. On the other side of the Flat, there were further rock quarries at St Kilda, Musselburgh and Tainui.

With the invention of fired brick around 3500 BCE kilns could be heated to above 1000 degrees Celsius. The invention of the Hoffmann kiln in 1858 allowed heat to be maintained at a constant temperature, 24 hours a day, and the volume and quality of brick production expanded dramatically. Kiln-making was a highly specialised job, but improved kilns, new technologies and the introduction of electricity reduced the importance of the kiln-maker's skill, just as improved efficiencies reduced the importance of other forms of craft knowledge.

Although itinerant brickmakers still worked in New Zealand, using fairly primitive methods, brickmaking had been almost completely mechanised in the mid-nineteenth century. The Shiels built a transverse-arch Hoffmann kiln to fire the bricks, a process that required a constant temperature of 900 degrees Celsius for around 48 hours. The transverse-arch kiln comprised a series of brick chambers built side by side, each spanned by its own brick arch, and the heat from a single fire kiln moved through the complex over a period of around 16 days. Managing the kilns was highly skilled and onerous work, often requiring 12-hour shifts.

The works usually shut for five or six weeks in winter for general maintenance and cleaning, a filthy job. The bricks proved to be excellent, and the field operated until the clay deposits were exhausted. In 1925, the Sheils shifted to Abbotsford, but the clay there proved inferior and in the 1930s this business went bankrupt.

Hocken Collections, *Otago Witness*, 19 September 1900, p. 47.

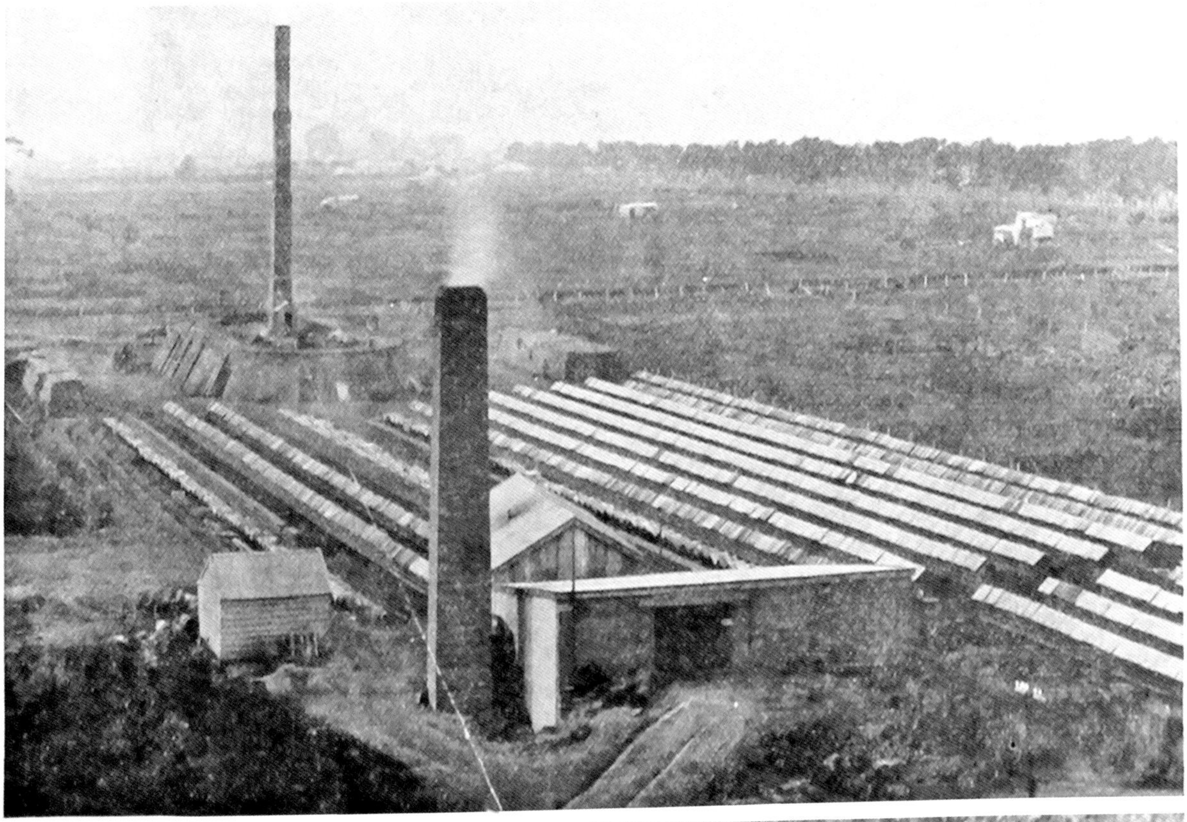

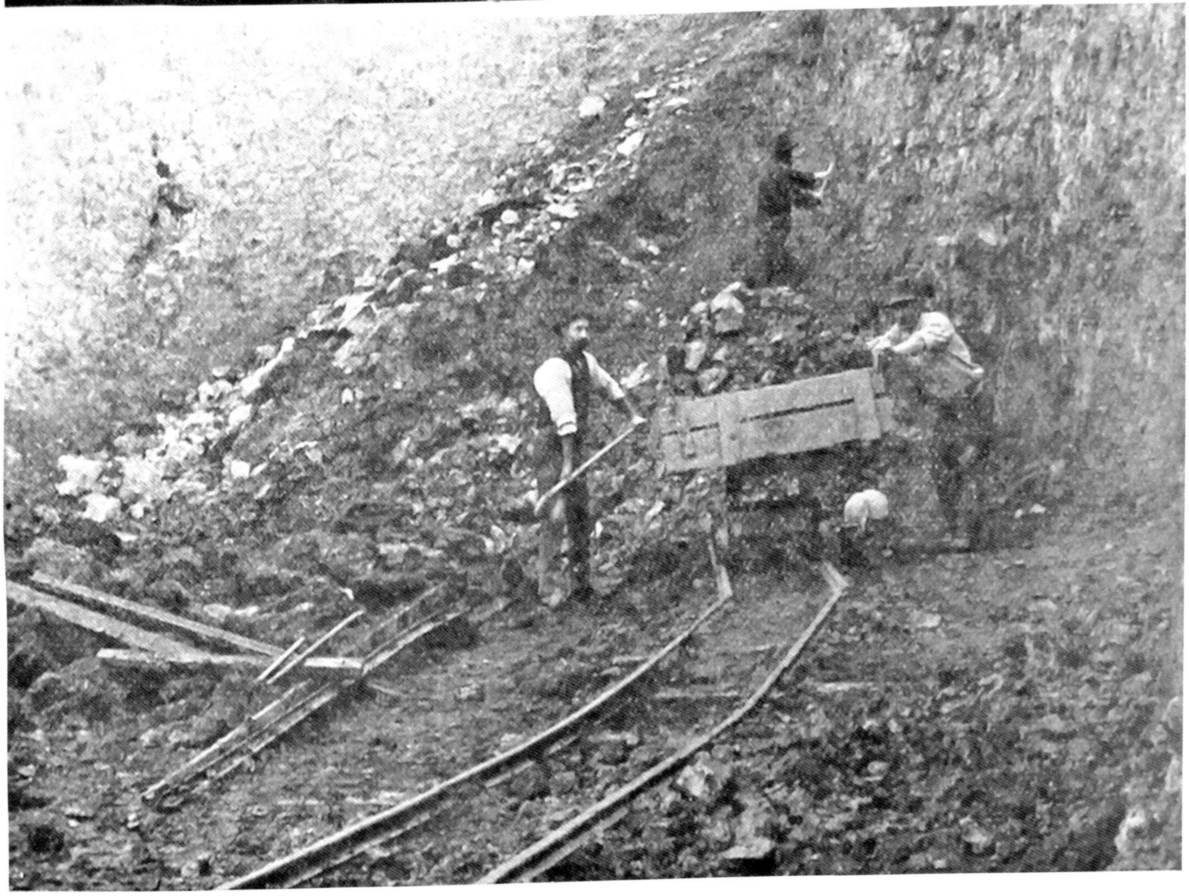

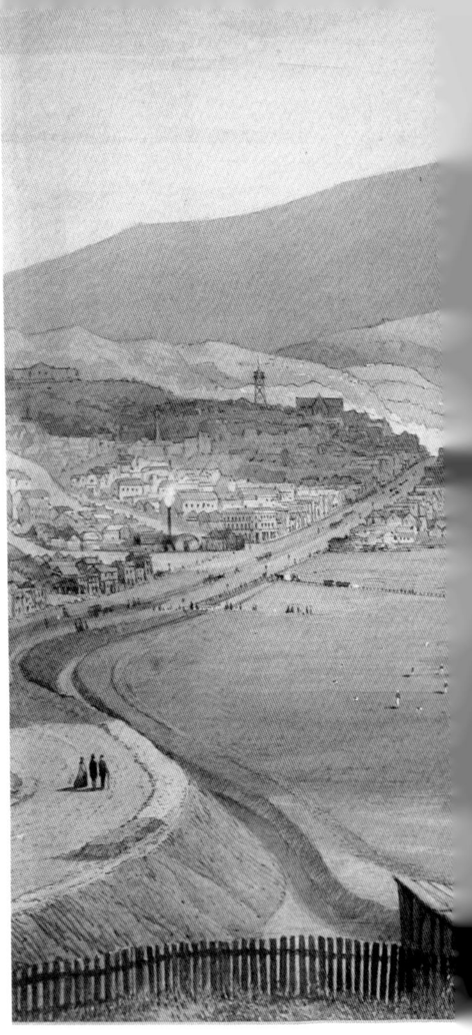

Dunedin 1865 from Southern Cemetery. House in foreground, watercolour, signed and dated by the artist, George O'Brien.

A native of County Clare, Ireland, O'Brien was probably born around 1821; he trained as an architectural draughtsman and migrated to Victoria in 1850. He appears to have worked in Melbourne as an architect between 1854–58 but was already painting watercolours. In 1863 he crossed the ditch to Dunedin, tried his hand at various jobs, but became famous for his luminous paintings. He was also famous because alone among his contemporaries he often painted man-made objects – buildings, townscapes, and even breweries and gasworks.

In this painting, as he did so often, O'Brien has altered the topography to achieve his perspective of the Flat and the City, with the newly reclaimed land west of the Exchange now sprouting a veritable profusion of new industries. I especially like the juxtaposition of the house, with its brightly coloured washing blowing in the wind, the marshy Flat, the dense concentration of industries between the Oval and the Exchange, and the backdrop of the suburban town set against Mt Cargill and Signal Hill.

Toitū Otago Settlers Museum, cat. no. 1988/78/1.

To the east of Kensington, another concentration of industries developed on the reclaimed land between the Oval and the Stock Exchange. As can be seen in George O'Brien's watercolour, even as early as 1866–67 industries dominated this area. The large area of reclaimed land lying between Princes Street South and Crawford Street had not then been built on, but by 1900 it had become home to a host of major industries and warehouses. The last area to be reclaimed had been seized by government for railway yards and running sheds. Thirty-five years later, walking towards the city from the Oval, one almost at once saw the tramsheds and three of the city's most substantial engineering works: Shacklock's, Reid & Gray's, and Cossens & Black's. George Methven, who had worked for Reid & Gray before going out on his own account in 1886 to manufacture agricultural implements in his back yard in Goodall Street, Caversham, had recently rented land from the Harbour Board and erected a one-storeyed brick factory on Crawford Street. If he could not boast of the largest establishment, he made it clear he had the most modern machines and appliances for engineering.

Most of the industries named in the map on page 65 would have been visible to our fictional pedestrian in 1900. To the north, on the lower reaches of Stafford and Rattray streets, one could see Farra Brothers, tinsmiths

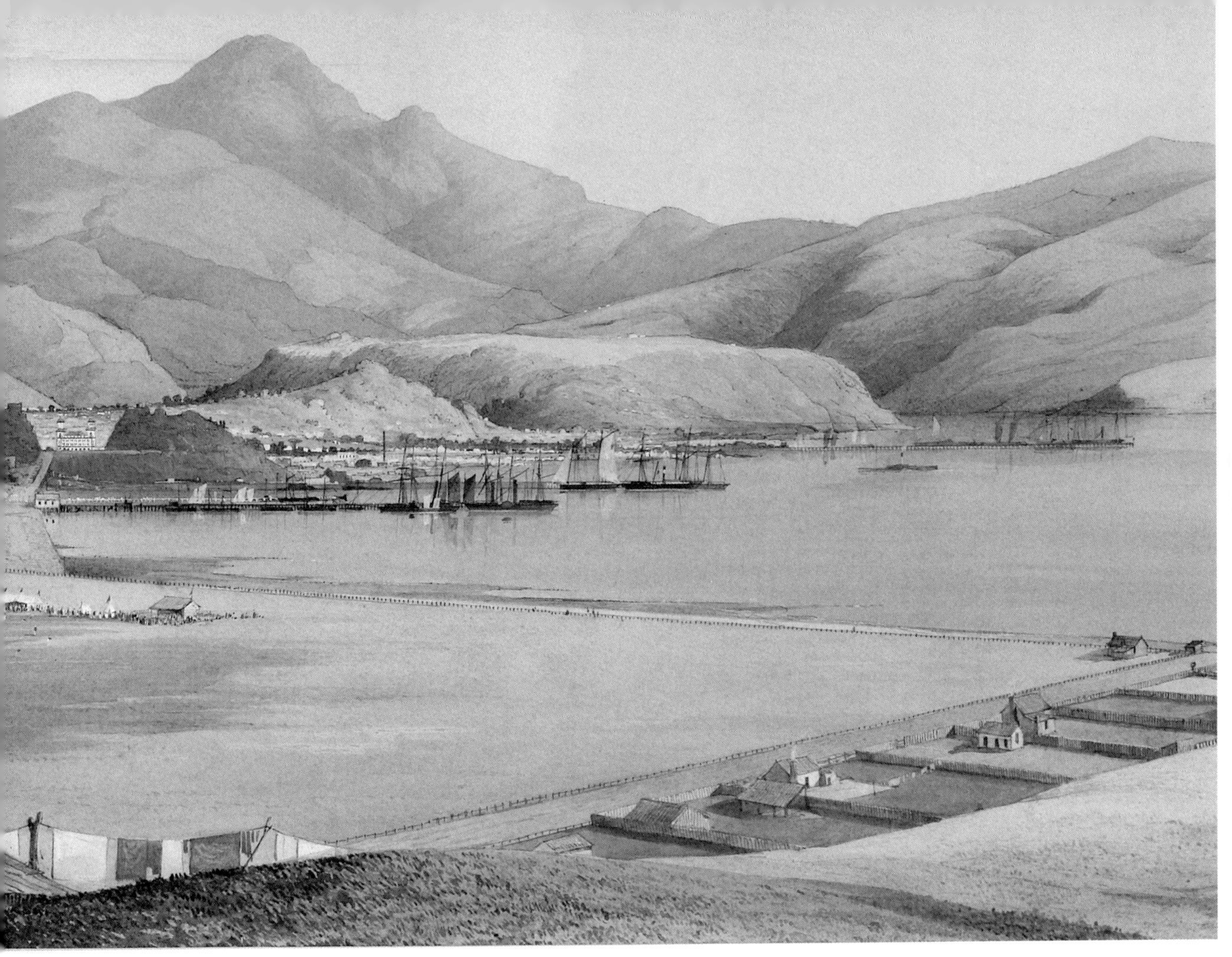

and japanners, various clothing and apparel factories owned by Ross & Glendining, one of Irvine & Stevenson's two stores and Speight's brewery. To the east and south of the Stock Exchange, then the city's transport hub, warehouses, bond houses, forwarding and express agencies, banks, and the head offices of three of New Zealand's four insurance companies could be found. Proximity to the city's wharves made this area especially attractive. The inconvenience of having the railway obstruct access to the port was doubtless softened by the close proximity of both.

Several other engineering workshops were situated east of the Octagon, mainly along Cumberland and Great King streets: the Otago Foundry, R. Sparrow's Dunedin Iron Works, McQueen & Co, J. McGregor & Co, and Barningham & Co, which specialised in manufacturing the 'Zealandia' coal range. A. & T. Burt, then the city's pre-eminent firm with a workforce of over 300 men, was also on Cumberland Street, with a substantial warehouse, showrooms and offices. Among manifold activities, A. & T. Burt briefly led the world in manufacturing hydraulic dredges and elevating machinery that was exported to Russia, Southeast Asia and the United States. Other engineering shops specialised in making and repairing ships, notably Stevenson and Cook and a branch of A. & T. Burt at Port Chalmers. All of these men were Scots.

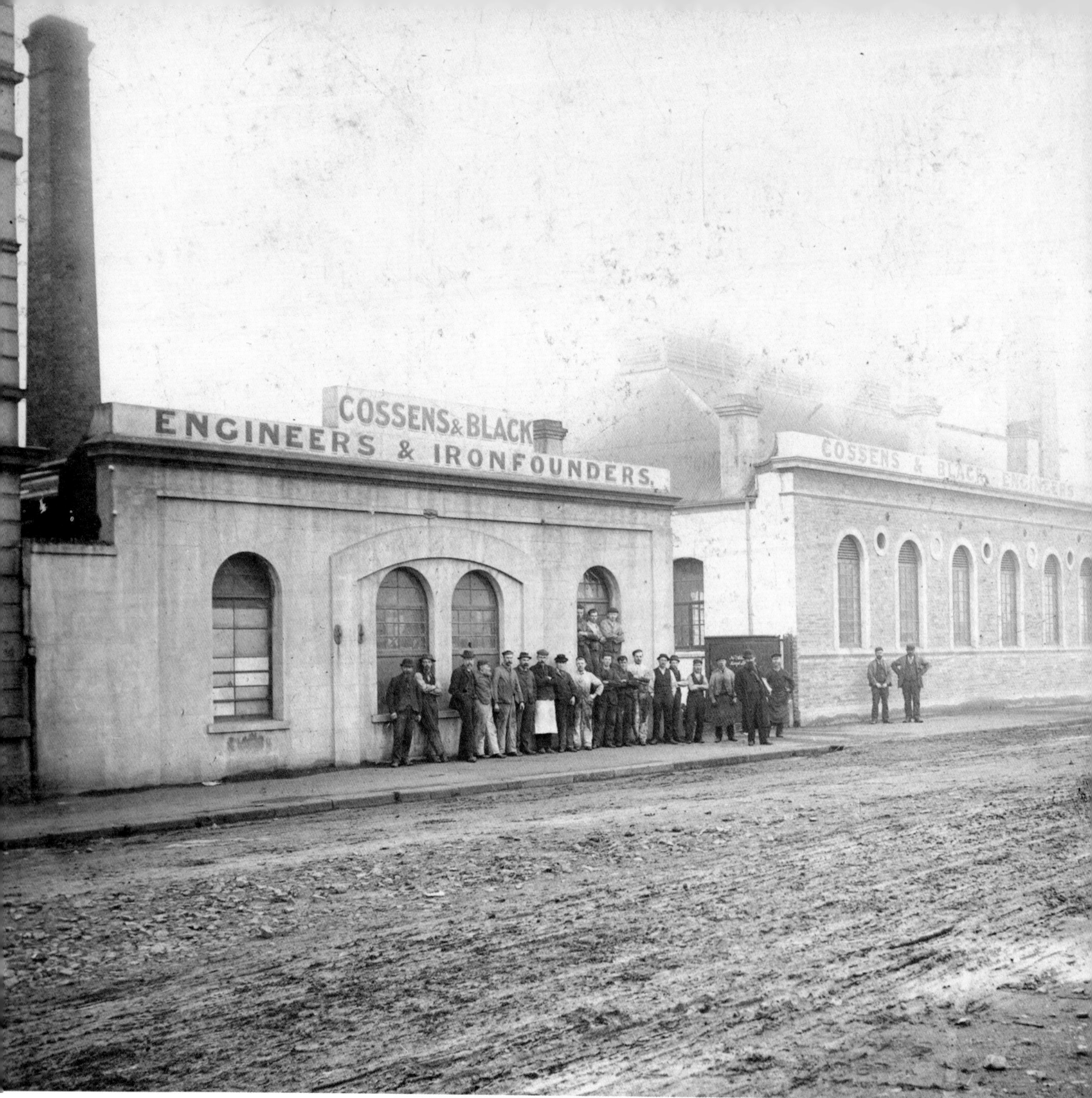

ABOVE **Cossens & Black, Engineers & Ironfounders,** were just to the south of Shacklock's, but had their main office on Crawford Street. Alexander Black and Thomas Cossens established their foundry late in 1874. When Cossens died in 1891, Black took over. Thanks to the gold-dredging boom, at its height in 1900, the city's need for trams for the new electric tramway system, and the city's decision to build a large hydroelectric scheme at Waipori, Cossens & Black were flourishing. Their extensive establishment included an engineers' shop, a blacksmiths' shop, a boilermakers' shop and a moulding shop. They employed between 50 and 70 hands. Black, born in Scotland, was the son of a farm servant. Photograph by Armstrong & Greer.

Hocken Collections, Cossens & Black Photograph Album, Ms-2989-004, S12-528f.

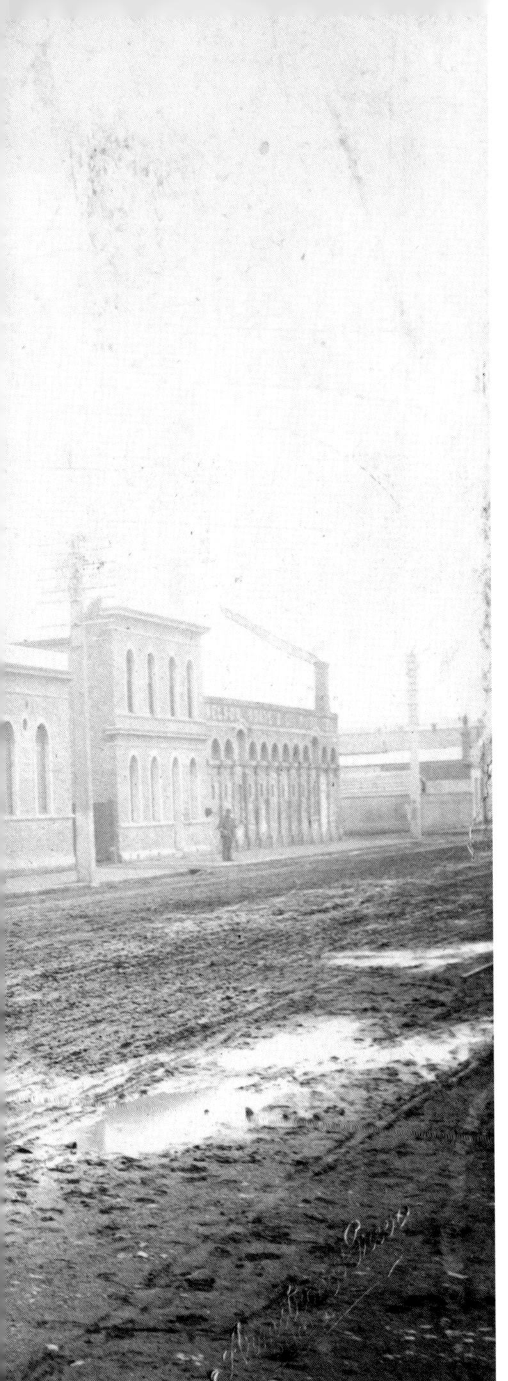

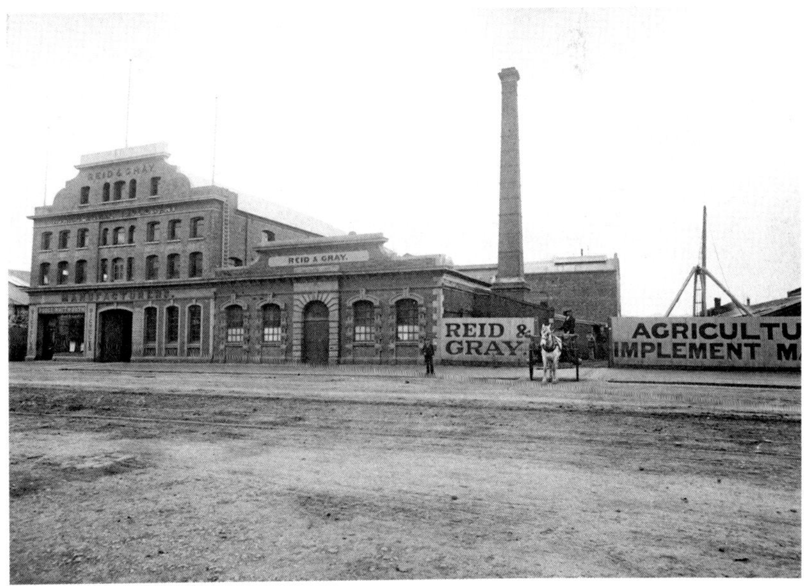

ABOVE, RIGHT **Reid & Gray's street frontage, Princes Street South and Crawford Street.** James Gray, great-grandson of George Gray of Uddington, Scotland, the first man to make an iron plough, arrived in 1863 to oversee a contract the firm had entered into. In 1868, he entered a partnership with Robert Reid and opened for business in Oamaru, on the eve of a vast expansion in arable farming and especially wheat production. Initially they imported John Gray ploughs from Scotland, but before long began manufacturing their own. Reid & Gray, Engineers and Iron-founders, moved to Dunedin from Oamaru in 1873. By 1900, their workshops occupied an area of one and a half acres and their blacksmiths' shop was the biggest in the city, with 23 forges and four furnaces. They were especially proud of their moulding shop, where they manufactured cast-chilled shares, among other things, and the blacksmiths' department where most of the work in manufacturing ploughs was done. Their double-furrow plough swept the New Zealand market and won them an export market. In 1900, on the eve of a spectacular decade of growth, the Dunedin factory employed some 200 'operatives'. Their handsome warehouse on Crawford Street, which included their counting house, had opened in 1880 and was the coping stone for a network of eight branches. Photograph by Armstrong & Greer.

Hocken Collections, Reid & Gray Photograph Album, Ms-1165-063, S12-528a.

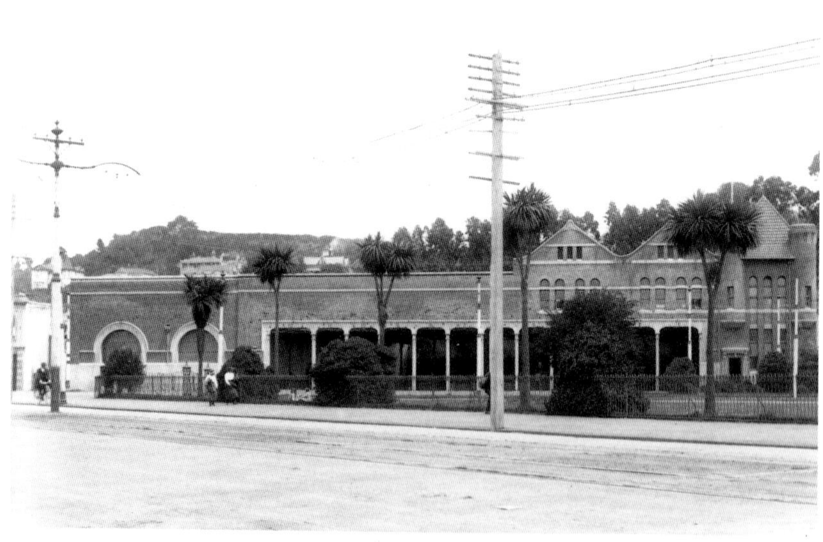

RIGHT **Dunedin City Corporation Tramsheds, Market Reserve, c. 1905.** After years of negotiations, the City bought all the trams. When Prime Minister Seddon opened the electric tramways in December1905, the first nine cars were comfortably housed in the brand new sheds at the Market Reserve. This substantial brick building, three storeys high, was built to hold 52 carriages. Well lit and well ventilated, the City boasted that it was 'one of the most efficiently equipped carhouses in the southern Hemisphere'. Even the examination pits were state of the art and under cover. The machine shop was equipped with the most modern tools from the United States. On the site of the old sheds on Crawford Street, the City erected a two-storeyed brick power house to transform coal into electricity, until the Waipori hydroelectric scheme began generating hydroelectricity in 1907. The Tramways employed around 200 men, 16 of whom were classified as 'officers'. Cleaners and oilers, a small workshop for maintenance, a maintenance and way department to maintain the tracks, and around 70 motormen and conductors comprised the workforce. DCC Archives.

I

Most of the newly established industries expanded and consolidated during the 'Long Depression' of 1878–95. Indeed, in the 1880s capital flowed into the development of manufacturing, especially the clothing and woollen industries. Young women began looking for work other than domestic service. When prosperity returned, fuelled locally by both refrigeration and the gold-dredging boom, these firms were ready to grab their chance. Others joined them. The Rutherfords established the New Zealand Wax Vesta Company's factory in the old Immigration Barracks in 1895. In 1902, they built a new factory on David Street, near Forbury Corner, and in 1903–04 the Barracks were dismantled. 'Spec' builders grabbed most of the timber. Alfred Morris also opened his dubbin factory in a large brick building on the corner of the Main South Road and David Street. While the Wax Vesta Company succeeded beyond anyone's expectations, having to expand its factory and workforce until it employed over 70 permanent staff, most of them women, Morris's business proved more precarious until he won a gold medal for his dubbin at the New Zealand International Exhibition in Christchurch in 1907. After that success, he obtained the contract to supply the Army. Although we do not know how many people Morris employed, for a few years it was all go.

Several smaller businesses also boomed across this period, helping to make Caversham township a flourishing and prosperous centre. Apart from George Methven's engineering jobbing shop, which he moved from his back yard to a larger site near the harbour around 1902, Frank Wilkinson, a qualified pharmacist who arrived from England in 1889, promptly established a dispensary at the corner of Goodall Street and the Main South Road. Boys with an interest in chemistry often did an apprenticeship in pharmacy. In the days before antibiotics, people suffered a wide variety of

OPPOSITE **Roslyn Mill, 1904.** Ross & Glendining's Roslyn Worsted and Woollen Mill dominated Kaikorai Valley (opposite, back view), across the Mornington ridge from the Flat. In 1904 it employed some 500 'hands', many of them young women, some of whom travelled from the Flat each day (one could walk the distance in about 35 minutes or take the tram and cable car). The train also made it comparatively simple to travel to Mosgiel, where the Mosgiel Woollen pains and ailments, and a pharmacist worth his salt became more popular than the local general practitioner. Wilkinson's success soon saw him buy the entire block between Catherine and Goodall streets and extend his premises. He took on his own son as an apprentice, a common practice at that time, and later bought other pharmacies. The firm was a manufacturing as well as a dispensing pharmacy. Robert Rutherford the grocer – a cousin of Wax Vesta Rutherford – also boomed beyond his wildest expectations and eventually bought all the shops on his block and erected a handsome two-storeyed brick building. Todd & Brown, tailors, moved into the upper floor. In these years, many successful owners lived upstairs from their business, but that was about to change.

It was not just industry that moved to the Flat in search of cheap land. Government played a key role, not least in constructing the Main Trunk Railway and locating the railway workshops at Hillside. Apart from the inevitable school, police station and post office – a handsome new brick one opening on Ruskin Terrace in 1901 – government also took advantage of cheap land to locate the Immigration Barracks, the Benevolent Home and the Industrial School on or overlooking the Flat. The impressive brick fortress known as 'the Benny' (and sometimes 'the poorhouse') dominated lower Caversham, but was run and owned by the Otago Hospital & Charitable Aid Board. On Sundays, the children from the Industrial School were marched down the hill and through the shopping centre to church. The

Mill was based. Both mills had hosiery departments, besides undertaking spinning and weaving, whereas in Britain separate firms usually specialised in these different lines of work. The original Roslyn mill began operations in 1880 and was substantially re-built and expanded in 1902–04. At the same time, the partners also built a large clothing factory in Stafford Street, just west of the Exchange, near their handsome three-storeyed warehouse on Princes Street, and soon began manufacturing hats as well. In 1909, they opened a substantial shoe- and boot-making factory at the western end of Princes Street, near the tramsheds. Apart from employing over 1000 persons in their various local factories and mills, Ross & Glendining also employed some 100 managers, warehousemen, clerks and heads of department in their head office and their seven branches. Photograh by Wrigglesworth and Binns.

Cyclopedia of Otago and Southland, p. 337.

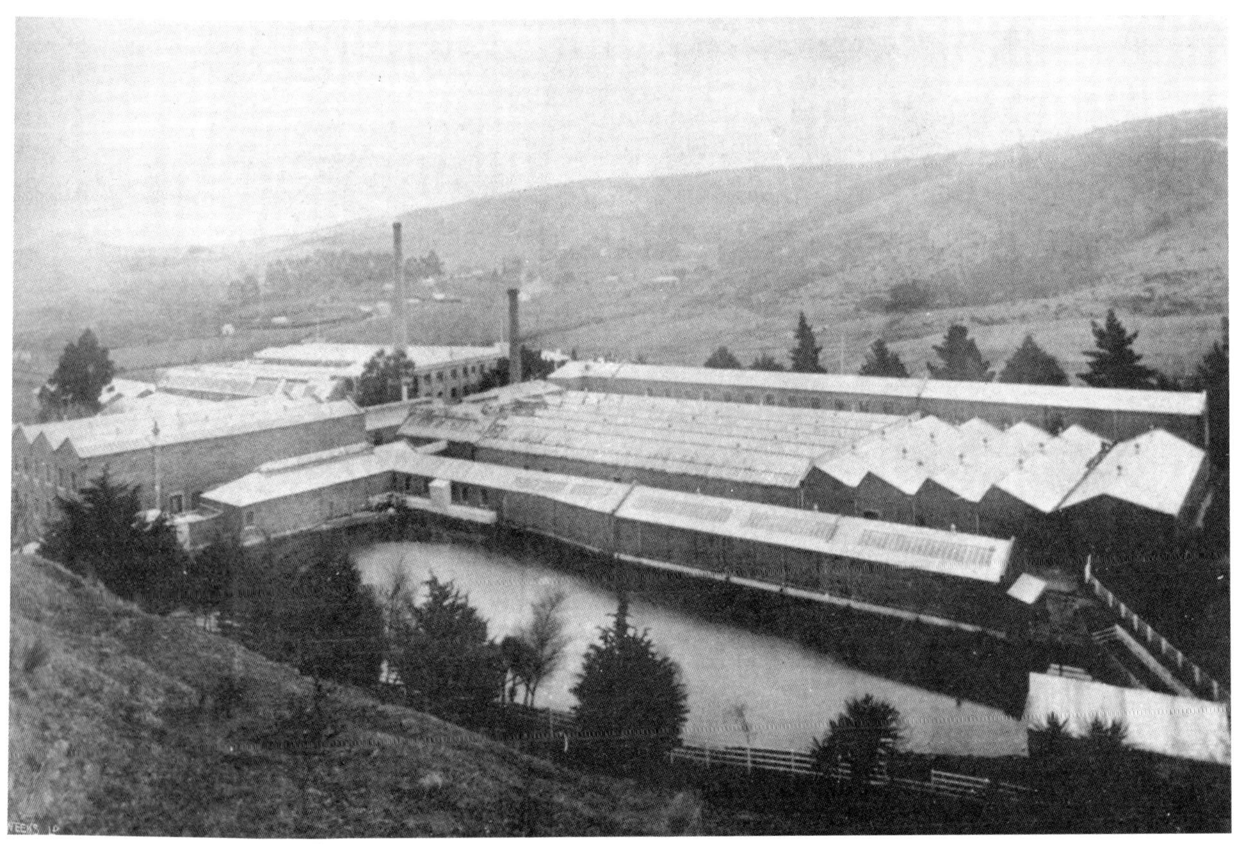

OPPOSITE **Sargood, Son & Ewen** began locally as a warehouse in 1862. During the Long Depression the firm expanded, opening a clothing factory and a bootmaking factory. Like many manufacturers in Victorian Otago, Sargood was especially attentive to the welfare of his workers. Not only was the factory designed according to the best modern standards, especially with regard to hygiene and ventilation, but the Company adopted policies to enhance the wellbeing of their workers. The firm's annual picnic and its Christmas party symbolised this paternal benevolence and familial ethos. These photographs, c. 1898, nicely capture the occasion and the spirit. Under the marquee Sir Frederick Sargood half turns to look at the camera, while Lady Sargood distributes prizes to the triumphant girls. Picnics in this period usually involved a train or ferry trip and invariably focused on a wide range of sporting contests.

Percy and Lucy Sargood – she was an Ormond from Hawkes Bay – lost their only son at Gallipoli, and as a memorial they gave half the cost of buying the building from the 1925 Exhibition at Logan Park for the Dunedin Public Art Gallery. They also gave many works from their private collection to the Dunedin and the National Art Galleries. Percy – made a Knight Bachelor in 1935 – also did good by stealth, supporting many young artists in several fields, as well as the YMCA, YWCA and the Scout movement.

Hocken Collections, Sargood, Son & Ewen Ms; image S12-528d&e.

Benny's residents were not marched to church, as various denominations (including the Salvation Army) regularly held services at the Home. Its extensive gardens and orchards provided employment for able-bodied men, but the old men incarcerated there were often to be seen heading for the Parkside Hotel and returning drunk.

II

Local society boasted two hierarchies: one, centred on large public organisations, was hierarchical, modelled on the army. The master and matron of 'the Benny', like the local headmaster or stationmaster, or the general manager of the Tramways or the Hillside Workshops, enjoyed considerable status in the local community. Then there were the businessmen, a broad category that included men working on their own account, even if only occasionally, and others like Wilkinson and Rutherford whose businesses proved remarkably stable and successful, often for three or four generations. And there were those whose success made them quite substantial employers, like the Shacklocks or the Wax Vesta Rutherfords. Frank Shacklock, one of three brothers who took control of the firm in 1900, lived next door to McCracken's store. Other substantial businessmen, with businesses in the city, lived in the area, such as John Wilson and J.J. Marlow (the latter one of several English Catholics who provided jobs in skilled and white-collar work for their Irish co-religionists). Wilson and Marlow were active in local politics.

The men who seized the business opportunities in this new society were overwhelmingly young, highly literate and numerate, and confident. Almost all of them were also immigrants. Many were fond of the contemporary saying: 'History in the past has enrolled the names of warriors. History in the future will perpetuate the names of the great leaders of industry and of social harmony.' The successful manufacturers saw themselves as the architects of a dynamic new society, a new industrial nation based on 'social harmony' rather than Britain's vast inequalities, a new society that would one day eclipse Britain as a workshop to the Pacific and the southern hemisphere more generally, if not the world. Dunedin's youthful industrialists, overwhelmingly Scottish by birth and well educated by the standards of the time, had a vision of Dunedin as New Zealand's industrial capital. The city was already the commercial and financial hub of the entire colony. In the buoyant 1900s, the engineering firms all established national markets and opened branches elsewhere; so, too, did Ross & Glendining, Hallensteins, and Sargood, Son & Ewen. Donaghy's and Methven's also expanded north. All achieved a national presence and some national dominance.

These entrepreneurs* were not only self-consciously the leaders of humanity but innovative and forward-looking: in a word, 'progressive'. Even though in most cases they had only recently left Britain, they regularly returned to the northern hemisphere to visit the leading firms in England and Germany, especially, and they usually made sure they visited the United States en route, to see for themselves the latest innovations and

* According to the *Oxford English Dictionary*, around 1885 'entrepreneur' came to mean a contractor acting as an intermediary between Capital and Labour. Our meaning derives from J.A. Schumpeter, *The Theory of Economic Development*, Harvard University Press, 1934, pp. 62–94.

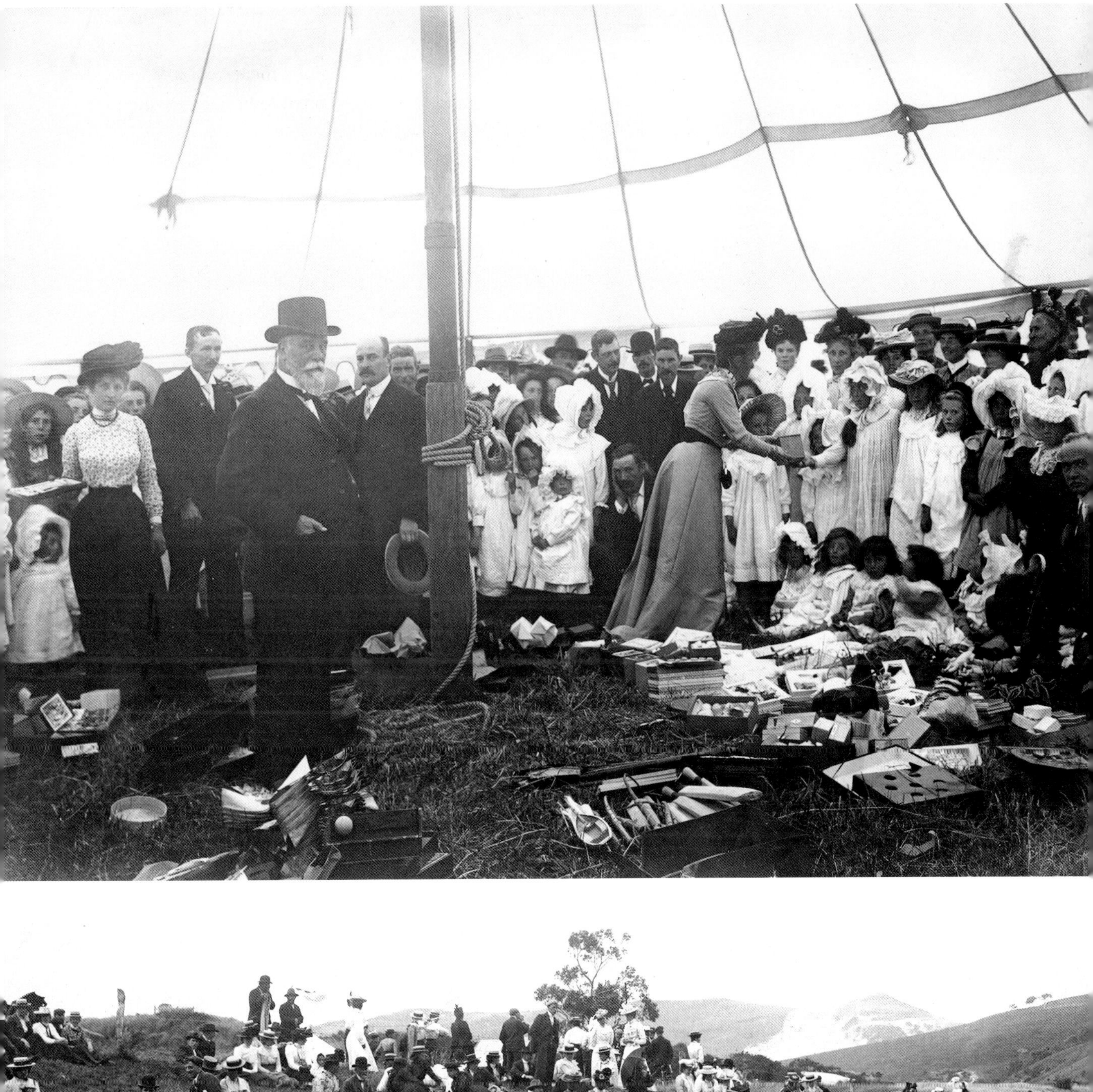

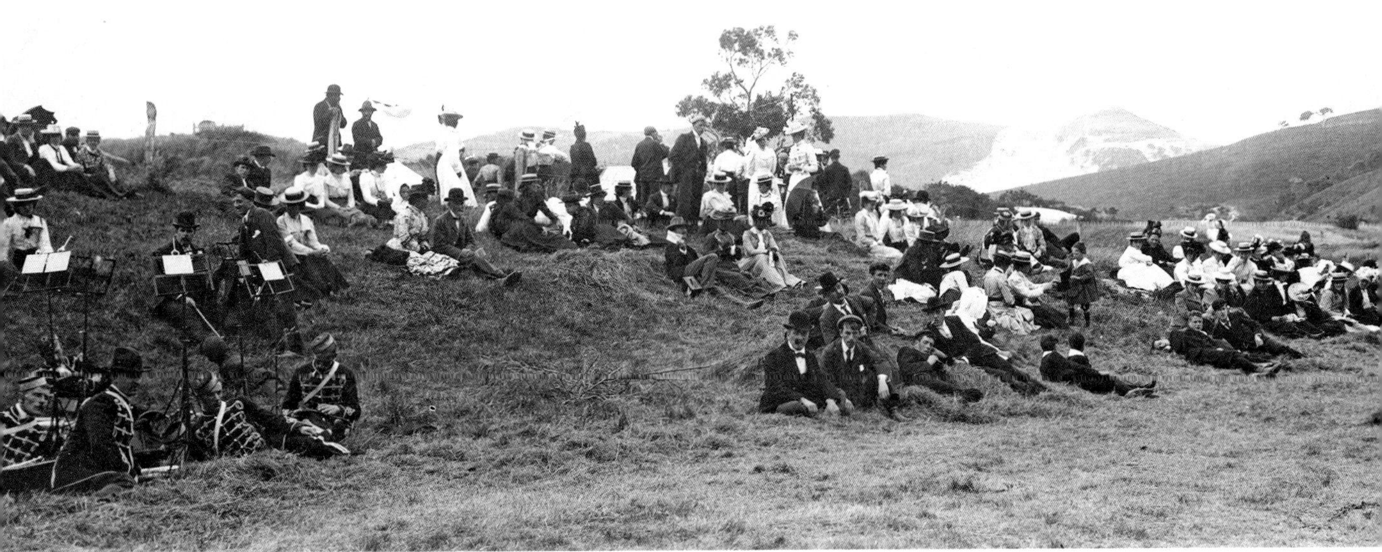

technologies. The largest of these firms, Ross & Glendining, kept one of the partners, Ross, in London permanently from 1870 until 1902, when he returned to New Zealand for the sake of his health. In the *Cyclopedia of New Zealand*, many contributors devoted much of their entry to accounts of their new machines, some even lovingly describing pieces of equipment and machinery as if they were stud breeds!

The British, of course, had become conscious of the momentous significance of the Industrial Revolution, which they named shortly after it was completed, so it is not surprising that immigrants from Britain's industrial districts would have been highly alert to the birth of what many thought might be an equally portentous industrial revolution here. They were wrong, as it turned out. The second industrial revolution, based on petroleum and plastics, was already underway in Germany and the United States; moreover, New Zealand's Liberal government (1891–1912) did not share this vision, preferring to develop instead the country's primary industries, especially dairying.

Many of Otago's manufacturers caught the spirit of Robert Owen, the famous Scottish mill-owner and socialist who made his mill and town, New Lanark, a model for all humanity. These men did not view their workers as mere 'hands'. Everyone followed Hallenstein's example, ensuring that their factories were well lit, well ventilated, and had well-appointed cloak rooms, comfortable dining rooms, and clean toilet arrangements. Ross & Glendining went further at the Roslyn Mill, by planting extensive gardens and making an amenity of Kaikorai Stream. Local employers also took a wider interest in the wellbeing of their 'hands'. In 1891, or thereabouts, Ross & Glendining established a savings bank to encourage thrift. They also established a benefit society to provide insurance against injury or illness. However, most of the Flat's working men joined one of four large friendly societies that offered insurance against loss of earnings due to illness or accident.

Nor were these manufacturers opposed to unions. The spirit of Robert Owen certainly worked on Robert Glendining, who oversaw the Dunedin end of the firm's activities from 1870 until 1902, and he had no difficulty in allowing the firm's 'hands' to join the appropriate trades union. It is not clear whose influence shaped Bendix Hallenstein's views, but he famously set the local scene by accepting the principal recommendations of the so-called Sweating Commission, including that a union be formed to better protect those who worked in the clothing industry. As one of those who paid the best wages and established the best conditions for his workers, he realised that he had nothing to fear from a well-run union helping to ensure that all workshops complied with certain minimum conditions. In the engineering trades, of course, the masters had generally served apprenticeships, worked as journeymen and joined their union, the Amalgamated Society of Engineers. This was a benefit society, among other things, and several retained its insurance long after becoming masters. It was the same in the building industry. Although after 1890 voting alignments and political loyalties in Dunedin were based on class, the culture of craft structured the organisation of work. John A. Millar, ex-leader of the Seamen's Union and now one of the three Members of the House of Representatives (MHRs) for Dunedin City, inaugurated Labour Day parades in 1890 to commemorate the founding of the Maritime Council. And on Labour Day, which Seddon made into a national holiday in 1898, unionised workingmen proudly paraded as members of their crafts.

These firms also became the centre for a lively social life. Most established libraries for their workers. Almost all fielded sports teams. Whereas the owners usually played a large part in setting up and running the libraries, which some saw as a means of fostering thrift as well as self improvement, few took part in the sports. Whether the owner or the staff took the initiative remains unclear, but Wolfenden & Russell fielded a cricket team, Rutherfords' Wax Vesta female employees had a marching team (between the wars), most of the engineering shops fielded both cricket and rugby teams and the principal workshops competed against each other. The Railway Workshops, on the other hand, usually only played other railway workshops in the region, the annual battle with Invercargill being the highlight of the year. Several of the firms organised floats to take part in civic processions, as when the Duke and Duchess of Cornwall visited in 1901. During the Boer War, the men at Cossens & Black organised a rifle club.

John Sutherland Ross, born in the hamlet of Girston in 1834, his parents refugees from the Highland Clearances, was apprenticed as a draper, worked in commerce, joined the Free Church and migrated to Dunedin in 1861. The following year he formed his famous partnership with Robert Glendining. After they went into manufacturing in 1878, initially in a modest way, Ross based himself in London and learned, among many other things, how to obtain the latest machinery. Over the next 22 years he became really expert. In 1903, after returning to Dunedin for reasons of health, he took off again to visit the leading manufacturers in the United States, England, Germany and Italy in order to equip the new mill with the most modern machinery. His account, published in the *Cyclopedia*, is as lyrical as the prosaic Ross ever became.

Every wheel in the new building is driven by new machinery. First to claim attention is a horizontal compound condensing steam engine, with positive Curliss gear, made by Messrs E.R. and F. Turner (Limited), Ipswich, England. The fly-wheel (9 feet 6 inches in diameter), which has seven grooves for 1½ inch ropes, connects with the main driving wheel on the first floor. In addition to this driving wheel there is another grooved wheel, 4 feet 6 inches in diameter, which transmits the power to a shaft 3½ inches in diameter, running the whole length of the building. The following figures in connection with the engine will prove interesting: Economical horsepower, 145; maximum horsepower, 200; diameter, high-pressure cylinder, 11 inches; diameter, low-pressure cylinder, 19 inches; stroke of engine, 28 inches; working pressure of steam, 120. The engine has a shaft governor for regulating the speed. The weight of the engine is 13½ tons. A Hornsby water-tube boiler has been placed in position. It has eighty tubes, each 17 feet long … From what has been said regarding the engine … anyone may judge that the company has secured the best that can be obtained for its requirements.

From *Cyclopedia of New Zealand*, pp. 341–42

III

The dramatic growth of industries in the late nineteenth century made Dunedin unique in New Zealand. In particular, the concentration of engineering workshops, textile and clothing factories, and brick and pottery manufacturers, not to mention the gasworks, the railways and the city's wharves, gave the area between the Hillside Workshops and the Exchange a distinct industrial character. Between 1895 and 1914, most of the larger firms also achieved national dominance in their principal lines of work, despite the absence of tariff protection. And those big firms also sustained many small ones. Nor did the masters and capitalists who drove industrialisation, or even their sons who took over from them, follow the example of so many English capitalists, who made good by investing their profits in land. These men – mostly Scots – took pride in their craft skills, got pleasure even in old age from supervising the education of their apprentices, and wanted no more than to produce excellent work and let the chips fall where they might. As it was, most of these firms, as we noted, achieved dominance within the New Zealand market. Many did better and exported to Australia, the Pacific Islands, and even to Britain. The engineering workshops that specialised in manufacturing gold-dredges briefly achieved global dominance. The vision of Dunedin as the Birmingham or Manchester of this 'Britain of the South Seas', as New Zealand was still widely known, was unchallenged.

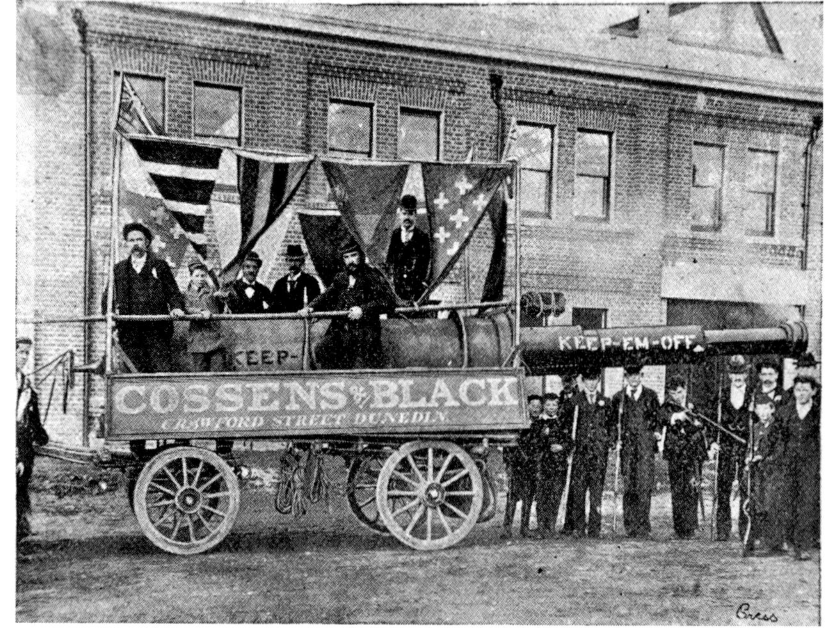

The Boer War (1899–1902) overlapped with the dredging boom. These employees of Cossens & Black in 1900 delighted to organise a float for the Boer War Procession. The company would have paid for the charabanc and given the men time off. The notes beside the photograph commented, 'Good Men Difficult to Get', '10 Dredges Built'. The photograph is by H. Wimpenny.

Hocken Collections, Cossens & Black Mss, Ms. 2989/004, S14-537.

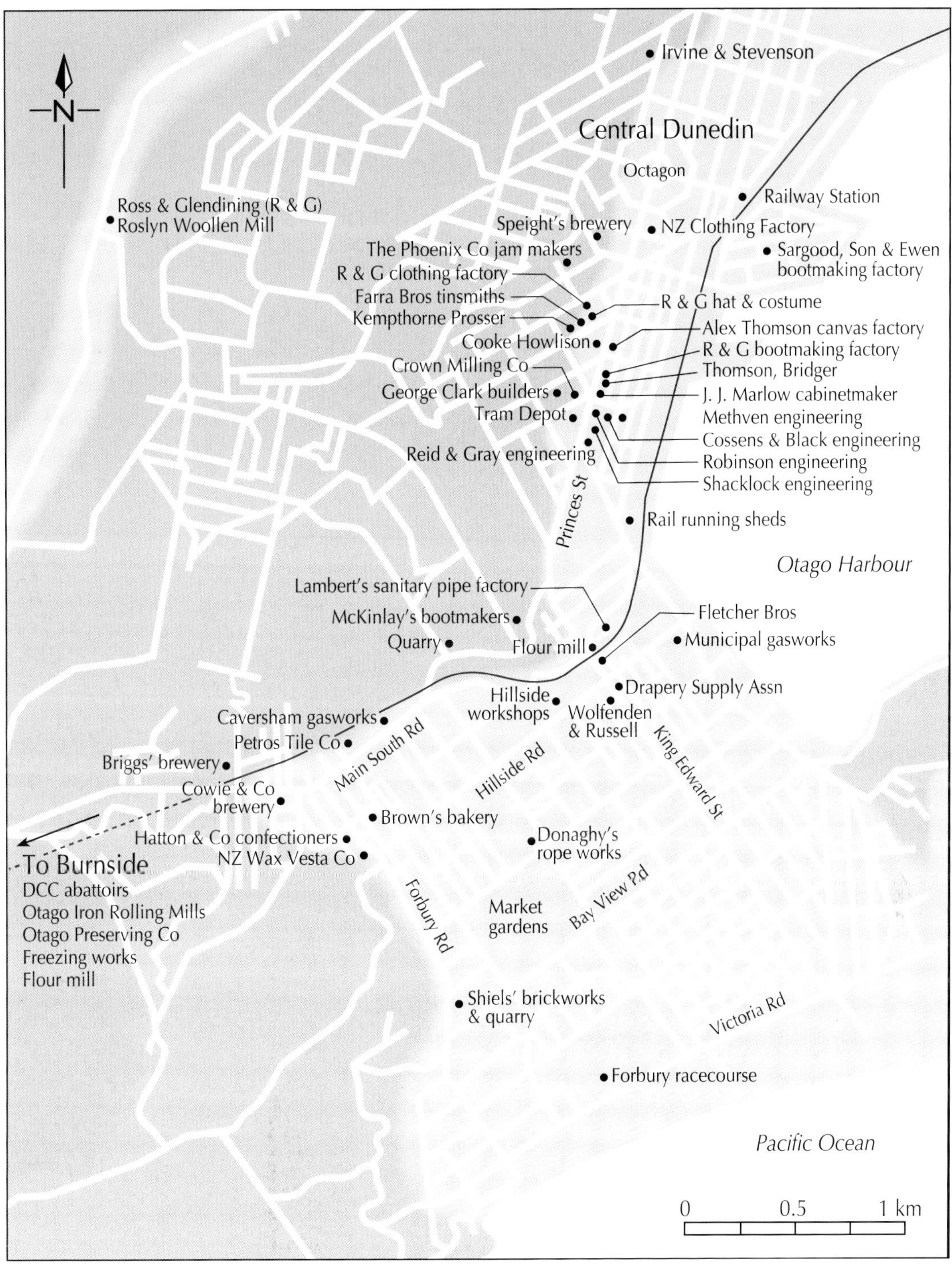
N
Irvine & Stevenson
Central Dunedin
Octagon
Railway Station
Ross & Glendining (R & G)
Roslyn Woollen Mill
Speight's brewery
NZ Clothing Factory
The Phoenix Co jam makers
Sargood, Son & Ewen
bootmaking factory
R & G clothing factory
Farra Bros tinsmiths
R & G hat & costume
Kempthorne Prosser
Cooke Howlison
Alex Thomson canvas factory
R & G bootmaking factory
Crown Milling Co
Thomson, Bridger
George Clark builders
J. J. Marlow cabinetmaker
Tram Depot
Methven engineering
Cossens & Black engineering
Reid & Gray engineering
Robinson engineering
Shacklock engineering
Princes St
Rail running sheds
Otago Harbour
Lambert's sanitary pipe factory
McKinlay's bootmakers
Fletcher Bros
Quarry
Flour mill
Municipal gasworks
Drapery Supply Assn
Hillside
workshops
Wolfenden
& Russell
Caversham gasworks
Petros Tile Co
Main South Rd
Hillside Rd
King Edward St
Briggs' brewery
Cowie & Co
brewery
Brown's bakery
Hatton & Co confectioners
Donaghy's
rope works
NZ Wax Vesta Co
To Burnside
DCC abattoirs
Otago Iron Rolling Mills
Otago Preserving Co
Freezing works
Flour mill
Forbury Rd
Market
gardens
Bay View Rd
Shiels' brickworks
& quarry
Victoria Rd
Forbury racecourse
Pacific Ocean
0
0.5
1 km

CHAPTER 3

Crafts, Jobs and Professions

WORK, IN OSCAR WILDE'S FAMOUS QUIP, may have been the ruin of Britain's drinking classes, but in the colony all classes drank and worked. As a growing army of Prohibitionists and Temperance supporters argued, increasingly with a battery of statistics culled from all over the drinking world, in New Zealand drink was too often the ruin of the working classes. Extreme Prohibitionists were inclined to claim that every social ill had its origin in alcohol, that peculiarly toxic food. Be that as it may, as the new century dawned, support for Temperance and even Prohibition grew.

In the Protestant world, especially, work had ceased to be merely one of the three unfortunate consequences of original sin. That devout Catholic from Ireland, Edmond Slattery, known as 'The Shiner', and a legend in his own lifetime, may have scorned work, but the great majority believed it morally worthwhile, a source of self worth and self discipline. This was especially true among those with schooling, which now included most working men and women. Literacy and numeracy, like craft skill, were esteemed in themselves and not just because they might command a premium in the marketplace. Few esteemed book learning above skill, however. Even Charles N. Baeyertz, whose monthly *Triad* preached cultural enlightenment to a large audience, affirmed the unity of head and hand. It is central, of course, to the performance of music, one of Baeyertz's passions, and one of the major leisure activities on the Flat, as in Dunedin. Those wonderful Romantics John Ruskin and William Morris recognised the unity of head and hand and made it central to their defence of manual labour. Together with David Ricardo's labour theory of value, picked up and elaborated as an indictment of capitalism by Karl Marx, the defence of manual labour provided the working people who migrated to New Zealand

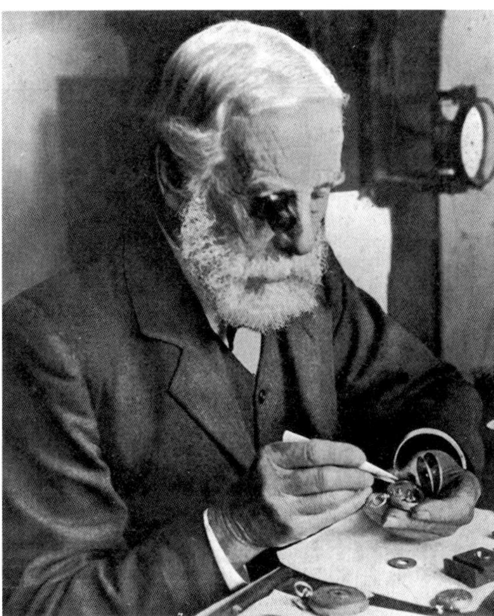

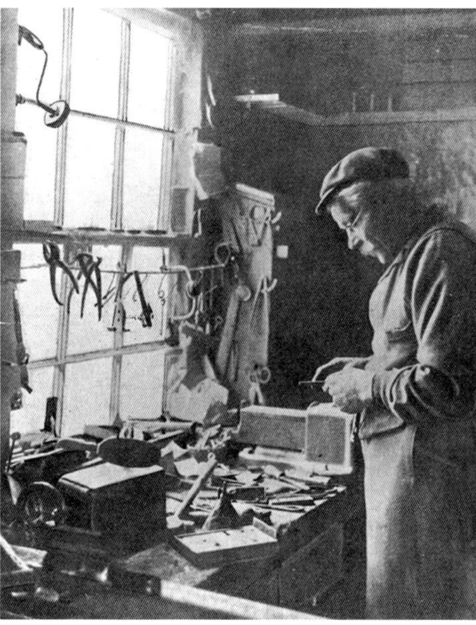

Some traditional crafts. The *Auckland Weekly News* published these photographs in August 1913, as part of a nostalgic look at historic survivals. In fact, each of these crafts was still flourishing, defying Sydney and Beatrice Webb, those famous English Fabians, who sneered that the survival of such an antiquated system showed New Zealand was lagging behind, rather than leading, the 'Mother Country'.

All skilled trades were still organised in the following way. After completing Standard 4 and increasingly Standard 6 at school, a lad or (less often) a lass was articled to a master of the trade who undertook to teach the youth the craft. The apprentice usually began by doing the simplest tasks in the workshop and was slowly brought on to more specialised tasks. Working only five hours a day over the seven years of the apprenticeship, the lad or lass would hone the relevant skills for at least 10,000 hours. Pay increased with experience. In New Zealand, unlike Britain, the apprentice rarely became part of the master's household and it was uncommon for a master to sue if the apprentice left.

Once out of their time, the apprentice was a journeyman or woman. Many young males still journeyed or tramped, often back to the 'Old Country', but most worked at their trade in one place. The key skills across all crafts were knowing how to do a job, what tools or other resources would be needed, and how long it would take. After some years, the journeyman or woman would often set up on their own account, either working from home or a workshop. In all trades, a master needed a wife to manage his household and his accounts. Often, she also ran the shop. (The sons and daughters of small employers were much more likely to choose a spouse from their own stratum than any other social group.) The master might also take on an apprentice and even hire one or more journeymen or women. Women rarely went out on their own account, unless widowed, usually forming a relationship with sisters or a close relation.

Successful masters had to be capitalists, their own buying and selling agent, technical expert, head of office, personnel manager, and even legal adviser! If they needed credit, they usually dealt not with a bank but with family, a friend, or a friendly society. In New Zealand's extensive handicraft sector, where there were usually as many masters as there were journeymen, this mode of production remained normal, as it still does in the decentralised building trades. The professions were organised on much the same lines, although they increasingly required some university qualification.

ABOVE LEFT **The watchmaker** represents the men of the handicraft trades where the progression from apprenticeship to journeyman to master working on his own account usually occurred as one aged. The master might take on an apprentice, but did not need one. His dress indicates his sense of status.

Hocken Collecions, Caversham Project, Ms-2690-199.

ABOVE **The tinsmith.** Many tinsmiths worked for New Zealand Railways at Hillside, while many others were employed by the gasworks, tramways and the engineering workshops to the east of the Oval. Others worked in smaller workshops such as Farra Brothers in Stafford Street, near the Exchange, or Hordern & White's, the coachmaker in Market Street, near the tramsheds. The trade was organised as a handicraft but opportunities for self-employment had shrunk as the industry became more capital intensive. However, many tinsmiths became plumbers or locksmiths, or specialised in making and repairing bicycles.

Hocken Collections, Caversham Project, Ms-2690-199.

OPPOSITE, TOP *The Carpenter at Work*, by David Con Hutton, chalk on paper, 1881. For centuries, carpenters worked with hand-held tools such as saws, hammers, chisels, augers, straight and moulding planes. Each tool had a number of specialised variations. The carpenter's tool chest was often his most prized possession. From the 1870s, an increasing proportion of all joinery was manufactured and sold pre-fabricated. D.C. Hutton's drawing of a carpenter captures nicely one of the unchanging jobs of the 'chippy', particularly important in a country where most houses were entirely made of wood.

Hocken Collections, Accession: A36 a

OPPOSITE **Fletcher Bros first joinery factory, near Lambert's in Kensington.** Prefabricated doors, door frames, sashes, architraves and stairs had largely captured the market in the 1870s and 1880s. All the builder had to do in building a house was construct the foundations and the frame. The extent to which local joiners bought the latest machines to manufacture joinery is unknown, but the youthful James Fletcher intended to capture the market in Otago and Southland, if not further afield. His factory had all the latest equipment, some of it manufactured by Cossens & Black. Here we see men in the planing department, where the timber was planed and dressed. Nothing if not innovative, Fletcher commissioned a series of photographs to advertise his firm, c. 1910.

Courtesy Fletcher Challenge Archives.

with a powerful philosophical sense of their own value. Colonisation made that value even more apparent.

The importance of skill and labour to building a new society was widely understood; the gulf that separated skill from labour in Britain simply never existed here. Indeed, the colonists and their children recognised the importance of the head to unskilled work and the importance of the hand to skilled work. The pre-eminent Victorian prophets of craftsmanship, Ruskin and Morris, delighted in the handmade over the machine-made. In Dunedin, David Con Hutton carried the banners in defence of craftsmanship. He understood the unity of head and hand and the intimate connection between problem finding and problem solving. A native of Dundee, Hutton graduated from England's famous National Art Training School in South Kensington, and in 1869 was appointed to teach drawing in Otago's schools. Before long, he set up the Dunedin School of Art, which affiliated in 1894 to South Kensington, and taught prospective artisans practical geometry, mechanical and architectural drawing, and skills essential to laying out work in the building and engineering trades. As the country's outstanding Pre-Raphaelite artist, he also taught freehand drawing and design, increasingly to young women. The success of the Technical Classes Association, established in 1888, owed much to his pioneering work and his partnership with Alexander Burt, the engineer, who chaired the Association for many years.

I

In this fledgling New World society, still largely confined by local product and local labour markets, small-scale businesses dominated the economy. Most of these businesses were owned and run by an artisan or mechanic working on his own, or less often her own, account, who would have their own workshop and be known as a 'master'. Male masters were invariably married. Every master had served an apprenticeship and then worked for some years as a journeyman or journeywoman (see caption, page 67). Skill legitimised authority. Craft skill had been acquired by slow learning and habit, although by 1900, thanks to Hutton's pioneering work, an increasing number recognised the potential value of technical education. From 1905 onwards, the state provided technical education and from 1908 it was free. Dunedin's King Edward Technical College opened in 1908.

In their workshops, masters also undertook responsibility for training their apprentices. Most masters employed at least one apprentice and sometimes as many as three or four journeymen, as well as (depending on the trade) one or two unskilled men. Each skilled man usually required roughly one unskilled to help: such unskilled men were usually known as 'skilled unskilled'. In trades now practised in factories, like bootmaking or engineering, the elite craft usually monopolised the training of apprentices. In bootmaking, for instance, this led to an over-supply of clickers and benchers in the 1880s and early 1890s and an under-supply of finishers. By contrast, in banks and insurance companies, stock and station companies and even warehouses, men who had risen through the organisation tended to end up running it. Whatever their origins or line of work, however, all believed in the importance of work both for the individual and for the society.

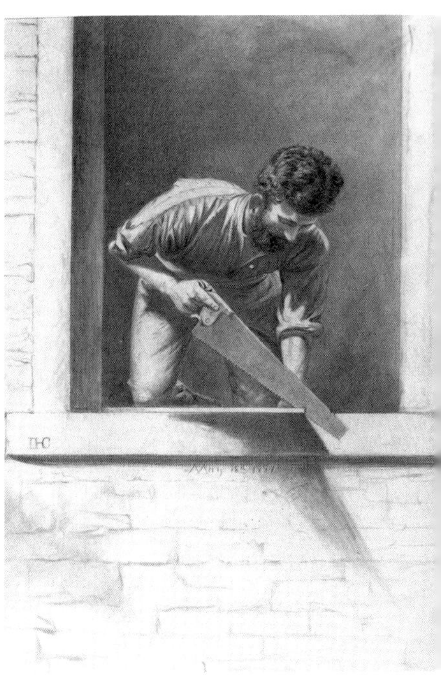

II

The nature of work and its meaning had not changed much with emigration to the colony. When you hired someone in colonial New Zealand to do a job, you assumed they knew how to do it, what tools they would need and how long it would take. In short, you assumed their authority and in return granted them considerable autonomy.

What workers expected had changed and changed dramatically, however. Not only had the four eights become normative – eight hours to work, eight to sleep, eight to play and eight shillings a day – but everyone expected to be able to find work. Almost all Pākehā also assumed that an able-bodied man would be able to earn enough to support his wife and children – which explains why the Long Depression (1878–95) proved so traumatic (as the Great Depression would a generation later). Even in the colonial period, New Zealanders were so touchy about unemployment that full employment became a major goal of public policy. Each winter, when seasonal

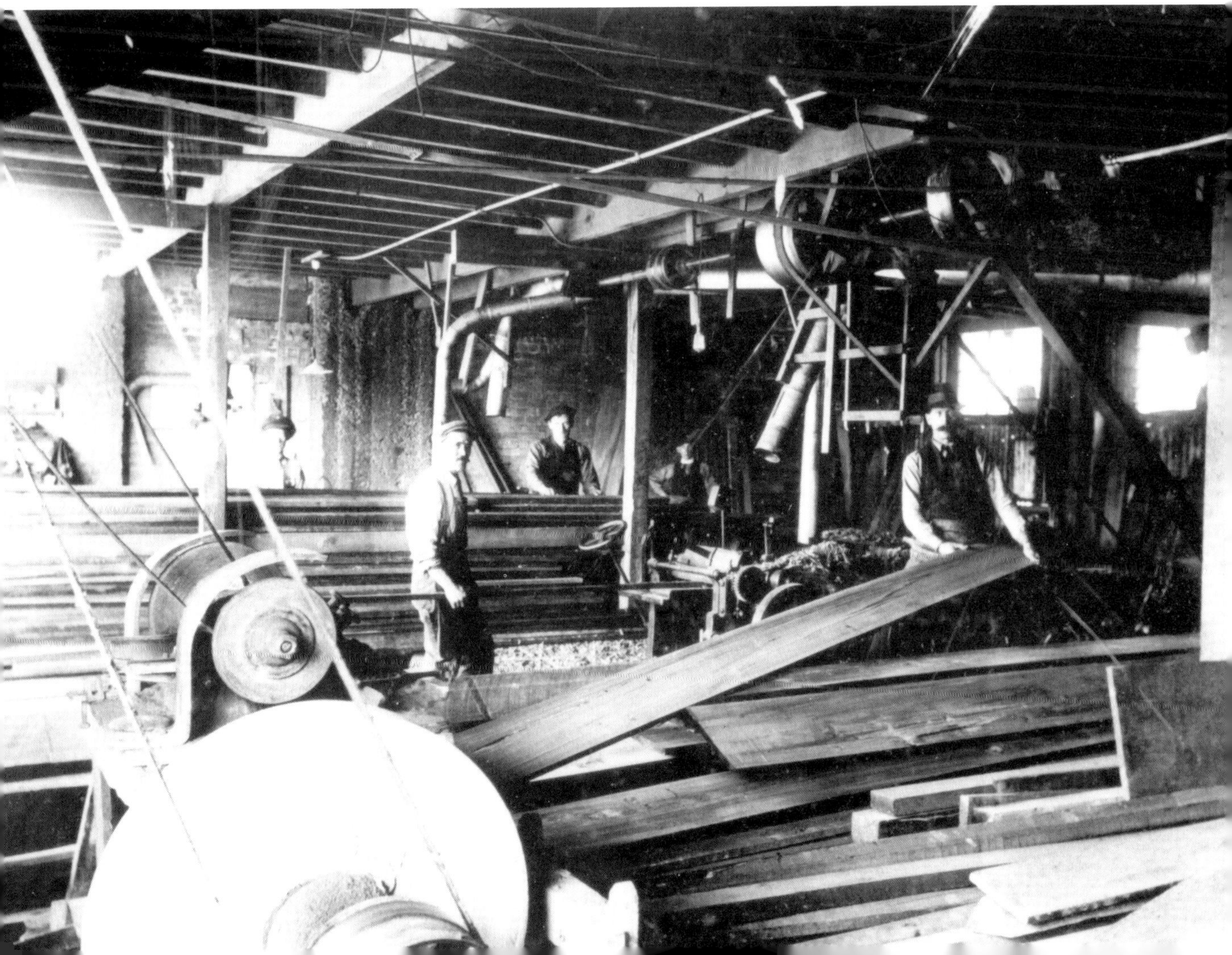

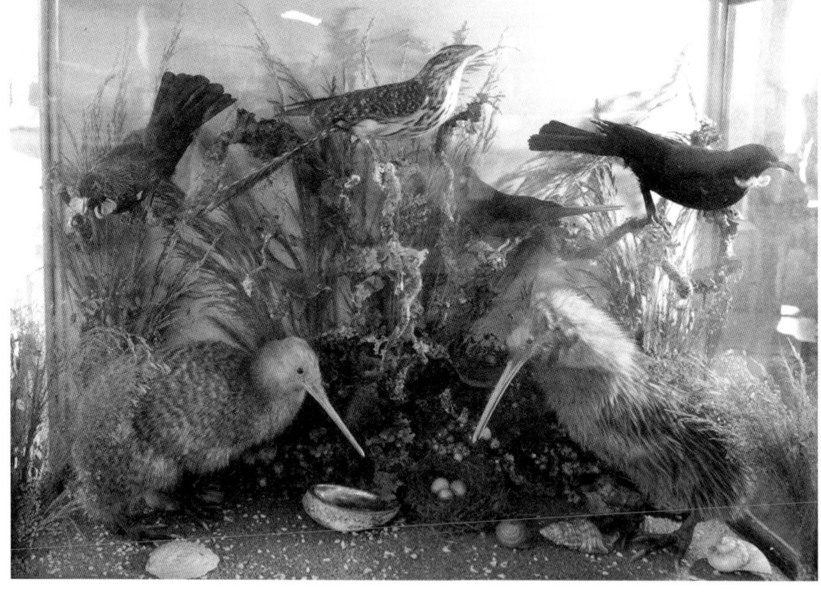

RIGHT **The taxidermist.** We have no photograph of taxidermist William Smyth or his workshop, which was tucked out of sight behind his home on the Main South Road (at the corner of Alexandra Street). He appears to have settled in Caversham late in the 1870s and took every opportunity to exhibit his stuffed birds at the industrial exhibitions of that decade. It seems he was also a Mason and a member of the Caledonian Bowling Club.

This photograph of a case of native birds represents his craft. A Methodist from Ireland, hard on himself and those around him, he migrated at the age of 34 with his wife and two boys. She left him and the boys in 1889. Preparing, stuffing or mounting the skins of animals and birds required not only obsessional attention to detail but knowledge of anatomy, sculpture, painting, tanning and, if you made your own cases, joinery.

Taxidermy was one of the few crafts left in which there could be no resiling from the pursuit of excellence. Nor were there any machines that simplified complex tasks, let alone took over menial ones. Technique and expression were inextricably one. Technical ingenuity and craftsmanship, the combination of manual and intellectual skill and knowledge, were essential. One imagines that Smyth's workshop was spotless and tidy. The skilled were famous for their general tidiness – a place for everything and everything in its place, as Samuel Smiles said.

Courtesy the *Otago Daily Times*.

unemployment was usually at its worst, deputations of the unemployed would march on the town hall and demand that the local body concerned do something. Such marches occured in the boroughs of Caversham and South Dunedin as well as in Dunedin City. Many councils tried to provide people with employment on public works, such as road building.

In some fields, the dramatic growth in knowledge changed the nature of work and what it meant. The growing knowledge of geology proved essential to the mining industry, just as advances in physics revolutionised engineering. The men of the metal trades – trades being revolutionised during the nineteenth century – often took an intense interest in physics and mathematics. Metal workers from Scotland and England knew how to make the six basic machines of classical antiquity and understood how they worked: the lever, pulley, screw, balance, wedge and wheel. By the 1860s, they also knew how to make and operate shapers, planers and lathes. Many understood the relationship between metallurgy, mechanics and other branches of knowledge. By migrating, however, they moved into a society that self-consciously looked as much to the United States as to Britain and thus they quickly became familiar with revolutionary new American techniques. The American system of interchangeable parts, which used small machines designed for one job, had finally produced the milling cutter, which, among other things, allowed high-speed machining without overheating. Ironically, in migrating from the centre of the Empire they became free to borrow from the whole world of engineering. An unknown number, of course, had migrated via America.

Knowledge of science and technical inventiveness were highly regarded. Arthur Beverly, the famous astronomer and clockmaker, was already a local legend because of his technical virtuosity and inventiveness. A few emulated Beverly and built workshops at home. When George Methven

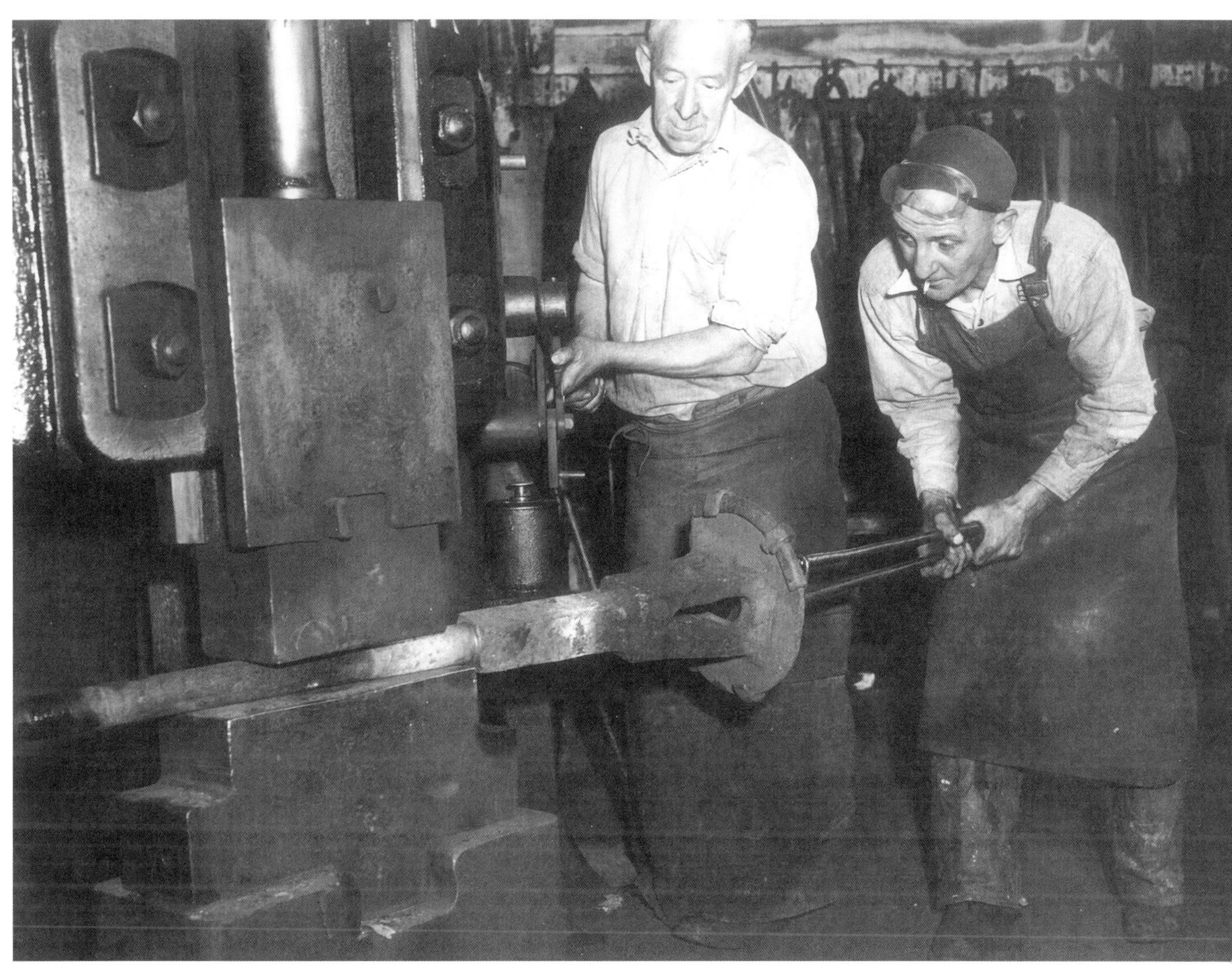

moved to larger premises in the city, he used his workshop in the back yard to design and build his own motor car. In the Railway Workshops, men also experimented with the different tools and the metals needed to make them. Joseph Mellor, like Beverly, was another local legend. He studied at the Technical School, earning his keep by working in various boot factories, including those of McKinlay and Sargood, and became fascinated by chemistry. In particular, ceramics interested him. It is hard to imagine that he did not spend time down at Lambert's, but after winning an Exhibition Scholarship in 1899 he proceeded to England, where he earned further distinction and became chemist to the Pottery Manufacturers' Federation and in 1905 director of the Federation's research laboratories. Over the next 20 years, he became the world's leading authority on inorganic chemistry and a world authority on ceramics. Around 1900, of course, some in the engineering trades were experimenting with electricity, telegraphy and radiotelegraphy.

ABOVE **A tradesman blacksmith and his striker** using his pneumatic hammer to forge a butt weld on a drawcar coupling. Men would down tools to watch outstanding blacksmiths and strikers at work, such was the brilliance of the choreography. It seems likely that physical coordination at this level prompted men and women to recognise the potential advantages of social cooperation. Mutualism and fraternity, in short, were fundamental to the culture of skill in many trades. Photographed by J.G. Duncan and J.F. Le Cren, 1952.

Archives NZ/Te Rua Mahara o te Kāwanatanga, Wellington, AAVK W3493, b3388.

OPPOSITE **The potter's craft**: making jam pots using a mould on a potter's wheel, Milton Pottery. The pots were then dried and turned smooth on a lathe. The third image shows the mould being removed from a more elaborate item, after air drying, before being glazed and fired.

There were two potteries on the Flat; a photograph of one of them survives, but there is no photograph of anyone working in either of them. Ironically, we have the fullest account of the work involved, written by Reg Lambert in his old age (next page).

Otago Witness, 25 September 1901, p. 40.

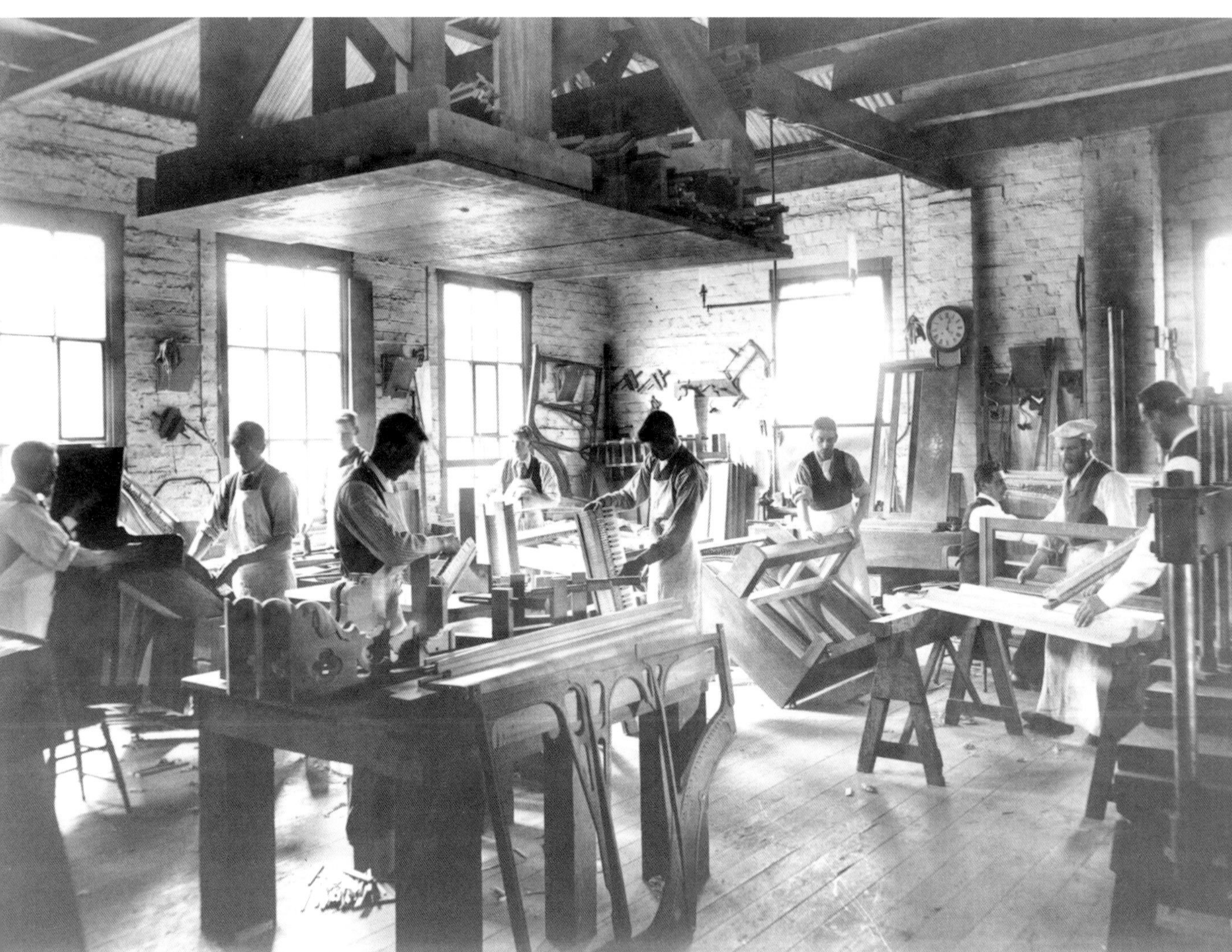

ABOVE **Howell's Piano Factory.** The dominance of local product markets meant that men and women in small skilled trades, such as piano makers (seen at work here), constituted a substantial proportion of the urban workforce. Although most pianos were imported, Dunedin boasted the country's three largest firms: Frederick Howell, Charles Begg & Co and the Dresden Pianoforte Manufacturing & Agency Co. In 1905, the London Organ and Piano Company was also operating. Begg's did not persist with manufacturing for long, although they maintained an extensive workshop in the cellar beneath their large showroom, for repairing and polishing. Dresden, by contrast, continued making both pianos and organs as well as importing a wide range of these and other instruments. Howells, by contrast, simply manufactured pianos. Such workshops were still much as they had been for the past 200 years.

Hocken Collections, c/n E6820/31, S09-350a.

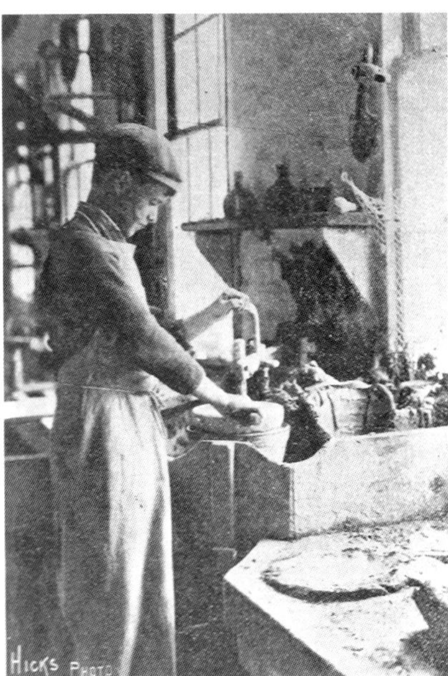

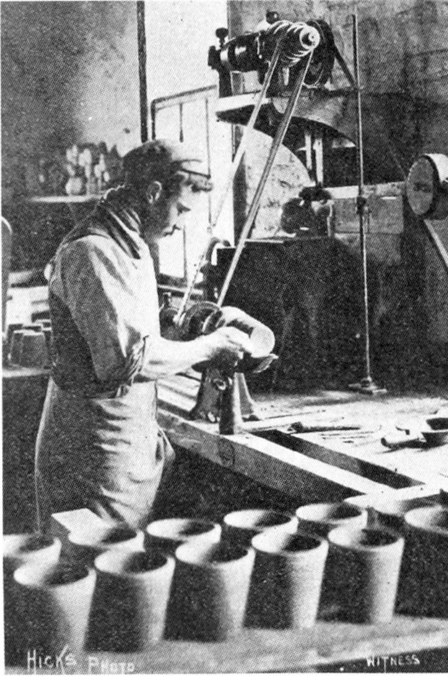

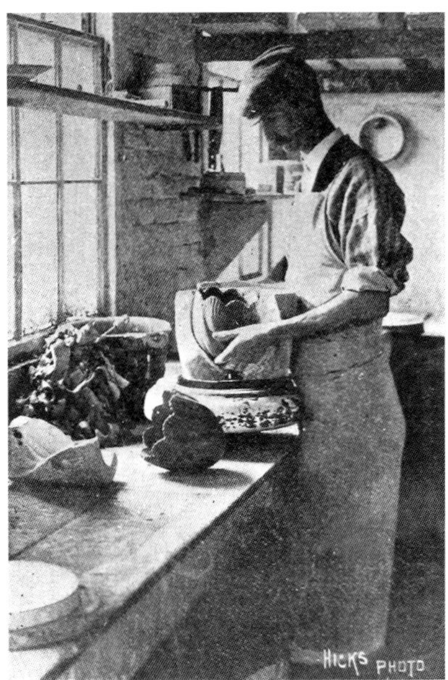

Humans have been making pottery for 30,000 years, glazing and colouring it for 3000 years. Most evidence suggests that potters invented the wheel. The clay had to be found, dug, and brought to the pottery, where it was watered, chopped and kneaded in a pug mill. 'The floor of the pug was always covered with water & slips were numerous. You would be carting a lump of clay and suddenly you would find yourself on your back in a mixture of clay and water.' The high water table in Kensington made this problem worse and rapidly rotted any floor joists touching the ground. Once the clay was the right consistency it had to be rolled, cut, flanged, wedged, dried, trimmed and then placed in the kiln. In the 1900s, puggers, wedgers, moulders, flangers, and setters were paid minimum award wages, towards the bottom of the range for skilled men. We do not know what the actual rates of pay were. In Dunedin, most skilled men making things from clay also had to work as labourers or go on to short-time.

By the 1880s, the potter's wheel remained much as it had always been, although it was now powered by a 10-horsepower steam engine fuelled by coal. For many items, however, the traditional wheel was used. There were several specialist kilns, including a downdraft kiln. The potter needed to be something of a chemist and a pyromaniac, for the kiln re-enacted the stupendous metamorphic processes that formed planet earth. As in ancient Greece, the kiln was still a chemistry laboratory.

While the mass production of bricks and tiles became normal, especially in larger brickfields, the potter's craft remained central to the production of jars, pipes, traps, chimney pots, limestone filters and butter churns, for instance. Production runs were much smaller for these items compared to those for sanitary pipes; small production runs helped to maintain the autonomy and authority of the craftsman. The need to identify which clays best suited which products, and to know how to achieve the optimal glaze and to manage the kiln to ensure a high-quality outcome, meant that the old-time potter remained central to the production process long after he had disappeared from brick-making. Master potters, like John Lambert and his son Reg, understood every aspect of the process.

Reg Lambert – one of the Lambert's founder's three sons who took over around 1900 – became acutely conscious of those skills as new technologies rendered them either redundant or less important in the 1910s and 1920s. In the 1950s he recalled some of the old methods, including his training. As a teenager – when he was paid 7/6d per week for six 12-hour days – he had learned to 'throw, engine drive, flange set, draw, repair & build kilns, had to put in several kiln crowns, no bricklayer in Dn knew how to do this … Of course at that time there were no text books regarding these matters and each individual had his own ideas & some I might tell you were pretty crude … Burning was a great art. The kiln would be lit on Monday, banked up Mon night. [Until Reg was around 16 or 17 his Dad did the burning but then] I took on, on Tuesday morning & fired all that day all that night until 12 o'clock on Wed & if not playing rugby … in the afternoon stayed on until 5 o'clock. … I usually took on Thursday night & stayed to the finish … It was far too much. It was hard work … Fancy now taking over 100 hours per kiln.' There were aspects of his training that he could not bear to write about. 'I would rather forget about some of the past …'

Hocken Collections, McSkimming Ms-90-059, AG-246-J.

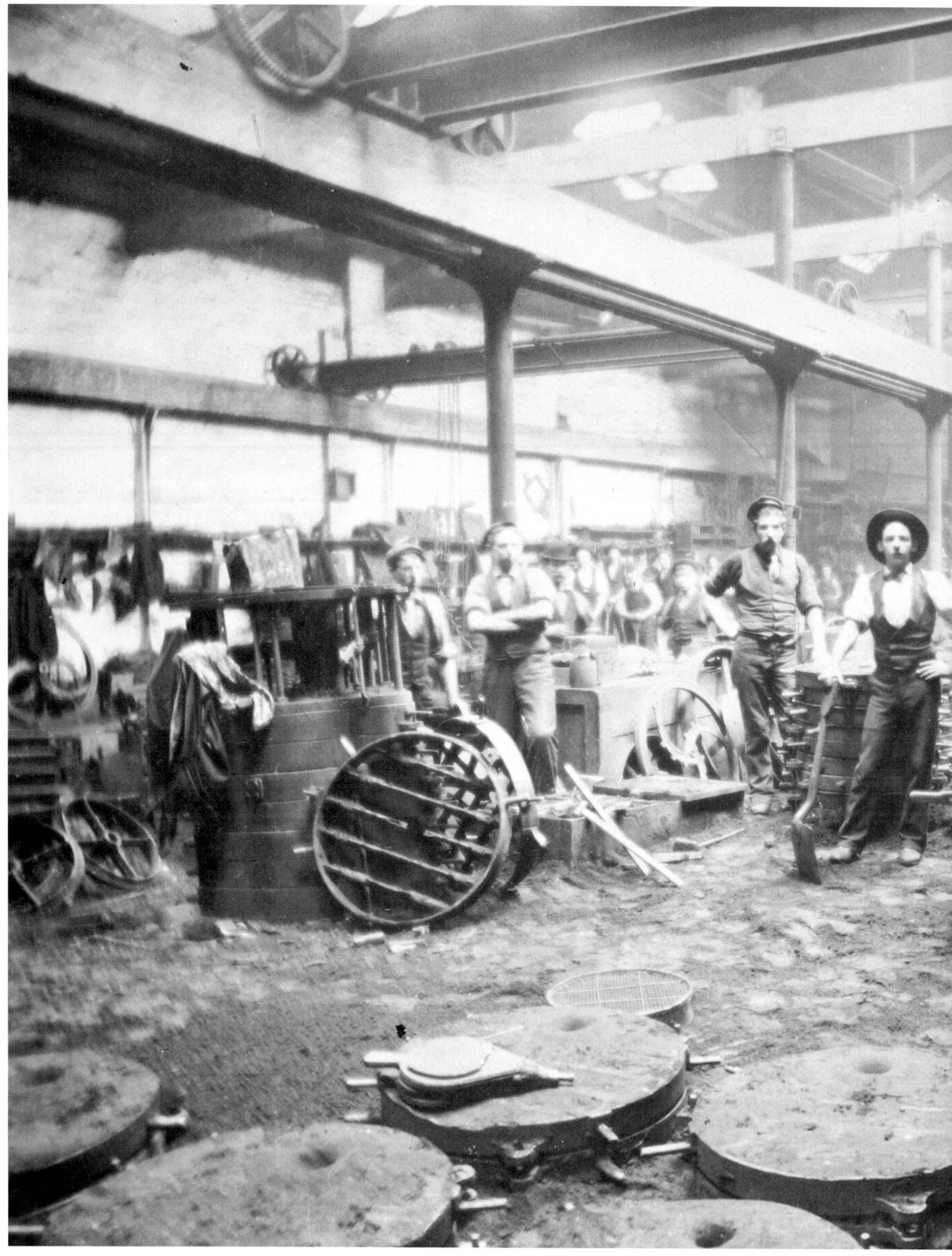

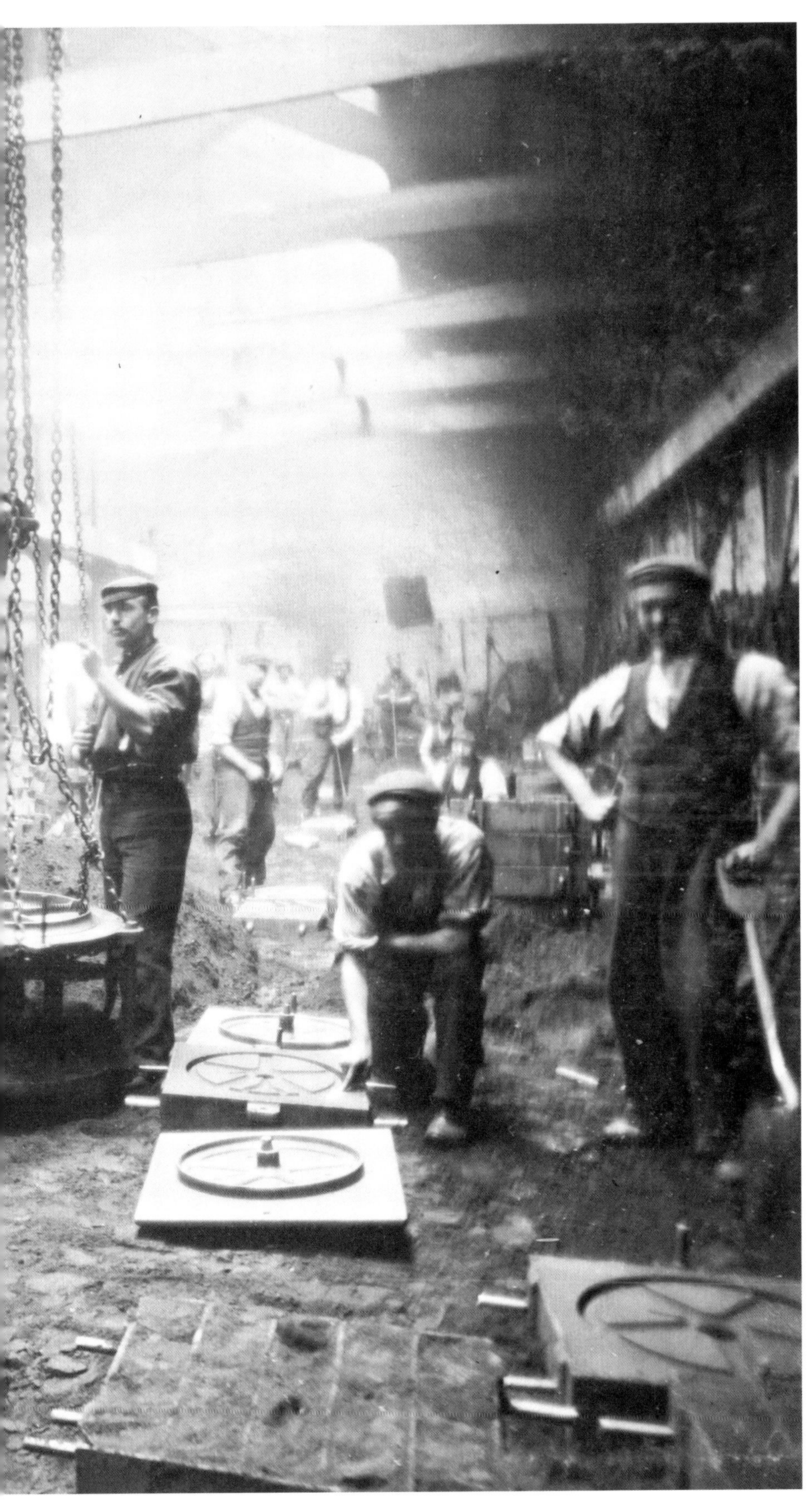

LEFT **The moulding shop in Reid & Gray's foundry.** The shop had scarcely changed since the Bronze Age, and most of the jobs, except that of pattern maker, were classified as semi-skilled. The pattern makers – highly skilled joiners – made the wooden moulds, around which the moulders packed the sand into which they poured molten metal to make the castings. Once the metal was set, the fettlers removed the castings from the mould and knocked off the roughest edges. Fettlers were considered less skilled than moulders, but neither served an apprenticeship. Men were also expected to tackle a wide range of jobs, whatever their specialty.

Hocken Collections, Reid & Gray Ltd, ARC-0695, Ms-1165-063, p. 08; image S13-593.

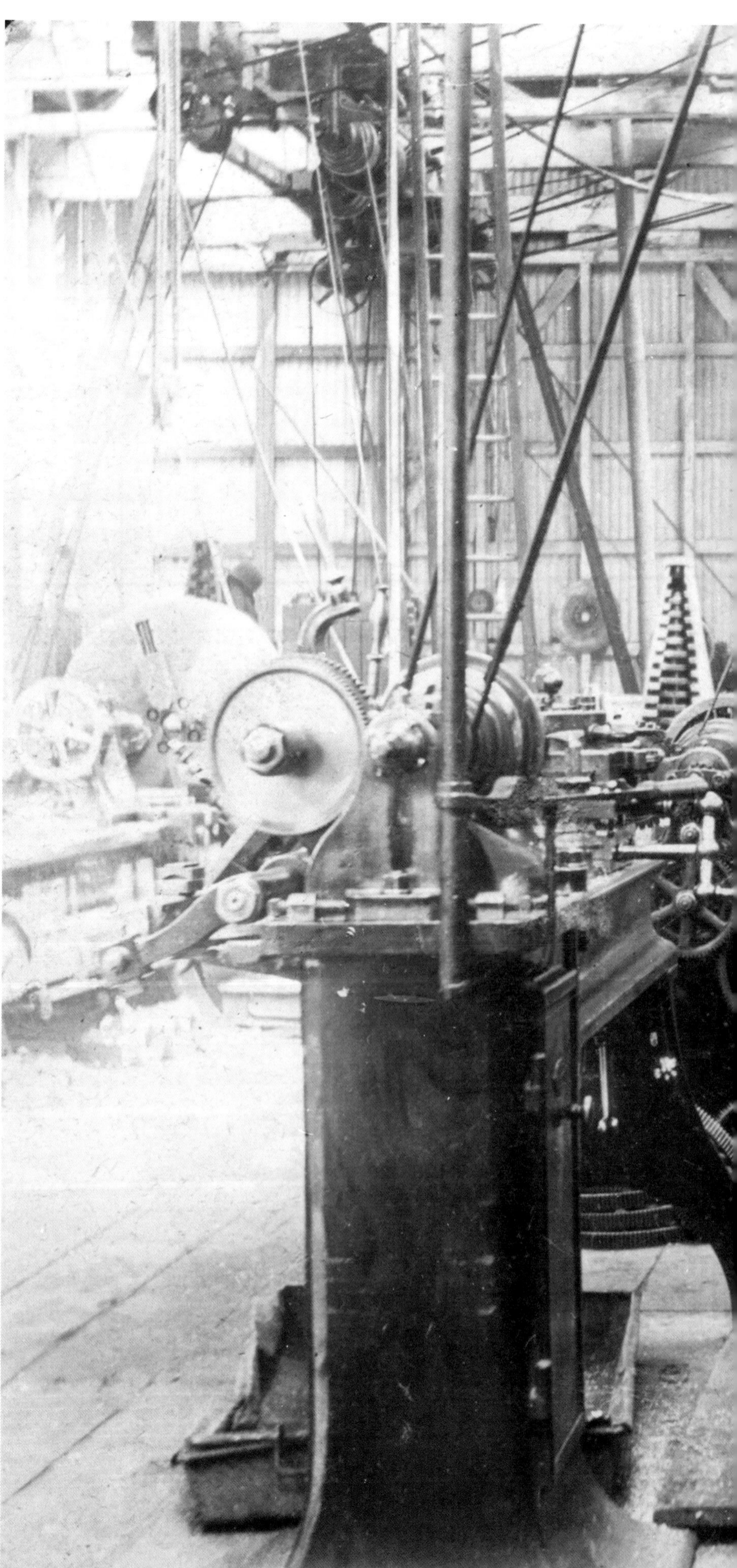

RIGHT **Hillside Railway Workshops.** An unidentified man standing alongside a lathe in the machine shop. William Williams, who transferred to Hillside in 1893, probably took this picture.

Alexander Turnbull Library, Jeavons Baillie Collection, PA Coll-296-1.

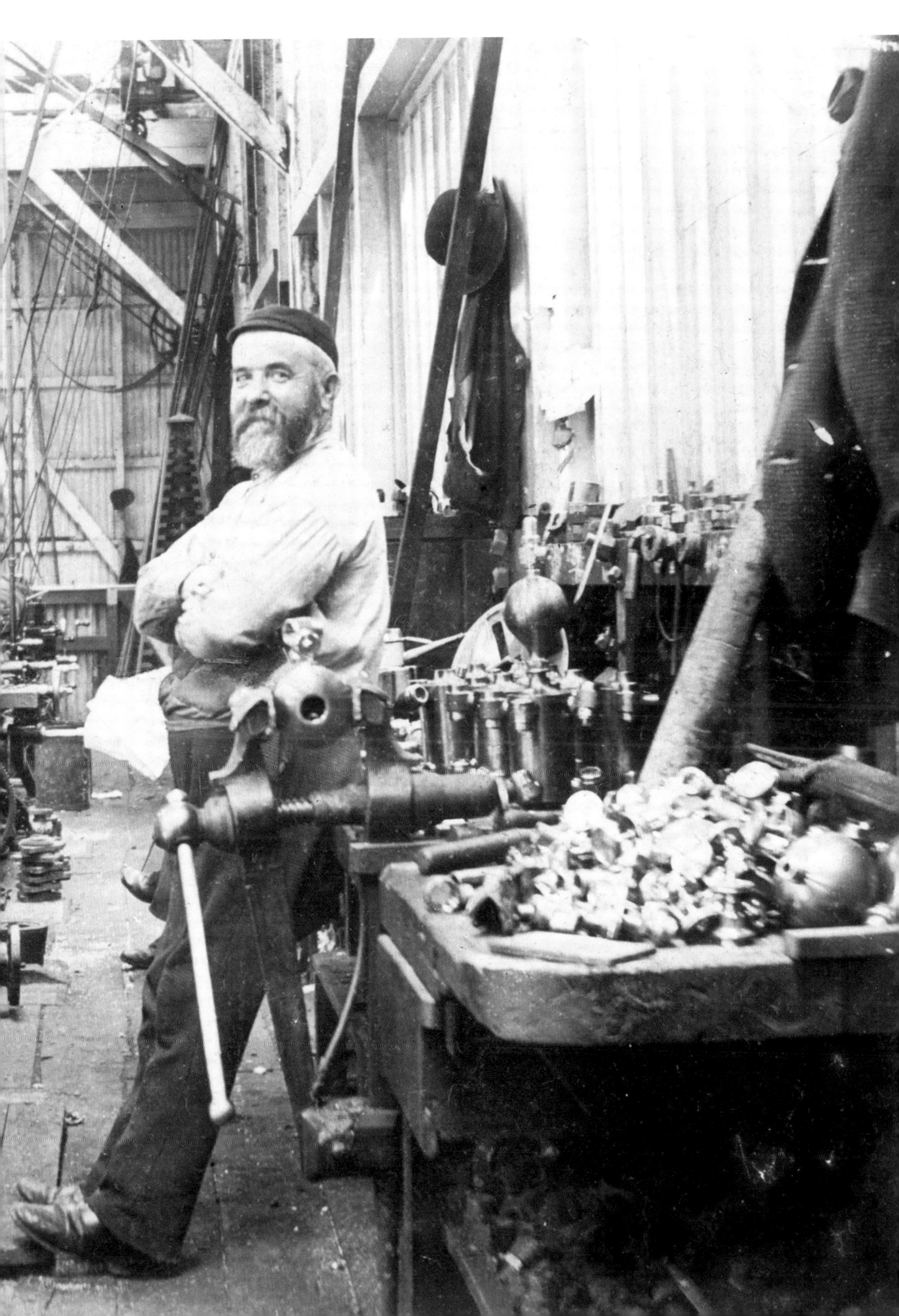

It is our good fortune that a passionate photographer of the industrial age and its technologies, Arthur Percy Godber (1875–1949), who was also fascinated by Māori art, documented the old Hillside workshops before their demolition, as well as life in the new workshops.

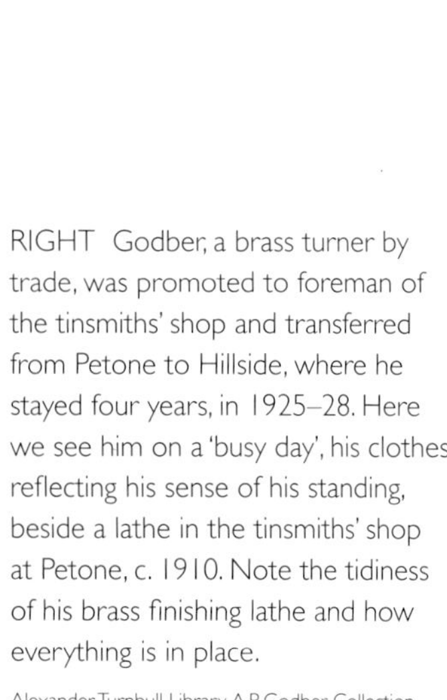

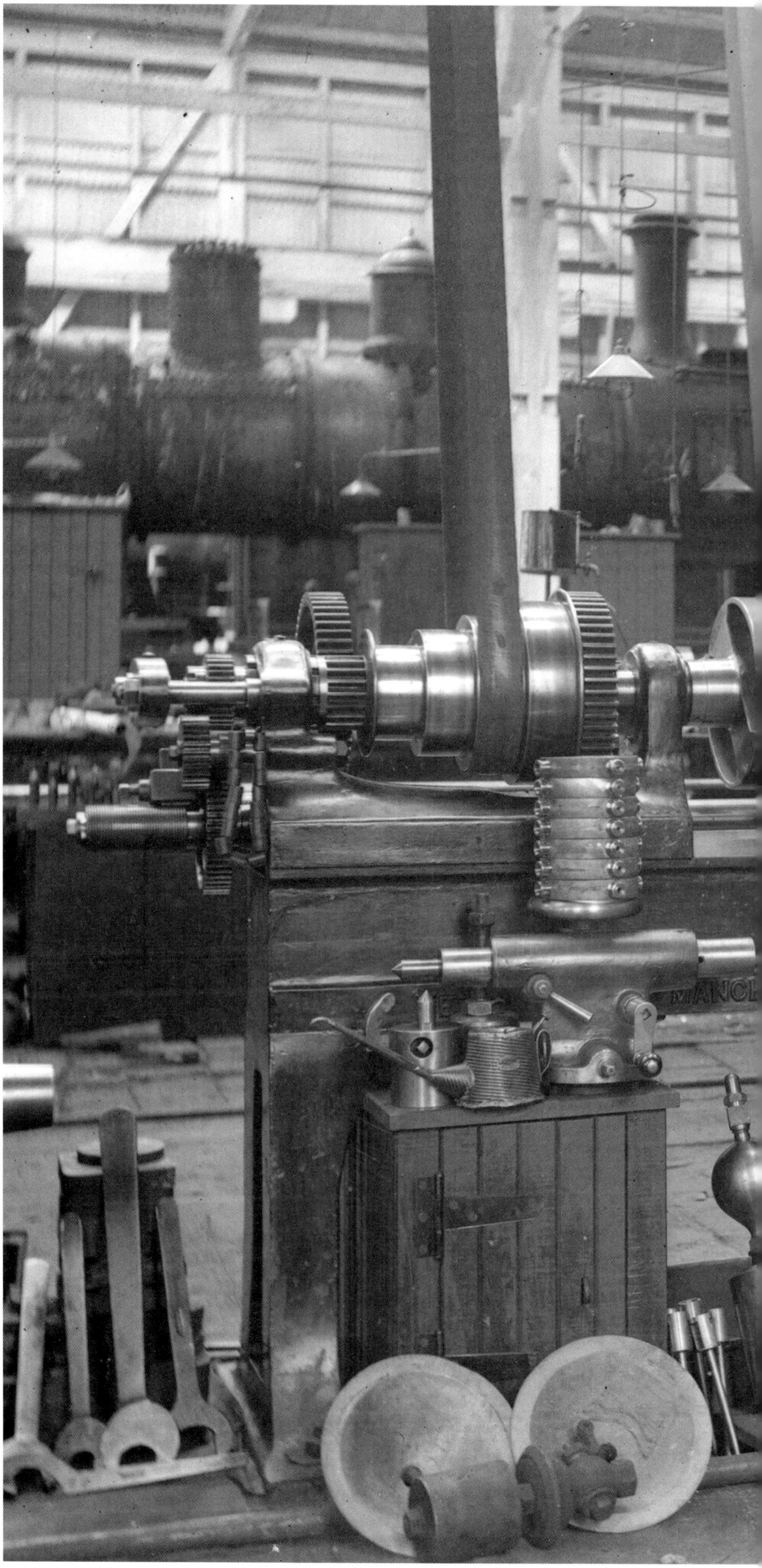

RIGHT Godber, a brass turner by trade, was promoted to foreman of the tinsmiths' shop and transferred from Petone to Hillside, where he stayed four years, in 1925–28. Here we see him on a 'busy day', his clothes reflecting his sense of his standing, beside a lathe in the tinsmiths' shop at Petone, c. 1910. Note the tidiness of his brass finishing lathe and how everything is in place.

Alexander Turnbull Library, A.P. Godber Collection, G-126-1/2-APG.

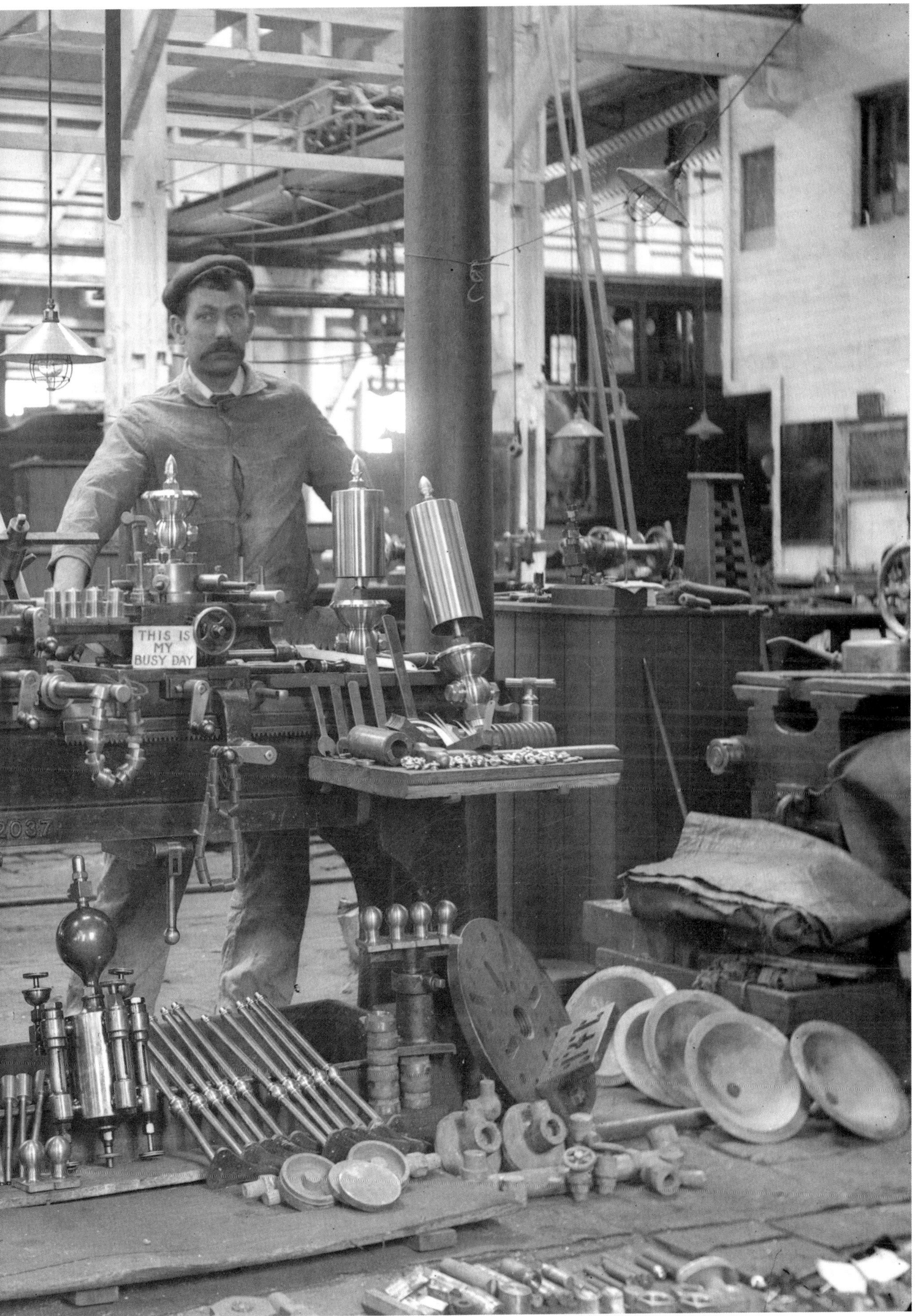
THIS IS
MY
BUSY DAY
2037

RIGHT **The tinsmiths' shop at Hillside** was a relatively peaceful refuge from the noise and dirt of both the foundry and other shops, especially the machine, smithing and boilermaking shops. This photograph reminds us, however, that even the largest industrial establishments often continued to be organised along pre-industrial lines as a congeries of separate shops, a sort of federation of equal communes, each under the supervision of leading hands and a foreman, all of whom had served apprenticeships and worked for many years as journeymen. The labourers – each skilled man had an assistant – were also highly skilled, and often learned to do all but the most complex work by watching and helping. The culture of craft was dominant and unquestioned. Only by moving to a small town where the union had no presence could an unskilled man easily pass himself off as skilled. Godber took this shot in 1926.

Alexander Turnbull Library, A.P. Godber Collection, APG-0936-1/2-G.

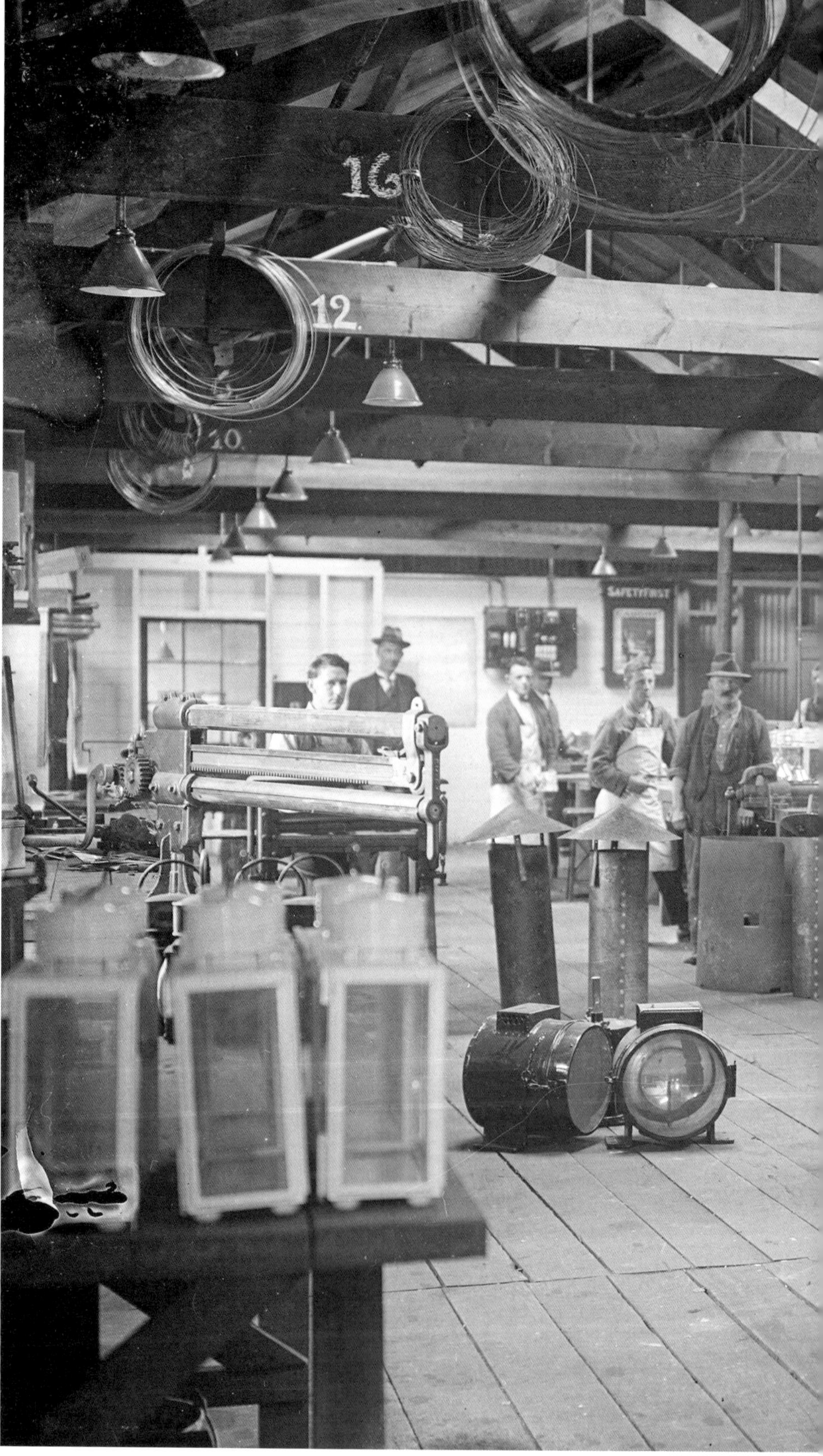

OVERLEAF
Albert Percy Godber's photograph of the old erecting shop, October 1925. The erecting shop was the kingdom of the fitters, the elite of the metal trades (although the erosion of their pay differential had annoyed them considerably and saw most break away from the Amalgamated Society of Railway Servants to form the Railway Tradesmen Association in 1925). In the erecting shop, the fitters not only assembled new locomotives and wagons but dismantled old ones for repair and maintenance. In the old workshops of the first industrial revolution, which had grown like topsy, the skilled tradesman was king: he

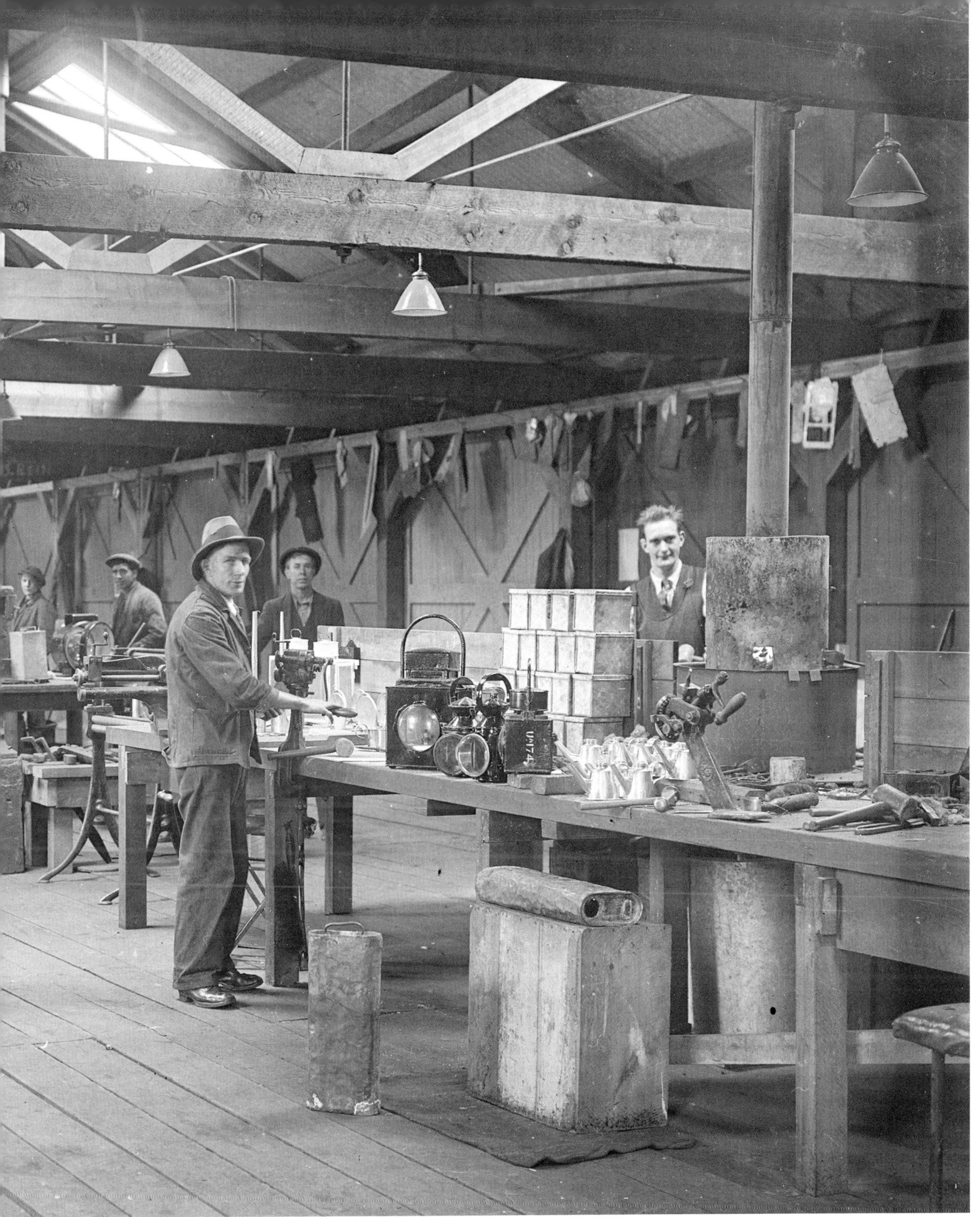

knew what had to be done, how to do it, what tools were needed, where to find them, and how long it would take. The old erecting shop was state-of-the-art when built in 1878, although dismantling engines was done outside, the men often breaking the ice as they lay on their backs to work beneath them. At Hillside, dismantling was done outside until 1918.

Electrification occurred 25 years later, making many of the shops even more dangerous because wires, shafts and cables were introduced willy-nilly, compounding the disorder and potential for disaster.

Alexander Turnbull Library, A.P. Godber Collection, APG-2025-1/2.

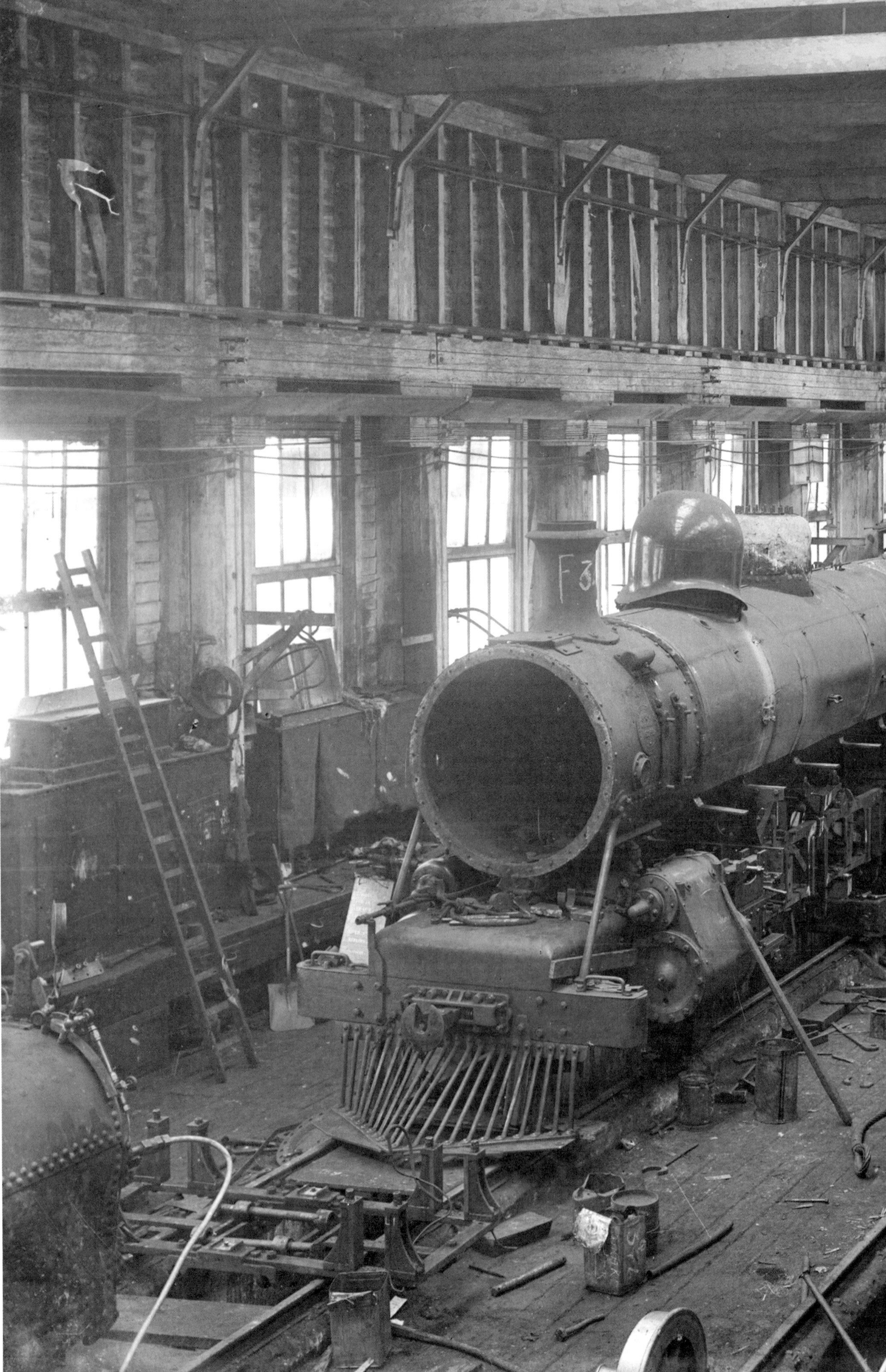
F 3

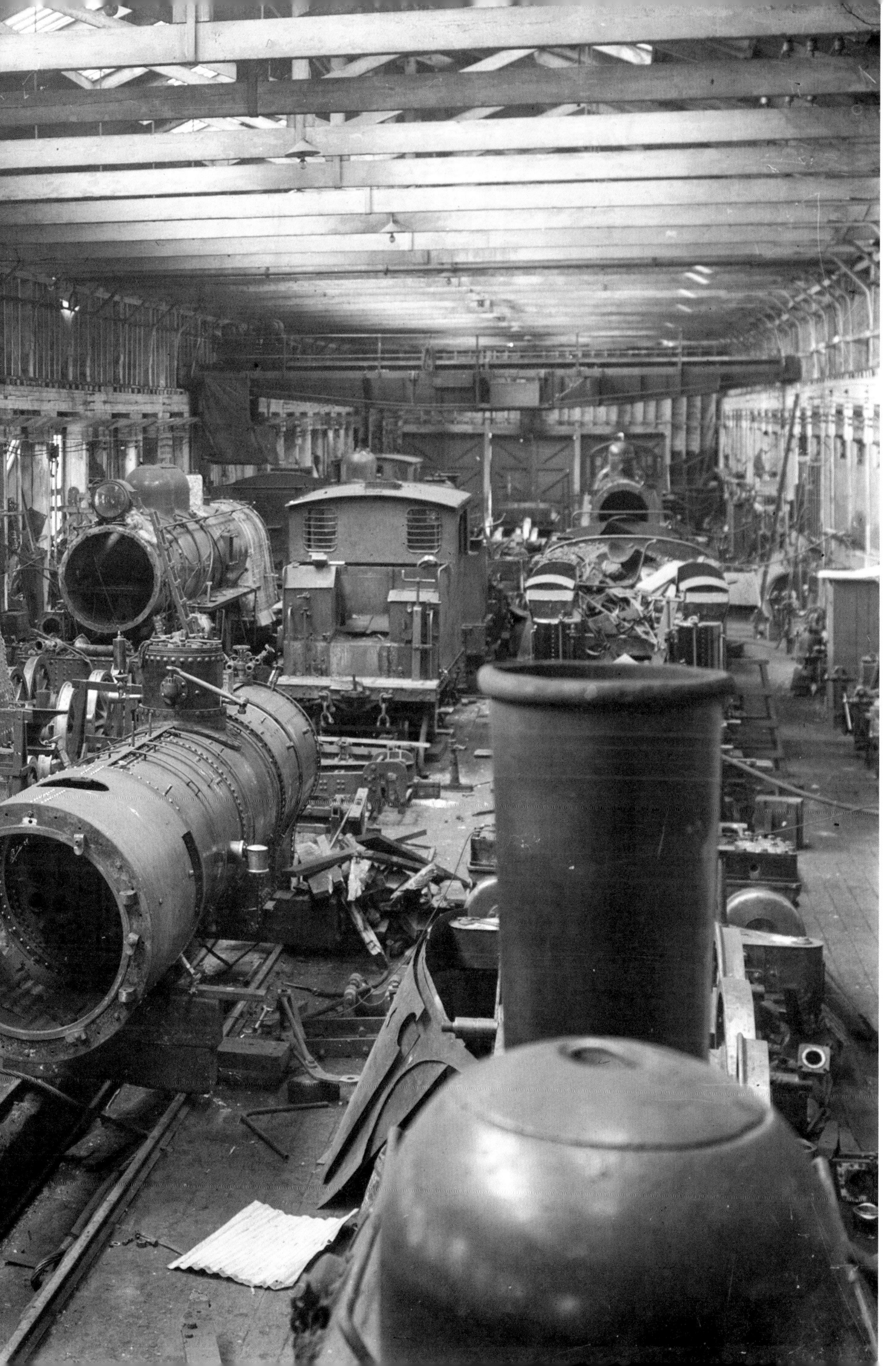

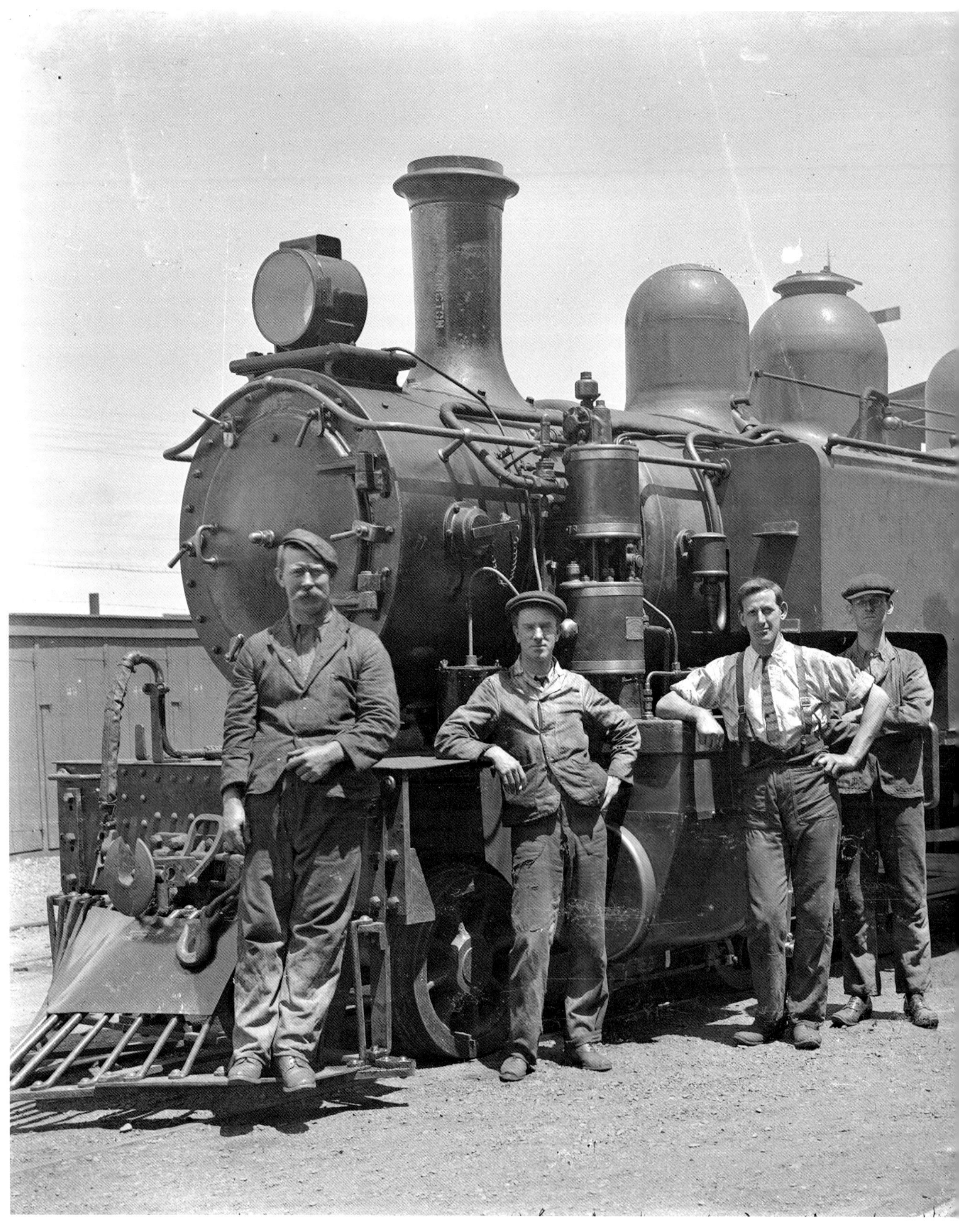

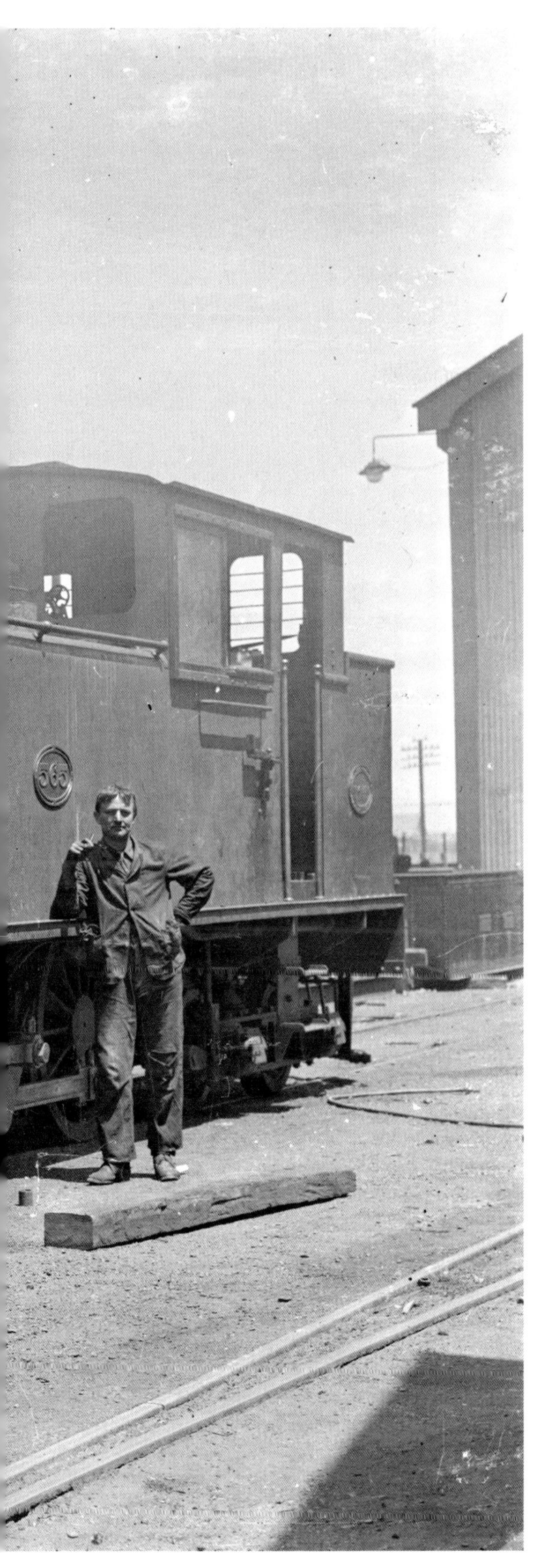

LEFT **NZR 565, 4-6-4T, began service in January 1914** and was written off in 1957. Hillside made its first locomotive in 1897, a Wa Class shunter, and in 1907 began specialising in making the Wf and then Wg Class locomotives. Between 1913 and 1919, Hillside built 53 Ww Class locomotives, including this one. The men at Hillside gained great satisfaction from building locomotives, although this was but a small part of their work. Constructing locomotives and rolling stock also allowed New Zealand Railways to retain its skilled workforce, then a common practice worldwide in both privately and state-owned systems. The men in the photograph are, from left: C. Strong, J.S. Cummings, K. McIntyre, C. Davis and R. Cunliffe.

Alexander Turnbull Library, A.P. Godber Collection, APG-0428-1/2-G.

OVERLEAF
Working men often resorted to fisticuffs if they wished to settle a dispute. Here Albert Percy Godber captures a fight outside the Petone Railway Workshops, 28 May 1915. Many working-class (or ex-working-class) men followed sports such as boxing, which involved strength, courage and physical skill. One's inclination to fight, especially fight 'dirty', or to take money to 'throw' a match, increasingly marked off the 'rough' from the 'respectable'.

Alexander Turnbull Library, A.P. Godber Collection, APG-0518-1/2-G.

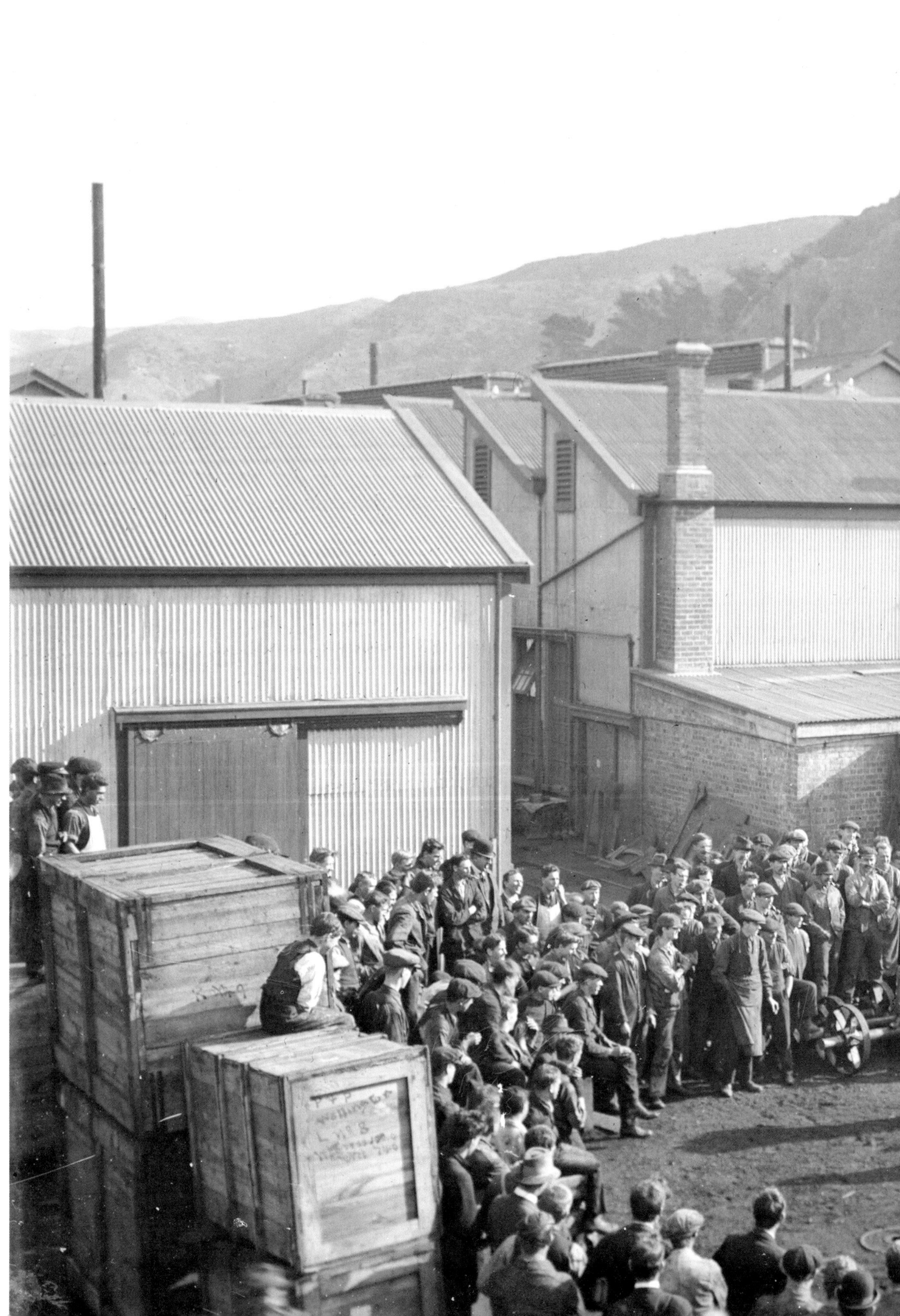

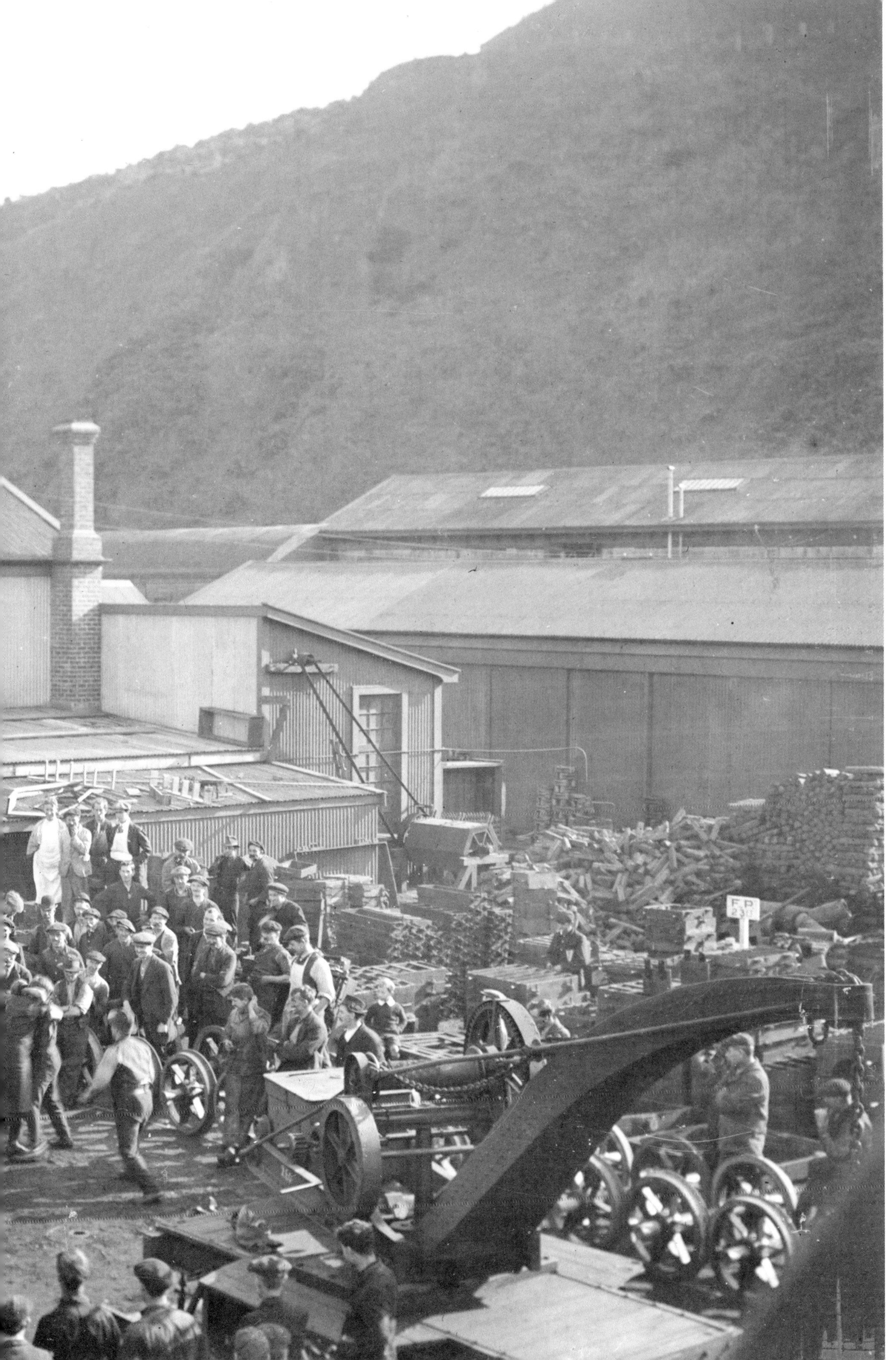
FP

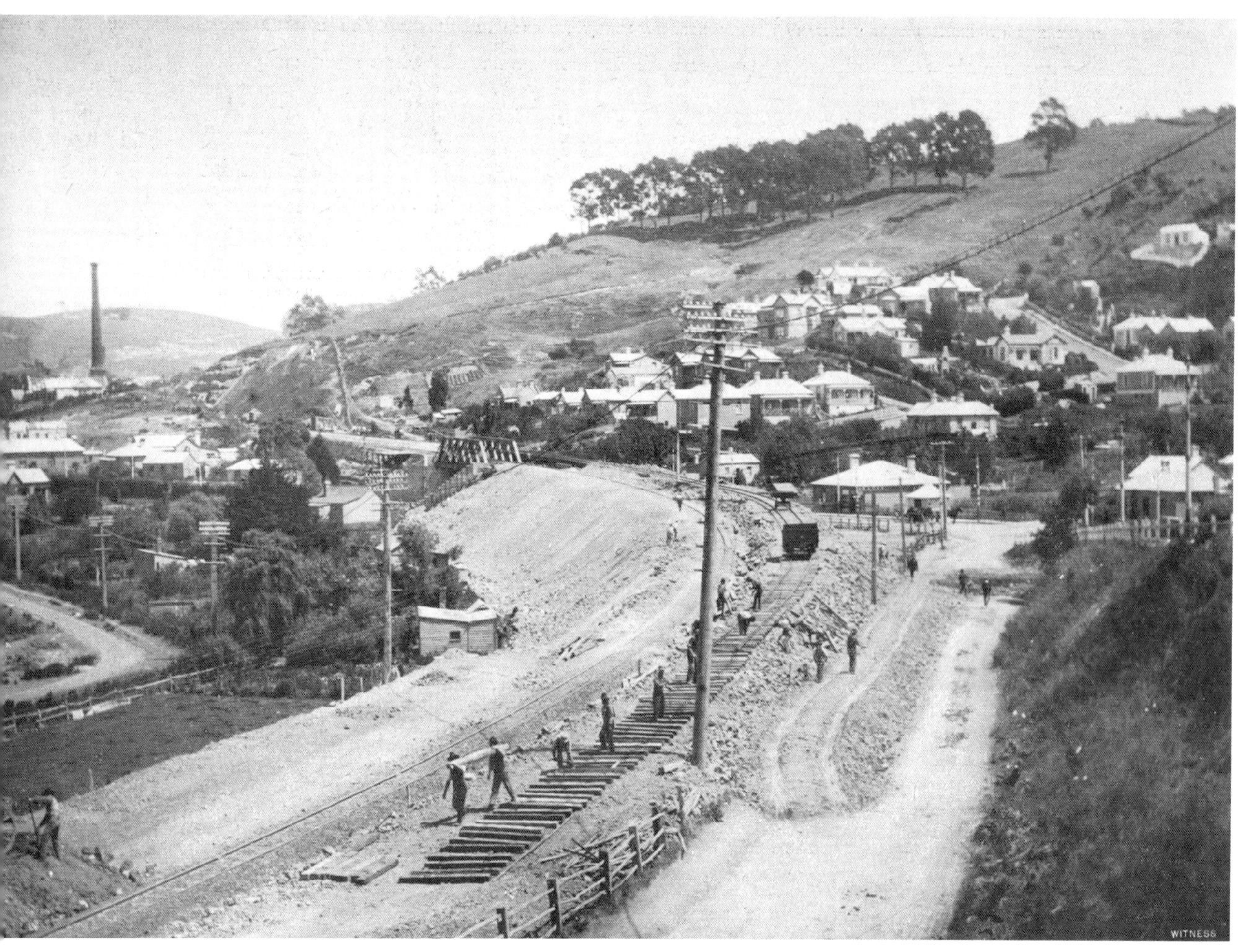

ABOVE **Laying a second railway line,** one of the many public works projects undertaken in southern Dunedin during the 1900s. This entailed construction of an embankment at the foot of the Glen, to remove several dangerous level crossings, and of a new tunnel beneath Lookout Point. The photograph above shows men carrying sleepers and rails on to the new embankment at Carisbrook, near Parkside, where the new railway bridge crosses the Main South Road. The chimney of the Caversham gasworks is visible in the left background. Photograph by Iles and Thomas.

Hocken Collections, *Otago Witness*, 6 February 1907, p. 43.

If you did not enter a skilled trade, chances were you looked for a labouring or unskilled job. Here there were three broad classes: 'skilled unskilled' work, labouring and navvying. Skilled unskilled work meant learning on the job, as sailors and shearers and miners did, or assisting a skilled man as a helper or improver. Alternatively, one became a labourer or a navvy. Labouring was a very broad church, whereas navvying meant physically demanding work with pick or shovel or sledge-hammer. Navvying was for young men in their prime.

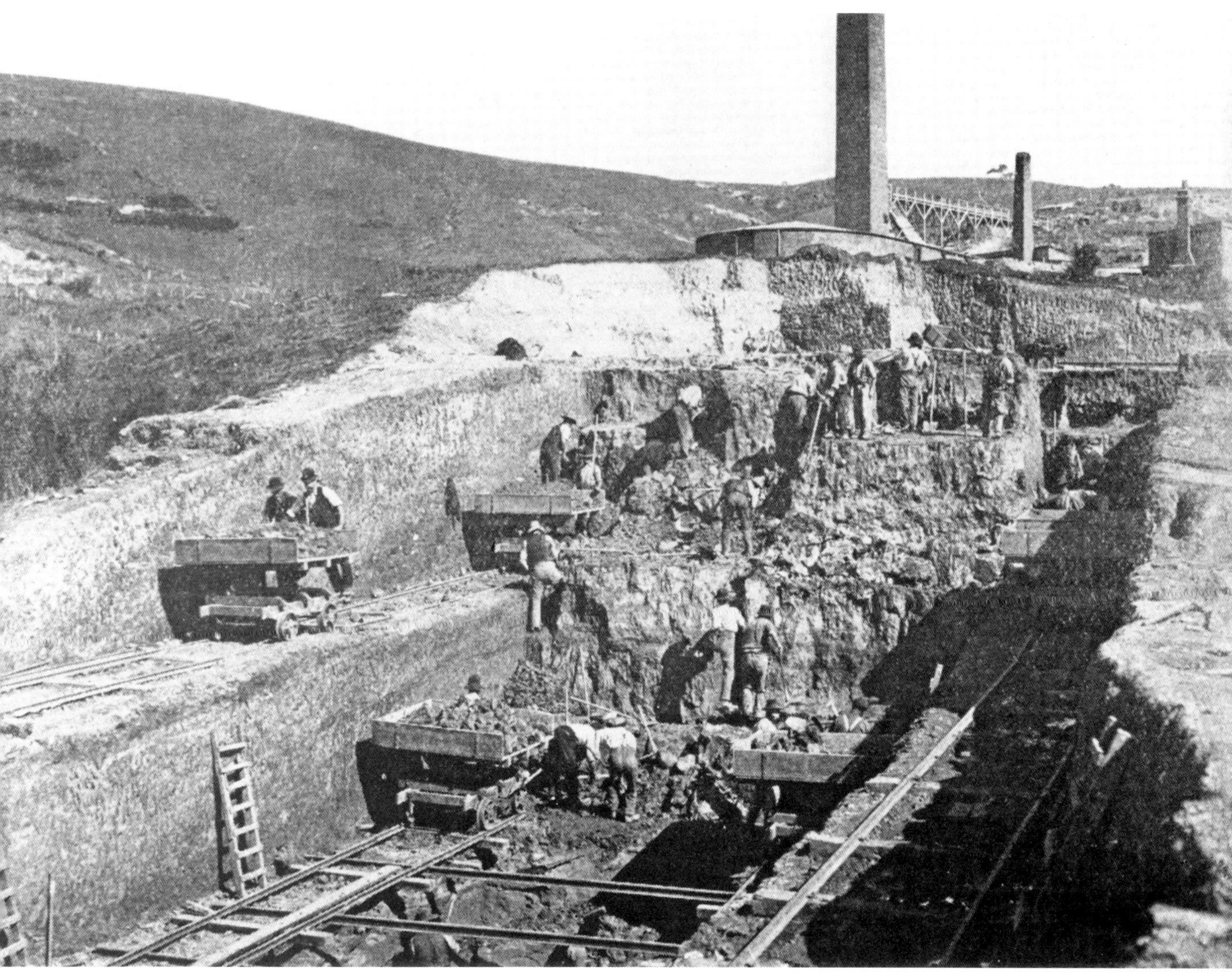

ABOVE **The excavations for the double-track tunnel** encroaching on Annand's Rockyside brickworks, above Caversham township. Photograph by Iles and Thomas.

Hocken Collections, *Otago Witness*, 15 May 1907, p. 40.

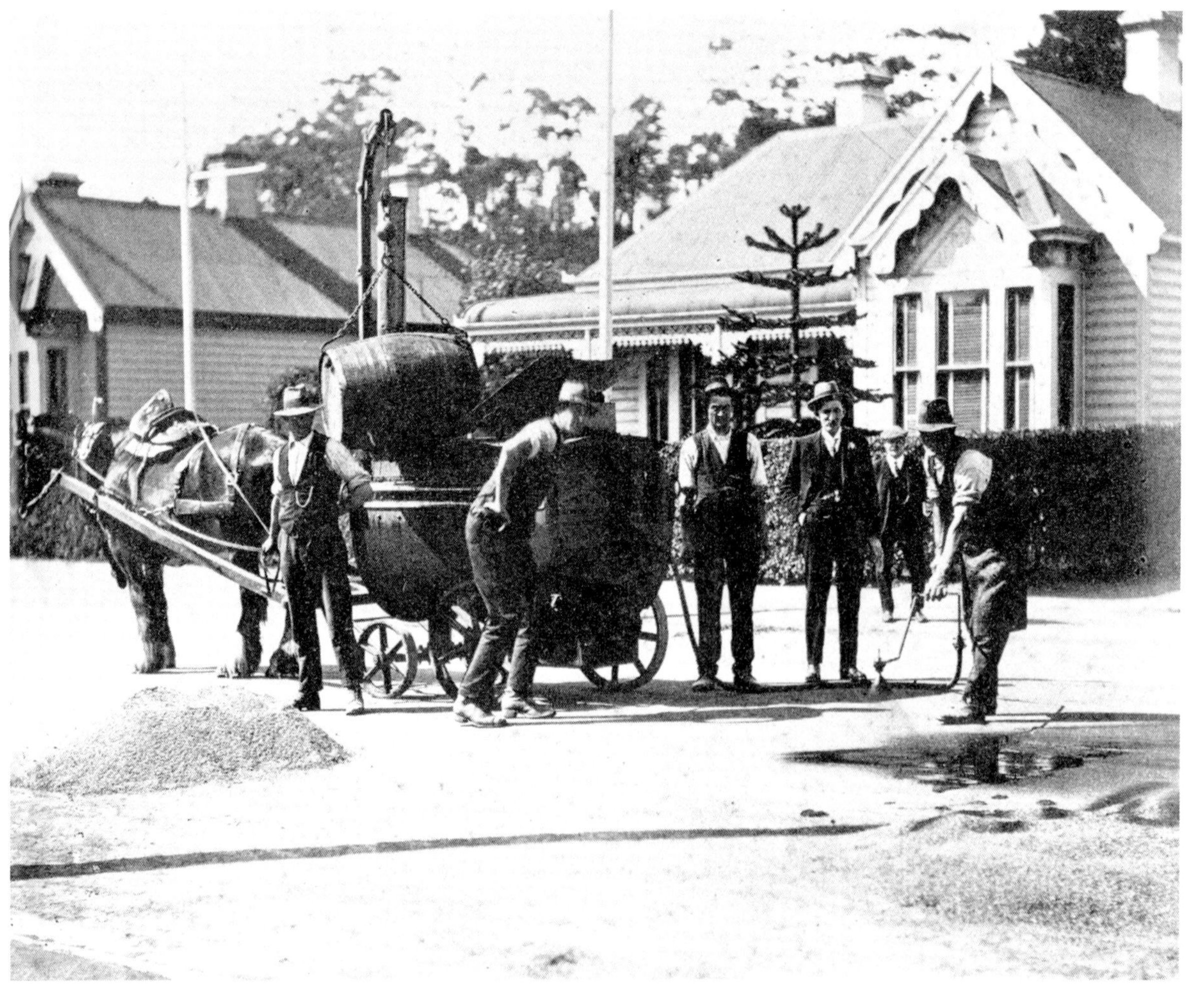

ABOVE **One of the City Council's new 'tar babies', probably in 1905** or thereabouts, designed to mechanise the process of sealing the streets with fine gravel sprayed with a tar seal. This process eliminated the problems caused by dust in summer and mud in winter, but proved costly (the handsome new villas in the background, together with their tidy hedges and an occasional ornamental tree, remind us that this was the most prosperous society in the world at this time). The inspector, with his trilby and fob-watch, clearly belongs to a different social class from the manual workers, but may have been promoted from the ranks. Under the impetus of a new city engineer from Australia, who declared Dunedin's streets worse than any other city's in Australia or New Zealand, even Sydney's, a sustained campaign was launched to pave all main thoroughfares. As the number of bicycles and motor vehicles multiplied, the need for tougher surfaces became clear. In 1908, Dunedin City began experimenting with Neuchatel asphalt and imposed a speed limit of four miles per hour for mechanically propelled vehicles turning corners. By the 1920s, the volumes of traffic forced the Council to embark on an entirely new programme of street improvements.
DCC Archives.

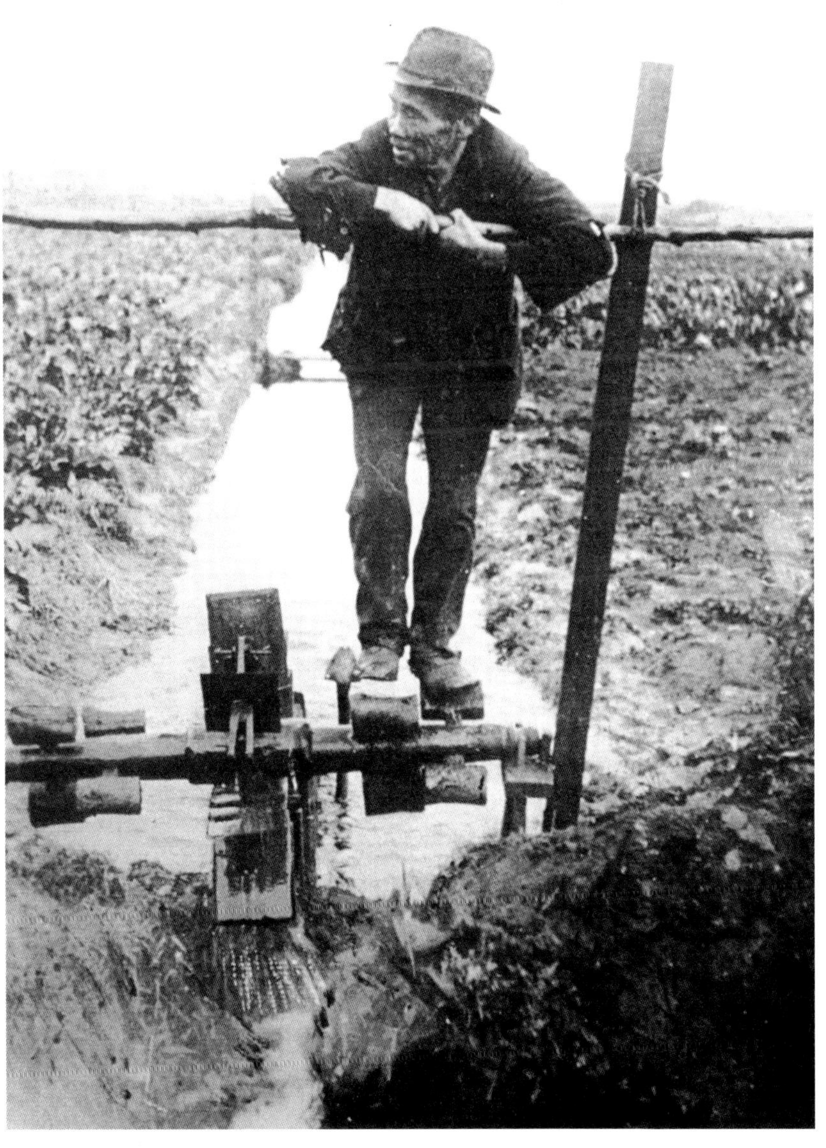

LEFT **The extensive market gardens and nurseries** that occupied the western reaches of the Flat from the 1860s until the 1930s provided an opportunity for Chinese workers to establish themselves within Dunedin. In the 1880s and 1890s some 40 Chinese men, mostly from Panyu county in Guangdong Province (near Canton), combined to rent land from James McIndoe and later from Charles and William Shiel. There they grew vegetables, which they hawked around the local streets and on City Rise. Some of the men lived in huts on the actual gardens, protecting them from larrikins and thieves, both being especially common on the Flat in the 1890s. Most stayed in an old wooden house in Caversham until around 1902, when they moved into the city.

Here a Chinese market gardener, name unknown, works at one of the irrigation control points. He is driving a waterwheel with his feet. Although classified by the *Census* as unskilled, like most farm workers, many of the men were highly skilled horticulturalists.

Courtesy of Kings High School, which was built on an extensive area of land long farmed by the Chinese.

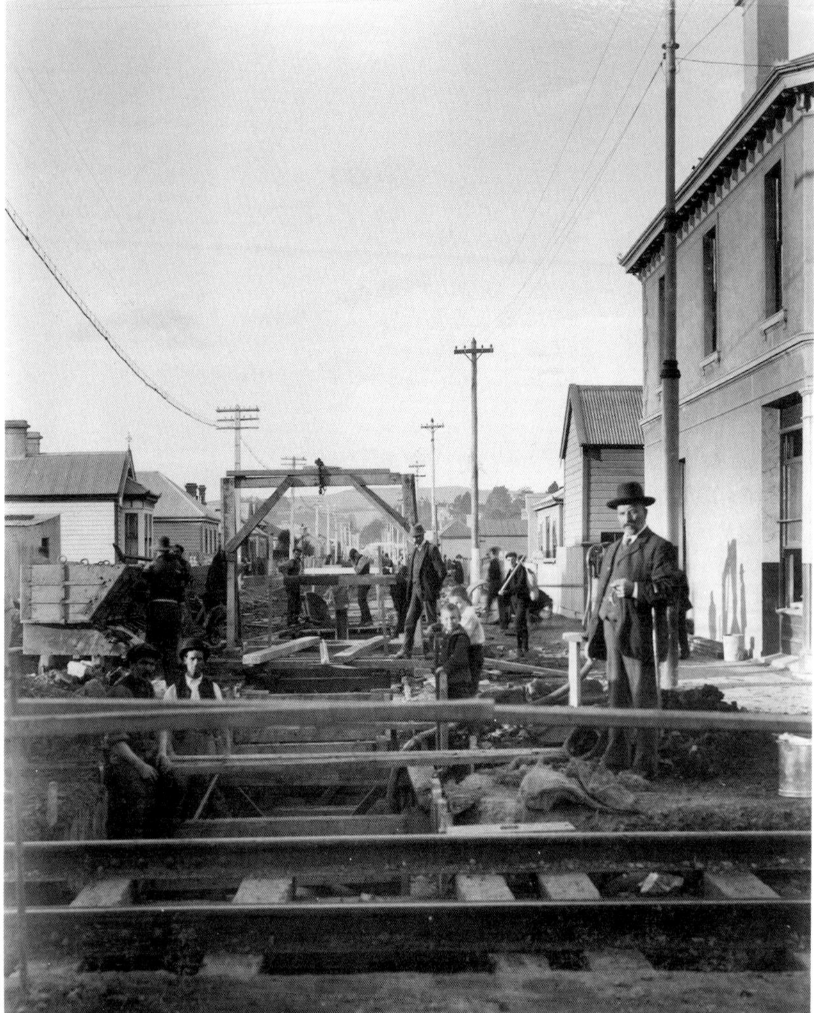

RIGHT **Installation of the foul sewer on Bay View Road.** A foreman or even perhaps a member of the Drainage and Sewerage Board looks at the gantry crane, as do some children who are taking a close interest. The travelling crane was probably being used to lower the pipes into the hand-dug trench. Gantry cranes, manufactured locally, were powerful machines consisting of a steel frame mounted on rail tracks, fitted with lifting gear that could move the heavy pipes both horizontally and vertically. Peebles & Halligan's Building, which housed a butcher's shop at street level, is to the right.

DCC Archives, Drainage & Sewerage Board Photograph Album, DD & SC 159/11.

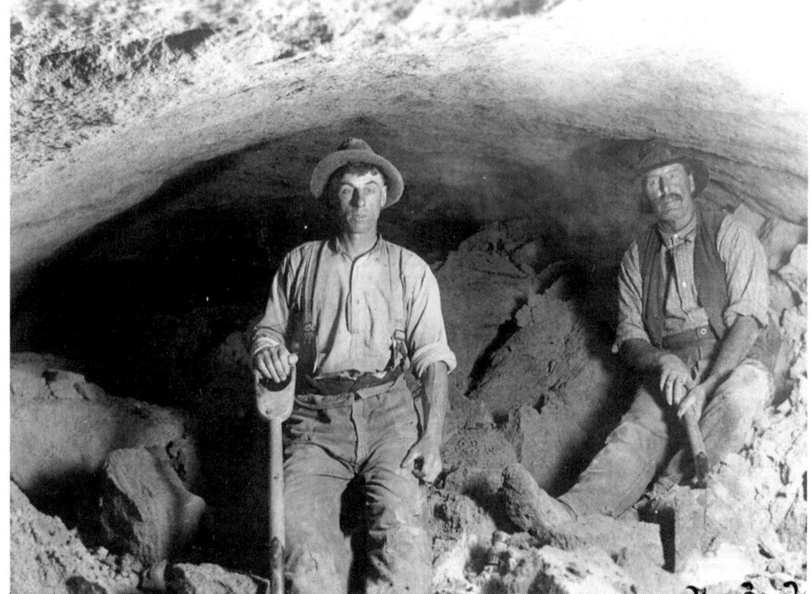

RIGHT AND FAR RIGHT **Tunnellers** still used pick and shovel, sometimes supplemented by dynamite, to dig through rock or even earth, and from mining knew how to ensure they were not buried underground by falls of rock or earth. The first pneumatic drills, known then as 'widow-makers', were trialled in the decade before World War I.

DCC Archives, Drainage & Sewerage Board Photograph Album, DD & SC 157.

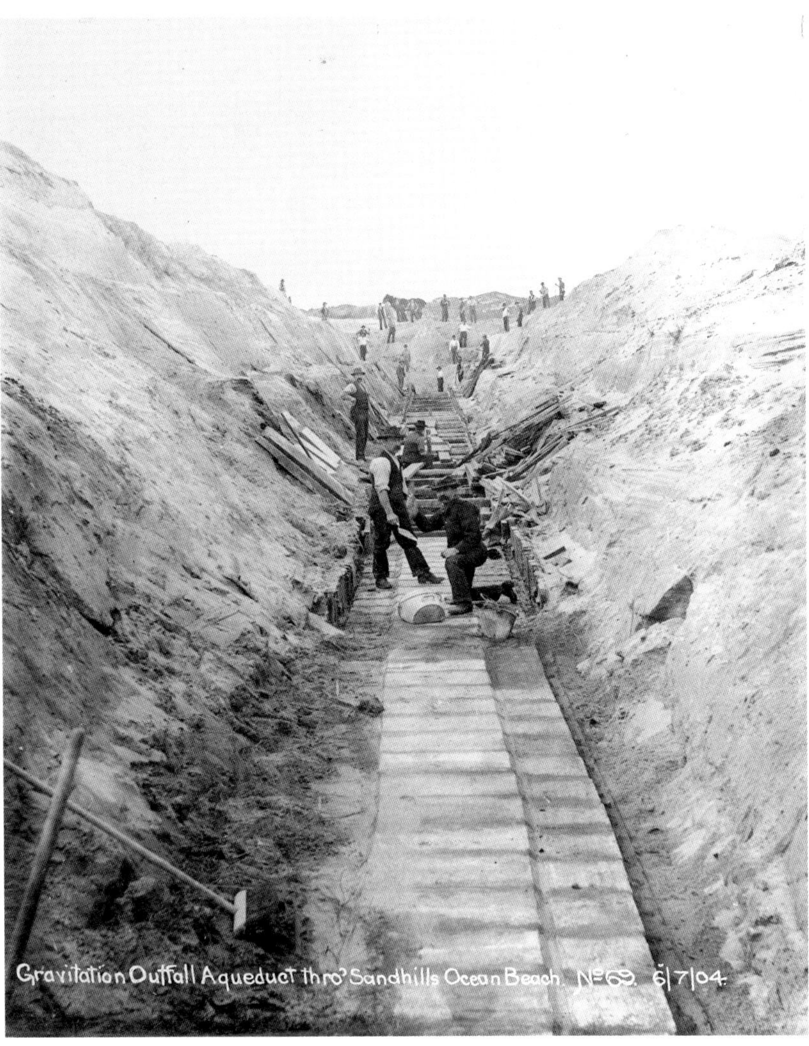

LEFT Building the stormwater gravity fall trench through the sandhills.

DCC Archives, Drainage & Sewerage Board Photograph Album, DD & SC 157.

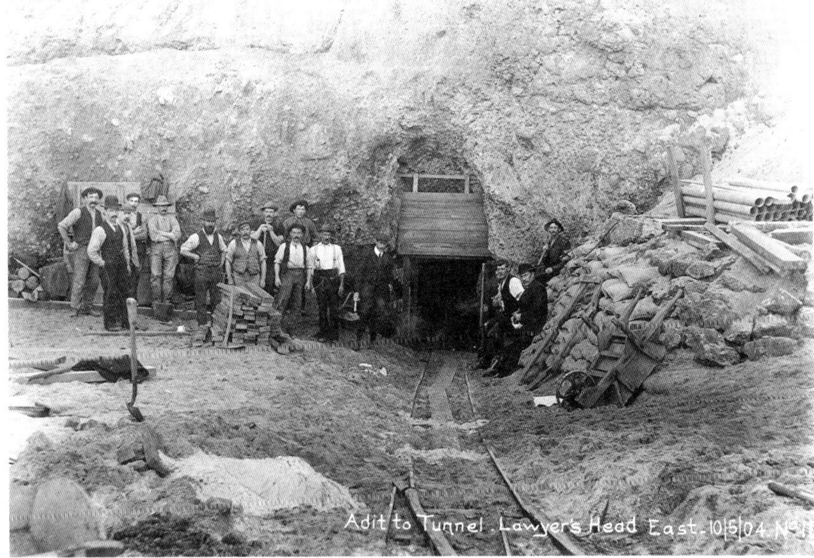

Between 1904 and 1910, the City Council reticulated clean water to most households on the Flat, and also launched an ambitious scheme to remove all stormwater and sewage from the Flat, other suburban boroughs and the city. Tunnelling was of the essence. Previously wastewater had been dispersed into the upper Otago Harbour. Under the new scheme the sewage was pumped 5 1/8 miles from the Botanic Gardens in the city, to a new pump station at Musselburgh, being fed by 40 miles of new sewers plus 26 miles of old sewers taken over from the City Council. A treatment station, built on a reserve beside the Musselburgh quarry, separated sewage from detritus. Work began on a gravitational outfall sewer tunnelled through the rock at Lawyers Head in 1907 and the first discharge of sewage took place in May 1908. Stormwater flowed separately through tunnels dug through the sandhills (themselves in large part the creation of the Ocean Beach Domain Board). The Dunedin Drainage and Sewerage Board commissioned W.H. Ombler to photograph this work in 1904.

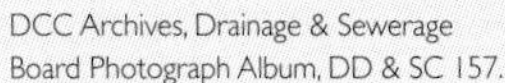

DCC Archives, Drainage & Sewerage Board Photograph Album, DD & SC 157.

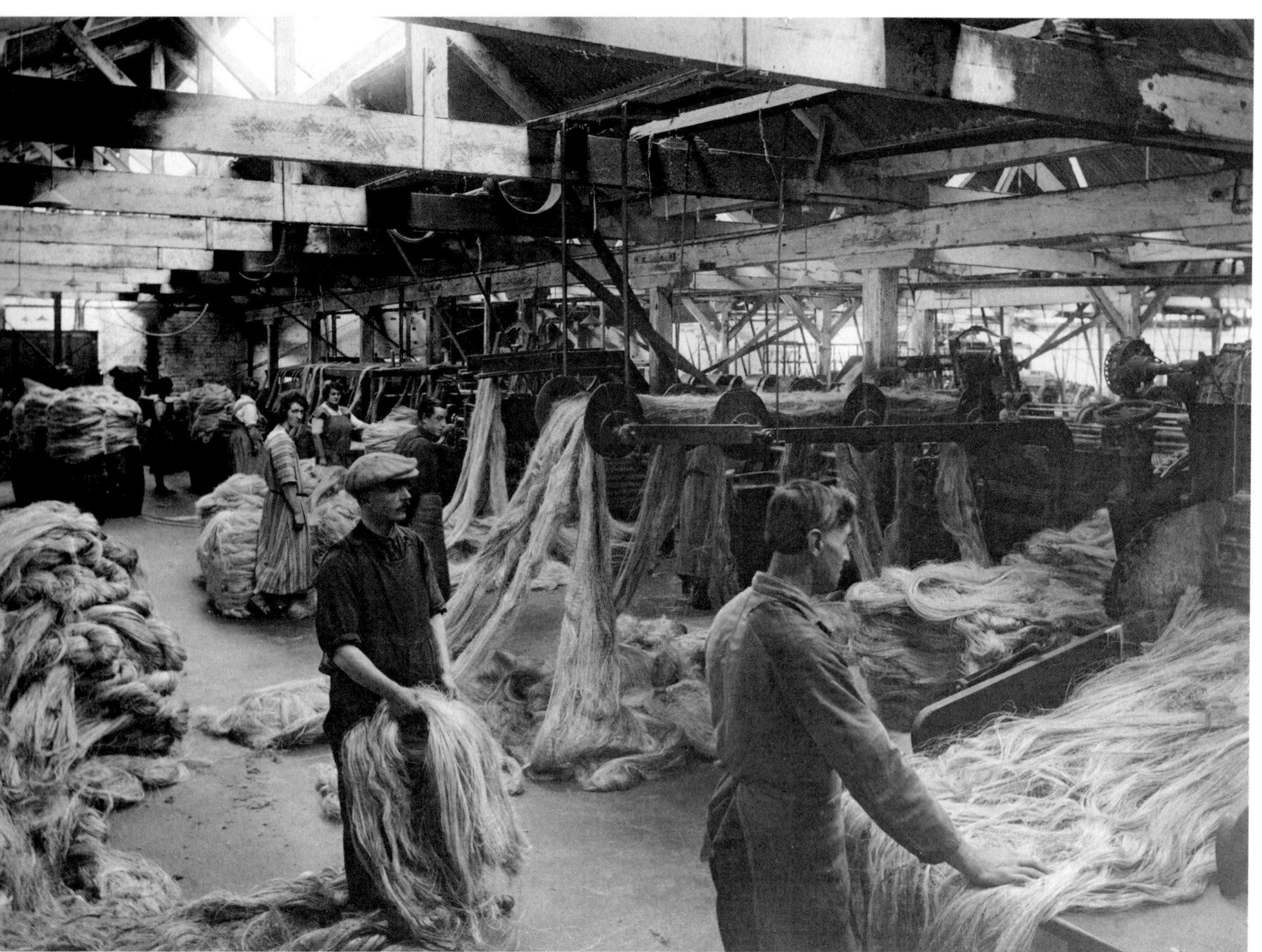

ABOVE **Donaghy's Rope & Twine Co Ltd's Dunedin factory.** In the preparing room, bales of hemp were opened, given a preliminary sorting, then teased into hanks that were wrapped around the cylinders that can be seen towards the back of the workroom. When women and men worked alongside each other, as they did in the preparing room at Donaghy's, the possibility that women might take men's jobs was always present if unspoken. Men were paid more than women on the assumption that they needed to support their families, even if they were single; while men tried to protect themselves by rigidly imposing separate spheres, the women often mocked their 'Lords & Masters'. Donaghy's historian claims that no women worked in the Dunedin factory until after World War I, but the photographic evidence says otherwise.

Hocken Collections, Donaghy Industries, Ms-ARC-0042; image S12-528h.

OPPOSITE, TOP **The men in the foreground are placing hanks of hard fibre – probably sisal – on a machine which carded the fibre in preparation for spooling.**

After the sisal had been carded, the hanks (known as the roving) were fed to the spooling machines, seen here. The spooling machine furthered the process of drawing and aligning the fibres preparatory to spooling. The roves can be seen on the spooling machines in the immediate foreground; the spools can be seen to the left. Each machine had to be watched in case the fibre jammed. After spooling, the fibre went across to the next process where it was woven into binder twine, or to the rope walk if destined to become rope. These machines had been invented in Britain during the Industrial Revolution for the woollen industry but required several adaptations to handle sisal.

Hocken Collections, S03-280d.

RIGHT **Men at work in the rope walk.** A simple machine wove strings of fibre into rope the length of the walk, many metres long. The ropemaker was king, his book of instructions for various twists and tensions being almost as sacred as the Bible. The last of the ropemakers, a tough old Scot, resisted mechanisation until the 1940s. Ropemaking then became a skill rather than an art, as the firm's historian remarked. Even before the ropemaking process was completely mechanised, the machines required constant supervision, not least because of the risk of fire, a particular hazard in rope factories. Until the mid-1890s, this walk was made of wood and several slats were left loose so that men could escape if a fire got out of control; the same exits were also used by any workers feeling in need of a quick trip to 'the Fitz' (the famous Fitzroy Hotel on Hillside Road, a major centre of gambling on almost anything, from rugby matches to pigeon races).

Hocken Collections, SO3-280c.

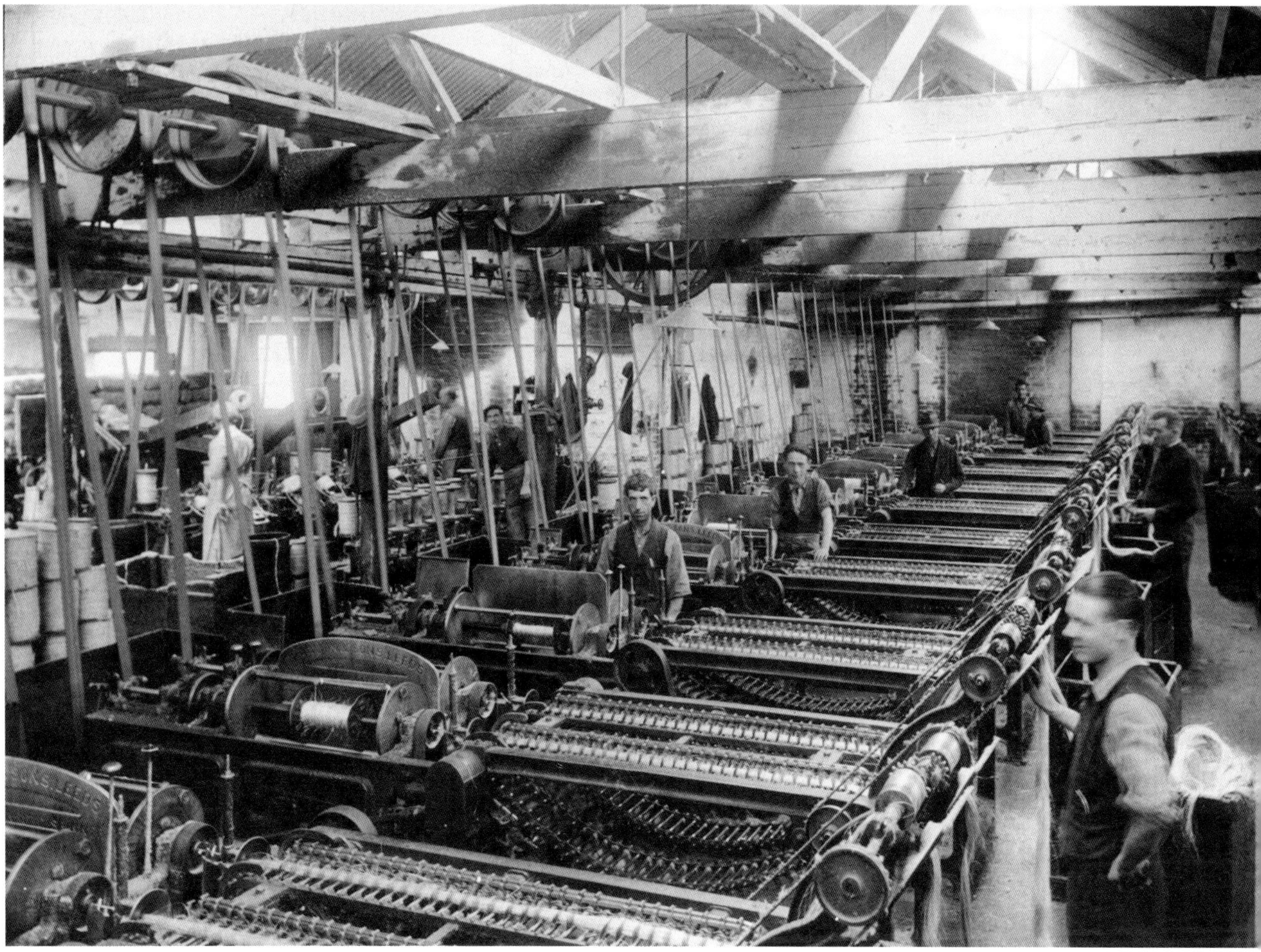

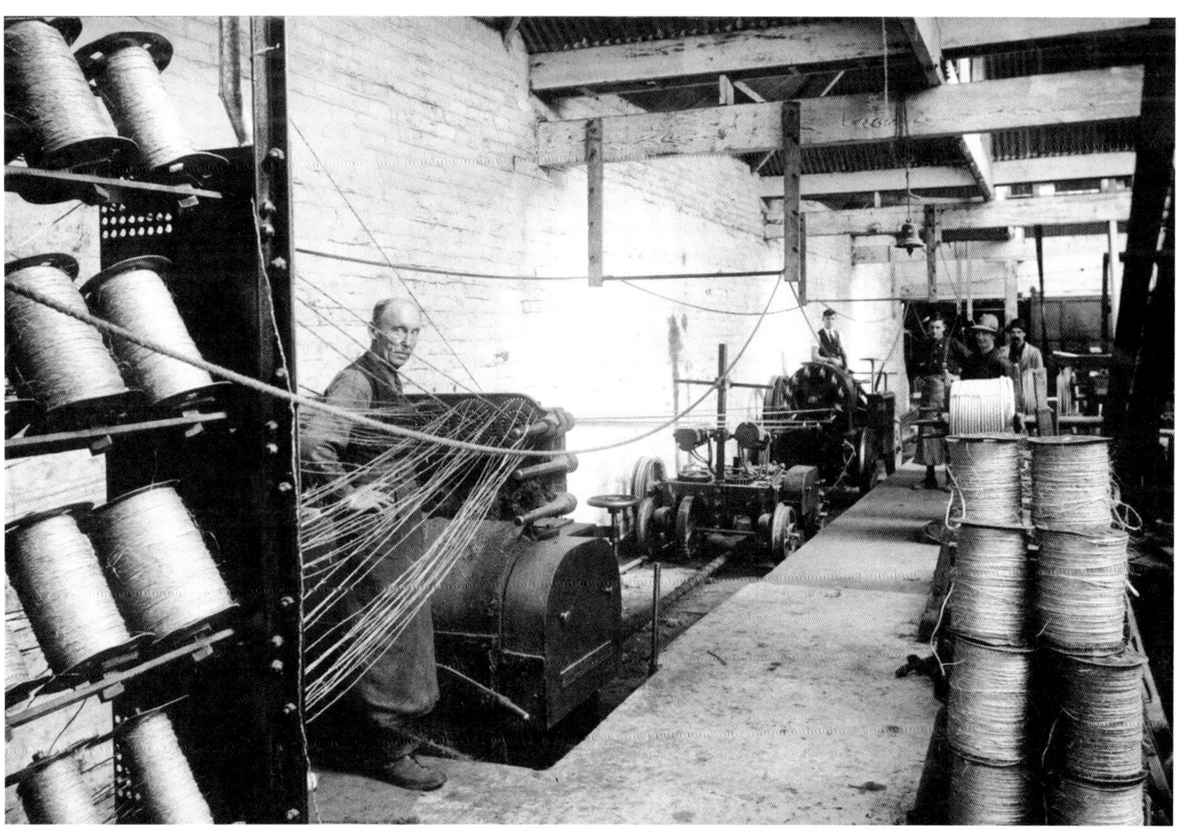

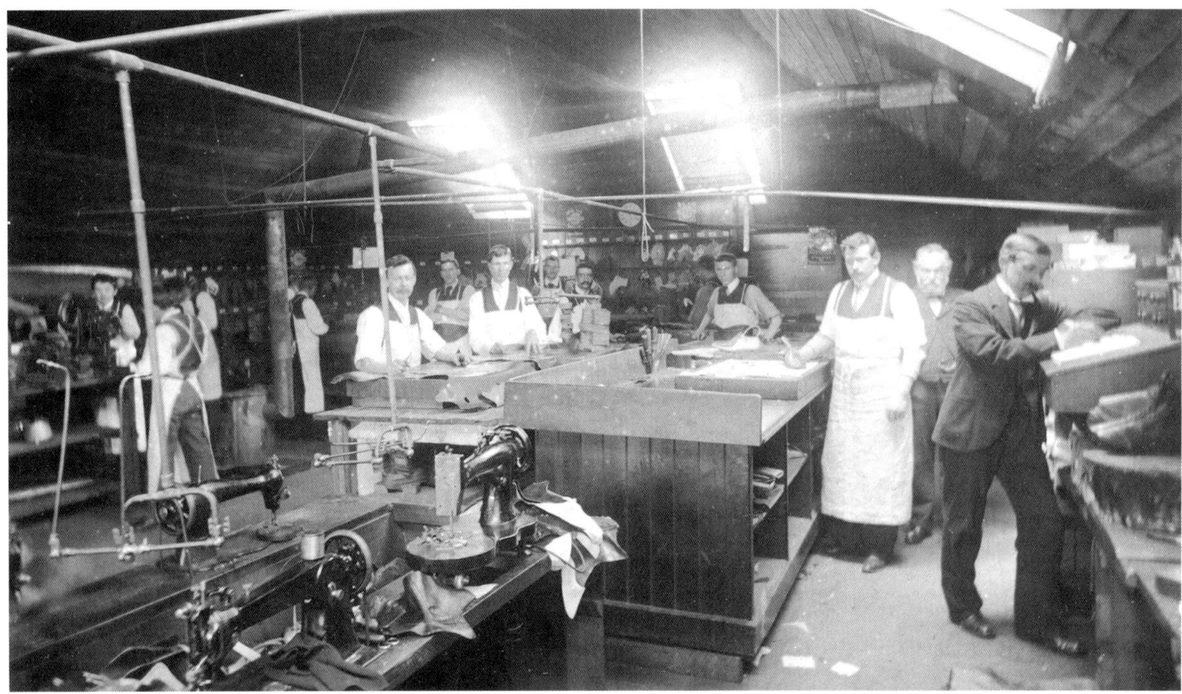

ABOVE **The clicking department at McKinlay's, where the clickers cut out the several pieces that would comprise leather footwear uppers.** Before it was mechanised, clicking was one of the most skilled jobs. This photograph suggests that clicking had not yet been mechanised at McKinlay's: the men would have used a hand-knife and a brass-bound pattern. The workshop has come to a stop while the photograph is taken. Robert McKinlay, dressed in suit and tie, stands just behind one of his sons, probably William, who is by the wooden clicking block, wearing an apron. Robert Jr, the foreman clicker, also wears a suit and tie and is working on his desk. The piping indicates that gas was used for lighting and heating the workrooms.

The sewing room was adjacent (foreground). Dr Jane Malthus, an expert on textiles and clothing, thinks the sewing machine at left foreground is probably a New Royal model, from a company based in Rockford, Illinois (originally, c. 1890, the Royal Sewing Machine Company, from late 1894 the Illinois Sewing Machine Company, and then, possibly about 1910, the Free Sewing Machine Company). From 1917, they produced electric models in conjunction with Westinghouse. The other sewing machine is different, and distinctive, but so far unidentified. Both machines are adapted in different ways for sewing footwear uppers.

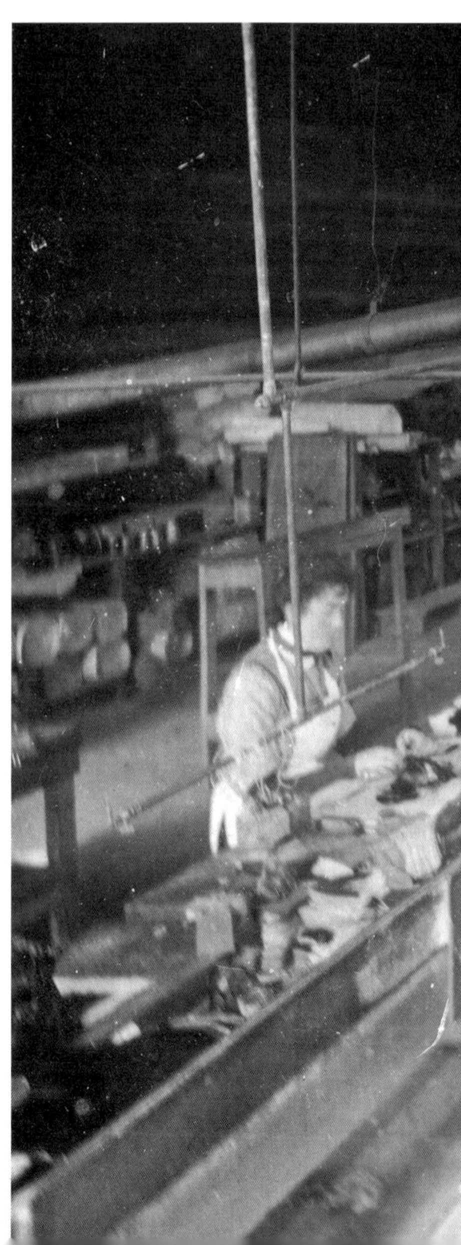

RIGHT **The machinery department at McKinlay's.** On the right, William McKinlay watches the women sewing uppers together, a process that had been mechanised during the 1860s (Isaac Singer, who patented his improvements to the sewing machine in 1851, initially manufactured sewing machines for industrial uses and others adapted them to shoemaking). The sewing machines were probably driven by under-bench pulleys powered by a gas (or possibly steam) boiler until electricity became available, sometime after the Waipori hydroelectric station came on stream in April 1907. Young women dominated the stitching or machine room (although a few were employed in the pump and cleaning departments). Most had probably worked there since leaving school at age 14 (Seddon's government increased the school-leaving age from 13 to 14 in 1901). Good hand–eye–foot coordination was essential for working with the machines.

McKinlay & Sons' shoe and bootmaking factory, c. 1906. Given that William, the third son, born in 1874, appears to be managing the factory in these pictures, his grandson, Bill, thinks they probably date from 1906 or even 1907. Less than 50 years earlier, a knife, needle and awl were the only tools a cobbler used and no difference existed between the shoes for left and right feet. By the 1870s, stitching uppers, women's work, had been mechanised by the sewing machine; next, binding was mechanised; and then the McKay stitcher mechanised bottoming. By 1890, every operation except cutting uppers had been taken over or quickened by machines. Because some 5000 patents had been taken out for mechanising aspects of shoemaking, in 1899 several sizeable American companies decided to share their patents and form the United Shoe Machinery Company, one of the first global corporations or 'trusts'. No industry other than textiles and clothing demonstrated so clearly the process of technological advance. Because they started in the 1870s and 1880s, Dunedin's various shoe and boot factories tended to have the most up-to-date machinery for stitching and bottoming.

One of Robert's innovative sons probably had these photographs taken for advertising purposes. They show five of McKinlay's workrooms and illustrate the process of manufacturing shoes and boots.

The sequence of photographs used here is courtesy McKinlays Footwear Ltd.

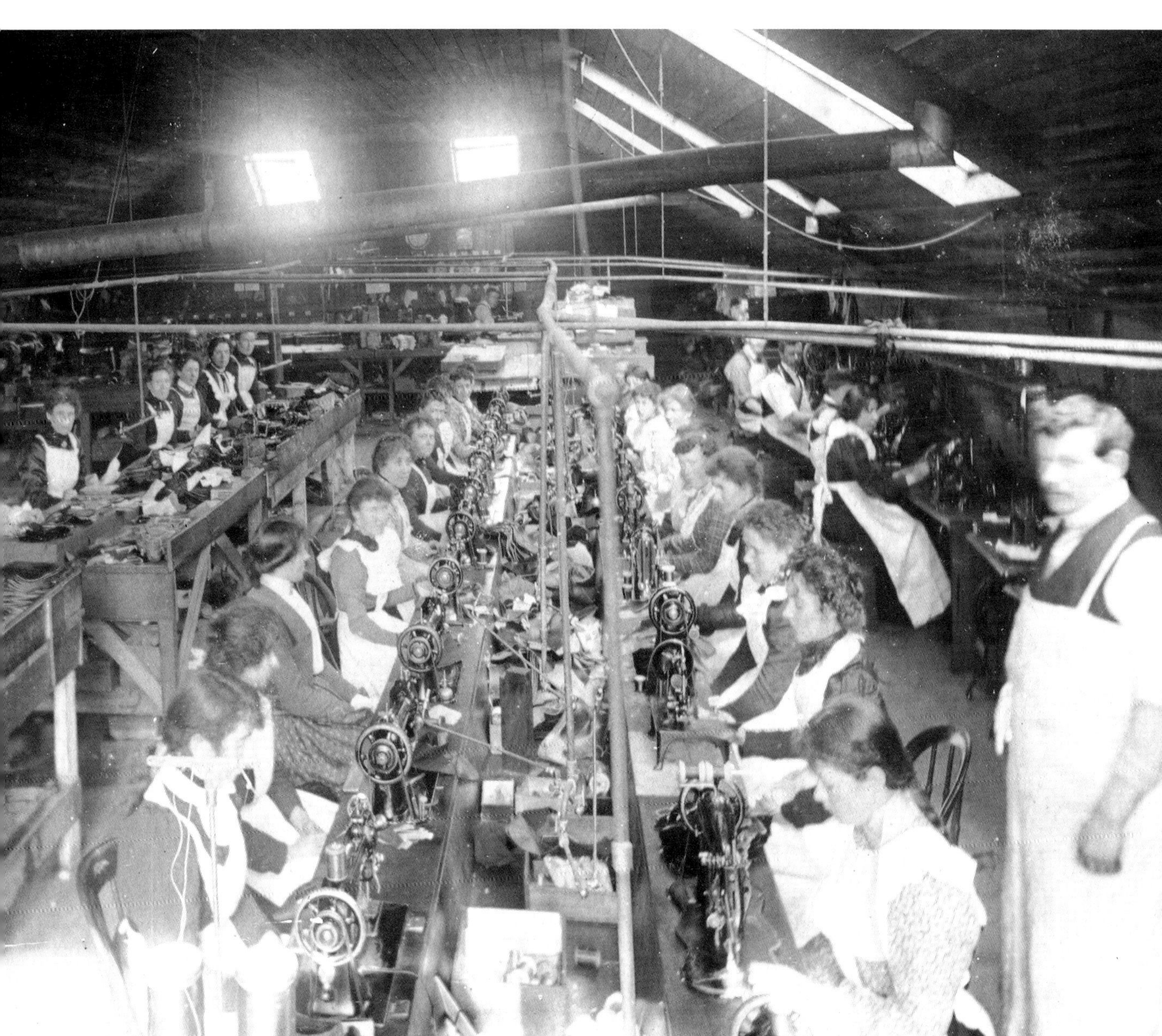

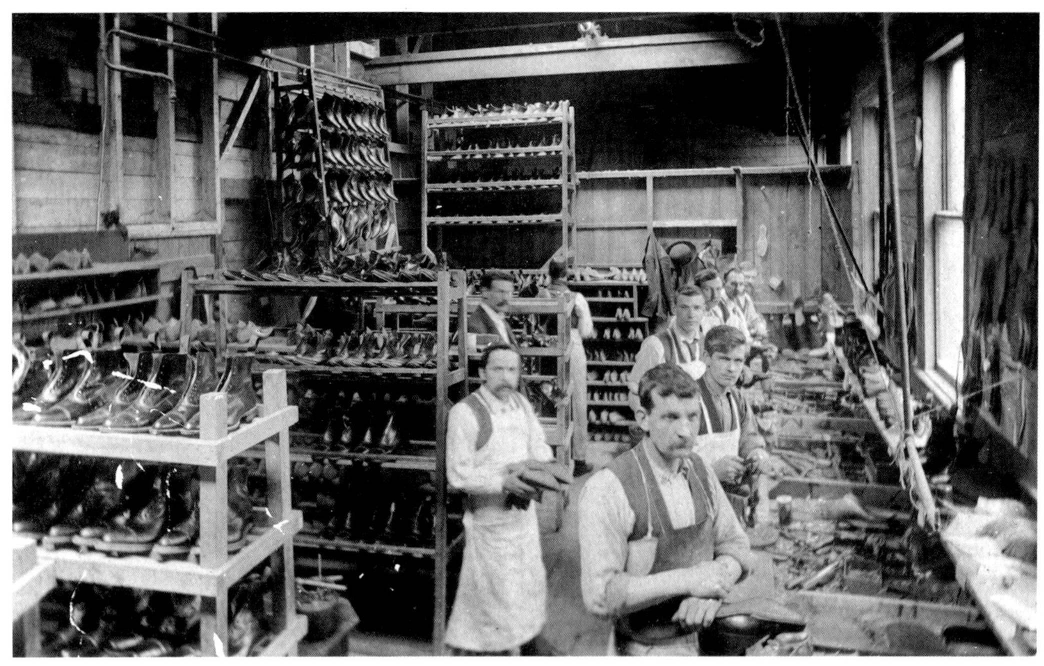

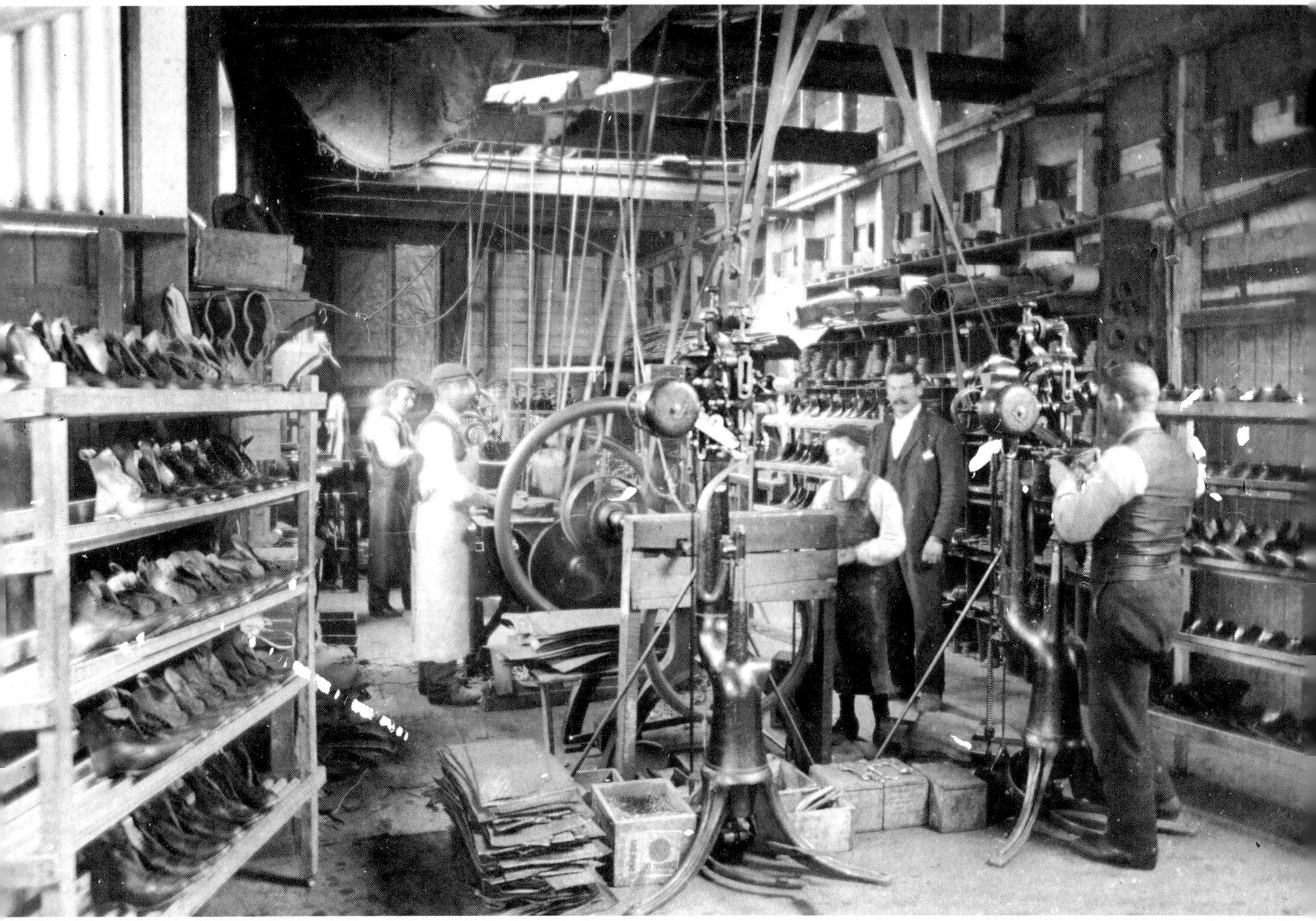

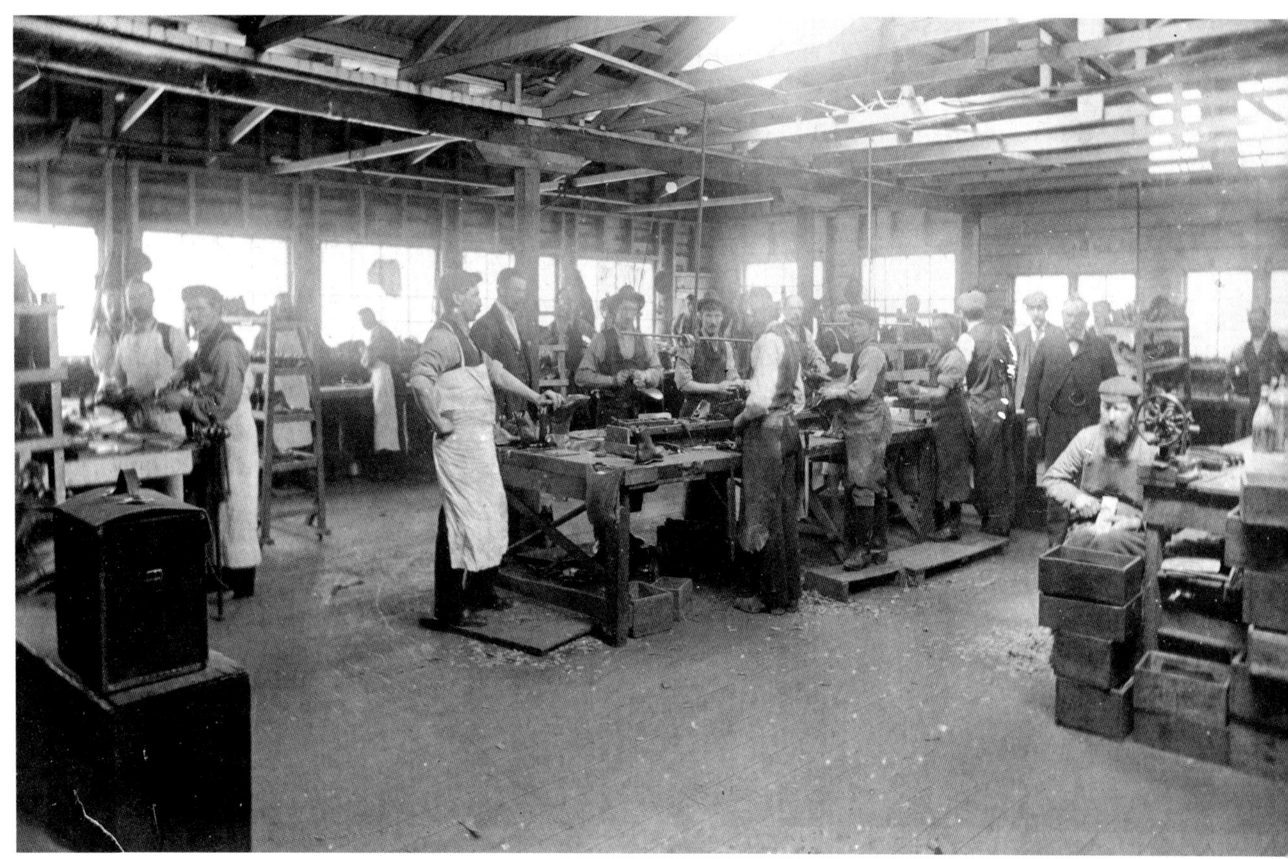

ABOVE LEFT **In this photograph, probably the third step in the process,** we can see the men of the 'rough stuff' department, although at McKinlay's insoling or assembling was also done in this workroom. These men have placed the upper on to a last – there were now specialised lasts for left and right feet – and are putting in the insole. This was no longer skilled work. One of the sons, possibly William, is watching from behind.

ABOVE **In the finishing room, the exclusive domain of males,** men still worked with sharp knives skiving, moulding and rolling stiffeners, channelling soles, trimming, scouring and setting edges, heels and bottoms. The man in the suit on the right is the founder, Robert McKinlay, and the man sitting in front is probably operating a machine that punched holes in the leather to facilitate screwing. William McKinlay, in the suit to the left, is watching the skivers. Judging from the extensive windows and skylights, this room had once been the dining room at the 1888–89 New Zealand and South Seas Exhibition.

LEFT **The assembling room.** Flanked by racks of boots and shoes in lasts, a pile of soling beads – premium leather from the backs of cattle beasts – stacked on the floor in the foreground is being mechanically compressed and flattened to make soles and heels. Once again, William McKinlay watches over the workroom. The man on the right is operating a machine that would be stitching or screwing pieces of leather together to make heels. Heels and soles would then be stitched to the upper. The most successful machines for this task had been patented by Charles Goodyear in 1871 and Lyman Reed Blake in 1888. The machine in the photograph is probably one or the other.

III

Because of the presence of two woollen mills and the various clothing factories, quite apart from the tailors and tailoresses, milliners and hat-makers, Dunedin's young women had enjoyed a range of opportunities denied to most of their colonial sisters. Bendix Hallenstein, who pioneered the department store locally in the 1870s, hired women to work in those departments specialising in products for women to buy. By 1900, when Dunedin boasted some 13 department stores, this had become a major source of employment for many young women. Indeed, if you chose to marry late, or not at all, the department store offered substantial career opportunities for women. In the late 1890s, several businesses – Hallenstein was again to the fore – began to hire young women as typewriters (as the occupation was then known), secretaries and filing clerks. Even the Post Office began hiring young women to operate the telephone exchange (the first in Caversham was situated in Wootton's hairdressing salon on the Main South Road). Except in the clothing industry, the number of women undertaking apprenticeships in the skilled trades fell away sharply.

In the early 1900s, countless married women were complaining that they could not find domestic help – every married woman needed such help when babies were born or when ill or injured – and factory owners were complaining that they could not get women workers. Although a powerful Victorian cultural convention required women to leave the paid workforce when they married, necessity compelled some to ignore this rule. Only government departments enforced it strictly. Married women usually could not manage full-time work away from home, however, and if forced to earn money preferred short-time, casual work or took it into their own home.

The white-blouse opportunities that opened in the aftermath of the Long Depression continued to expand. Thousands of young women found themselves able to escape domestic service or even factory work – the only choices for their mothers – and become teachers, nurses and, in a handful of instances, doctors. In these years a surprising number of young Catholic girls entered the Sisters of Mercy, which had only recently taken over the South Dunedin convent, orphanage and schools. An equally surprising number of young Protestant women became deaconesses or missionaries.

The largest racial minority, the Chinese, were excluded from many lines of work. Any attempt to enter industries with powerful unions inevitably provoked an explosive backlash. Dunedin's 'Big Bill' Belcher, national secretary of the Seamen's Union, was the foremost champion of racial exclusion, but the Liberals were generally supportive. The entrepreneurial frontier, unlike many labour markets, was more open, as both the Chinese market gardeners and later the 'Assyrian' hawkers discovered. There were other 'aliens' living on the Flat, including at least one West Indian and several Italians, but they were so few that nobody felt threatened by them.

It is more difficult to know to what extent religion became the basis for discrimination in the labour market. While it is unlikely that members of the Church of Christ or the Brethren would have hired atheists or Catholics,

The clothing industry, one of the first to mechanise, had long had a gendered division of labour and a distinctively hierarchical structure, depending on which market segment was being targeted. Male bespoke tailors and female bespoke dressmakers came at the top of the hierarchy. The mechanisation of the clothing industry made the industry even more complex and diverse than it had been. As factory production successfully won more of the market, Roslyn Mill owners Ross & Glendining, among others, opened their own clothing factory specialising in ladies' clothing. Rising real incomes and a newly fashion-conscious market made the venture successful. Before long, department stores created their own workshops, which competed with the factories.

By 1900, some 27 per cent of Dunedin's workforce were employed in the clothing industry, almost 80 per cent women. Independent workshops in the suburbs, like Todd & Brown tailors in Caversham, came under increasing pressure as the electric tramways created a unified urban shopping market. To make matters worse, the sewing machine (100,000 were imported between 1880 and 1900) made each woman potentially her own dressmaker.

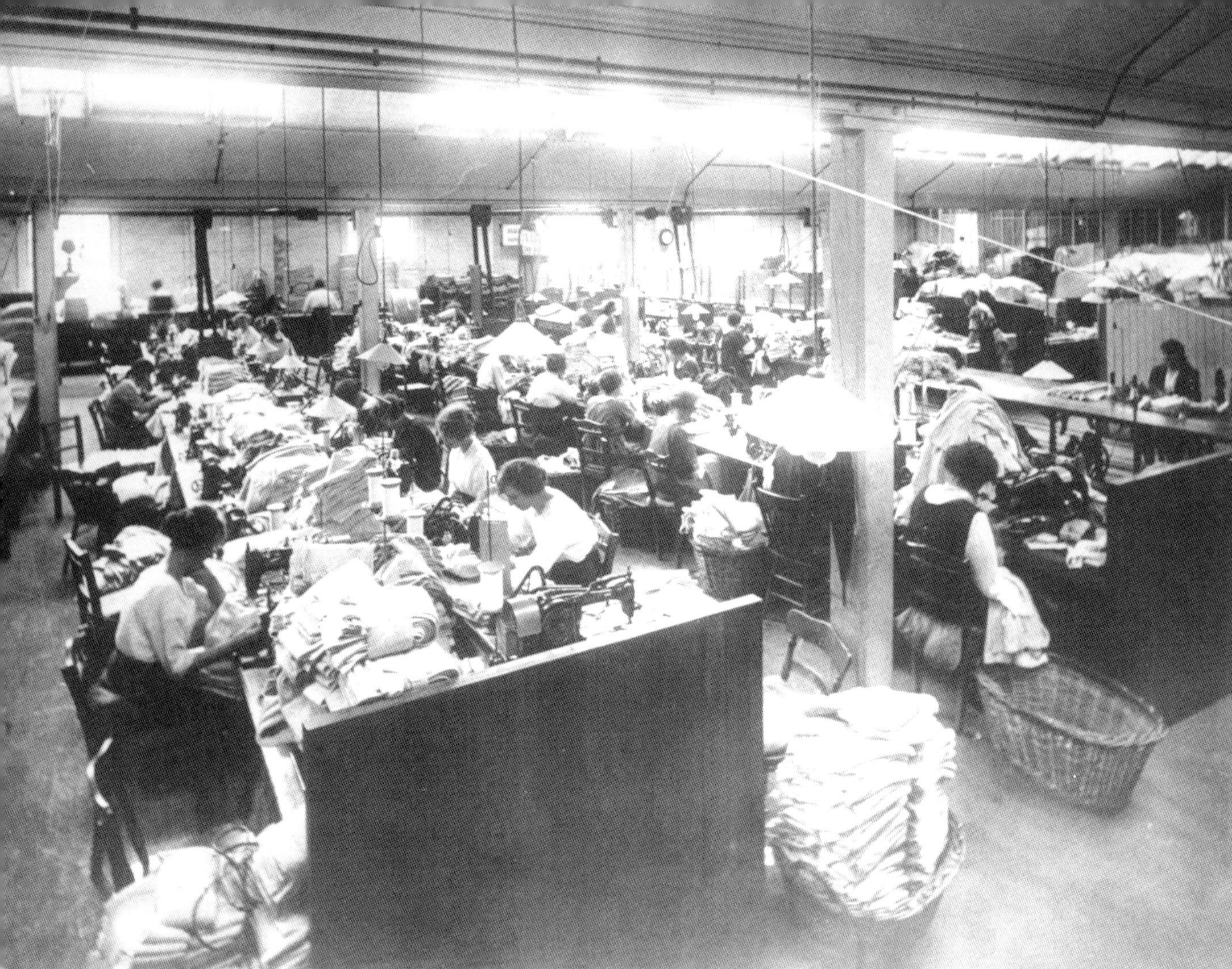

it was also unlikely that Catholic employers would have hired members of those Protestant sects. Deeply devout members of some denominations often preferred their fellow parishioners. Henry Shacklock, a leading Congregationalist, preferred fellow Congregationalists. William Ings, the Baptist nurseryman, preferred to hire Baptists. The church, of course, was not only a place of worship but the centre of a network of potential help and assistance. In other instances, nationality and religion doubtless combined. Grocer McCracken tended to recruit his customers and staff from Ulster's various Protestant communities; grocer Rutherford, immediately across the street, tended to recruit his customers from Scotland and England.

Firms may have selected on the basis of religion – there is much anecdote but little systematic evidence – but it certainly never happened across an industry, unlike in Britain. There were plenty of Catholic employers locally, of course: Petre, Lee Smith, and Marlow, all English Catholics, and the Shiels and Heggarty, Irish Catholics, ensured that there were ample opportunities for young Irish-Catholic lads not only to find work but undertake apprenticeships. Irish-Catholic lasses, by contrast, were more constrained by traditional expectations. Nor did the state, the largest employer in the country by far, discriminate on the basis of religion. When the well-known

ABOVE **Workroom at Ross & Glendining's Woollen and Worsted Mill, c. 1921.** Women are working on heavy-duty sewing machines designed for heavier weights of cloth, such as blankets (possibly they are edging blankets, but we cannot be sure). The work was organised on a batch system, where the garment or item was moved through a production line on which each worker executed a separate operation. These machines were electrically powered.

Hocken Photograph Collections, SO3–184d.

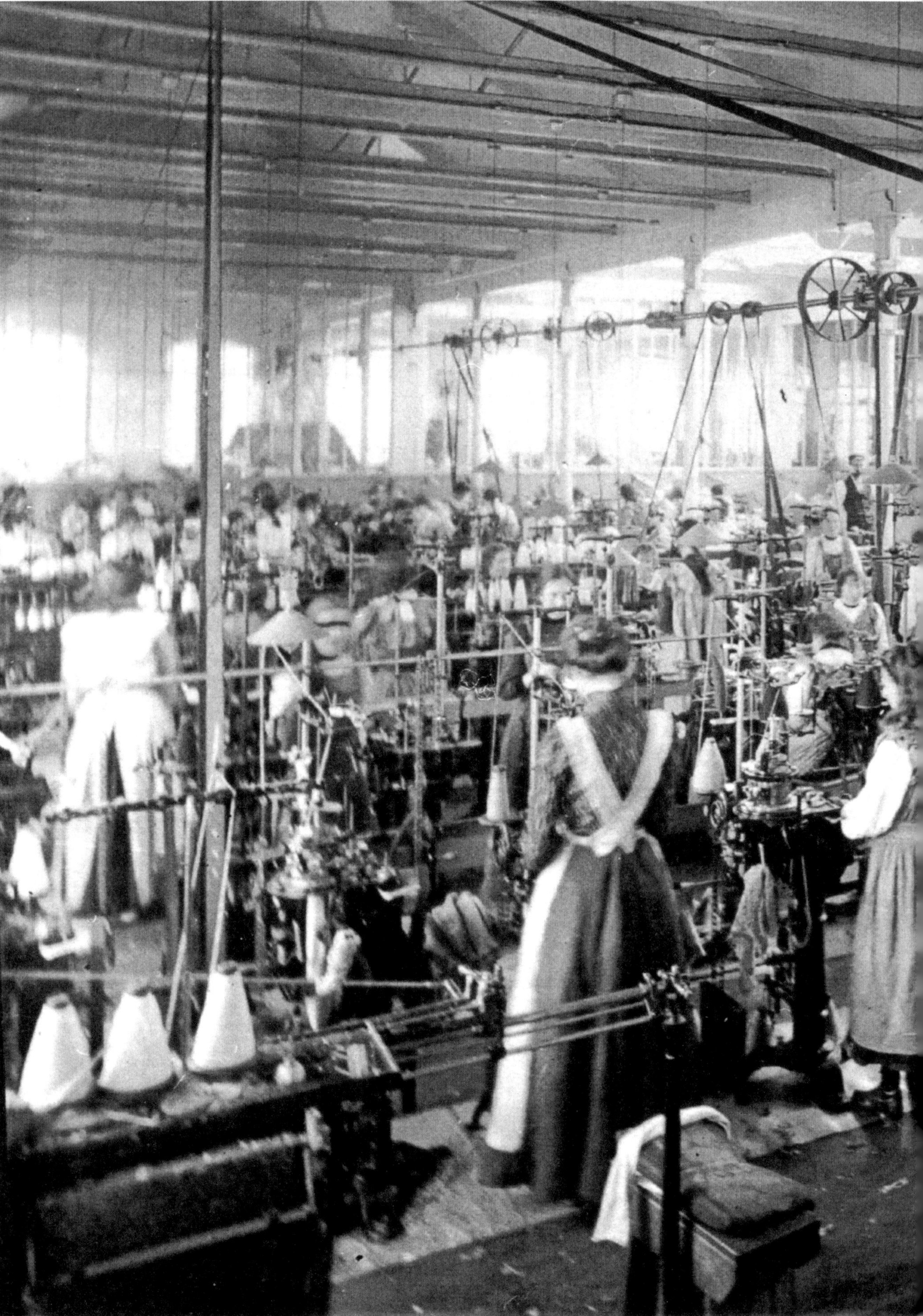

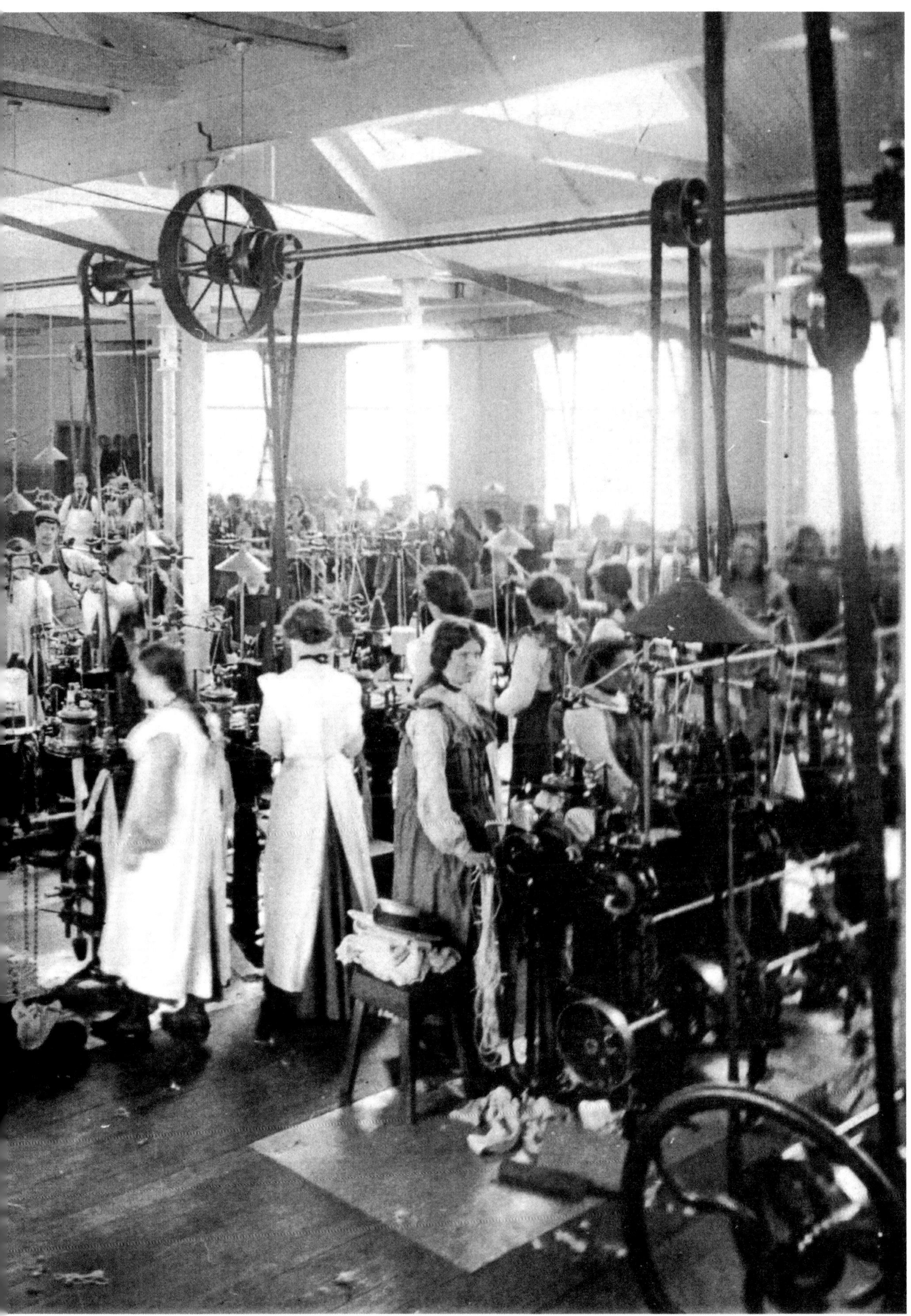

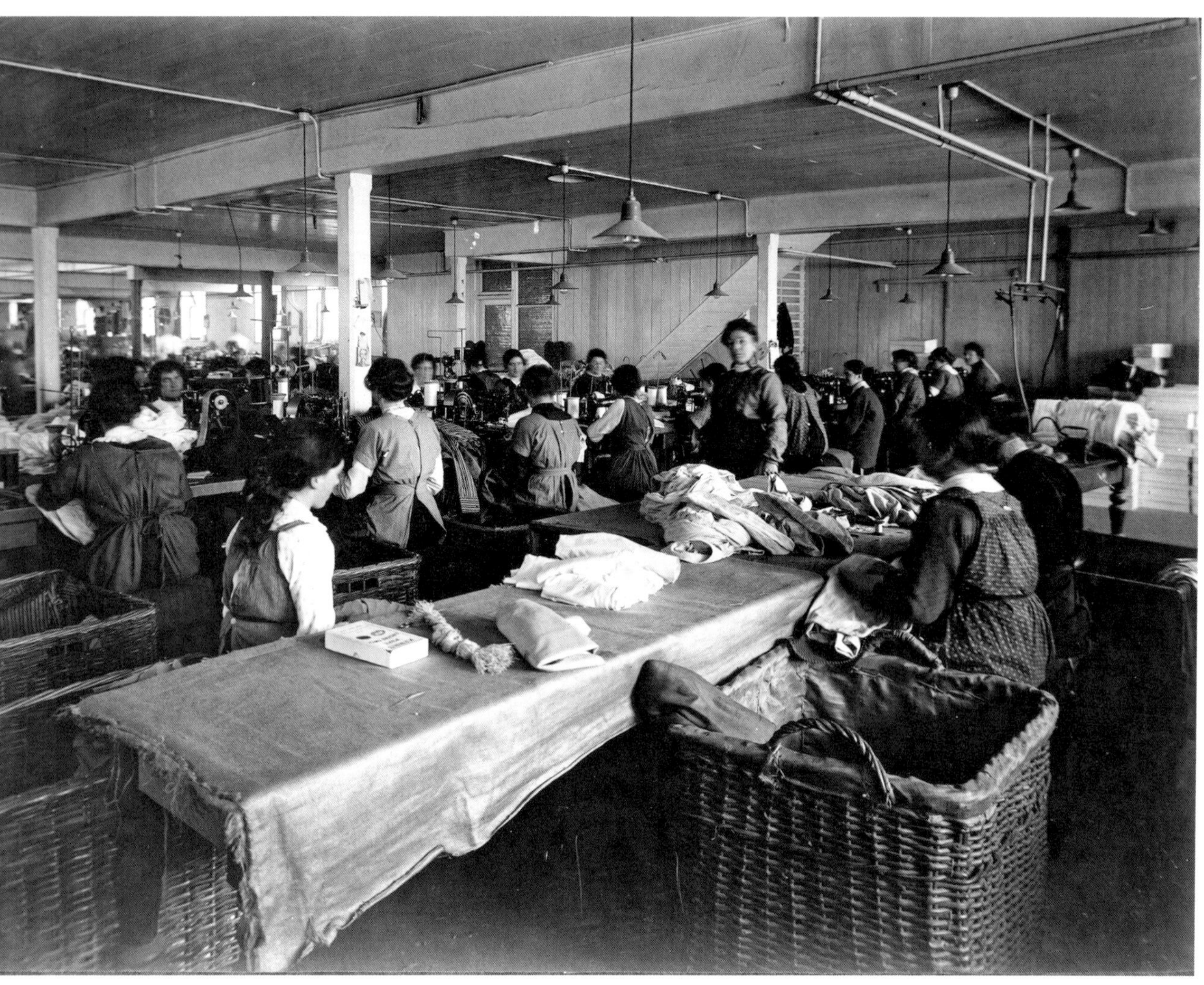

PREVIOUS PAGES
The Hosiery Workroom, Ross & Glendining's Roslyn Woollen Mill, c. 1910. These young women are supervising steam-driven circular knitting machines that are manufacturing hosiery. The workers had to stop the machine if a stitch was dropped or a mechanical fault occurred. Latch needles were especially prone to malfunction. It is not clear from the photograph whether an entire sock or stocking was being made. If it was, however, the worker then had to alter the settings when the toe and the heel were to be sewn.

At the Roslyn Woollen Mill in Kaikorai Valley, according to S.R.H. Jones's excellent history, *Doing Well and Doing Good* (2010), giant looms now wove the woollen yarn into cloth. Woollen rugs, blankets and garments were also manufactured. The hosiery department was expanded in 1901–02 in response to rising demand for high-quality hosiery. Demand kept rising. In the new worsted mill, which opened in 1904, the hosiery department occupied the first floor. The old Griswold knitting machines were replaced by new automatic circular knitting machines.

'Women working in the Roslyn Woollen Mill', Alexander Turnbull Library, Making New Zealand Collection, PAColl.3060, MNZ-0704-1/4F.

ABOVE **The shirt-making room in the New Zealand Clothing Company's factory in Dowling Street.** Each table in this workroom specialised in a different part of the work, and when each process had been completed the shirt was tossed into one of the big wicker baskets, as can be seen in the foreground, so that it could be moved to the next table. The seamstresses at their sewing machines dominate the workroom, while the women nearest the camera are either hand-stitching or labelling. They may also

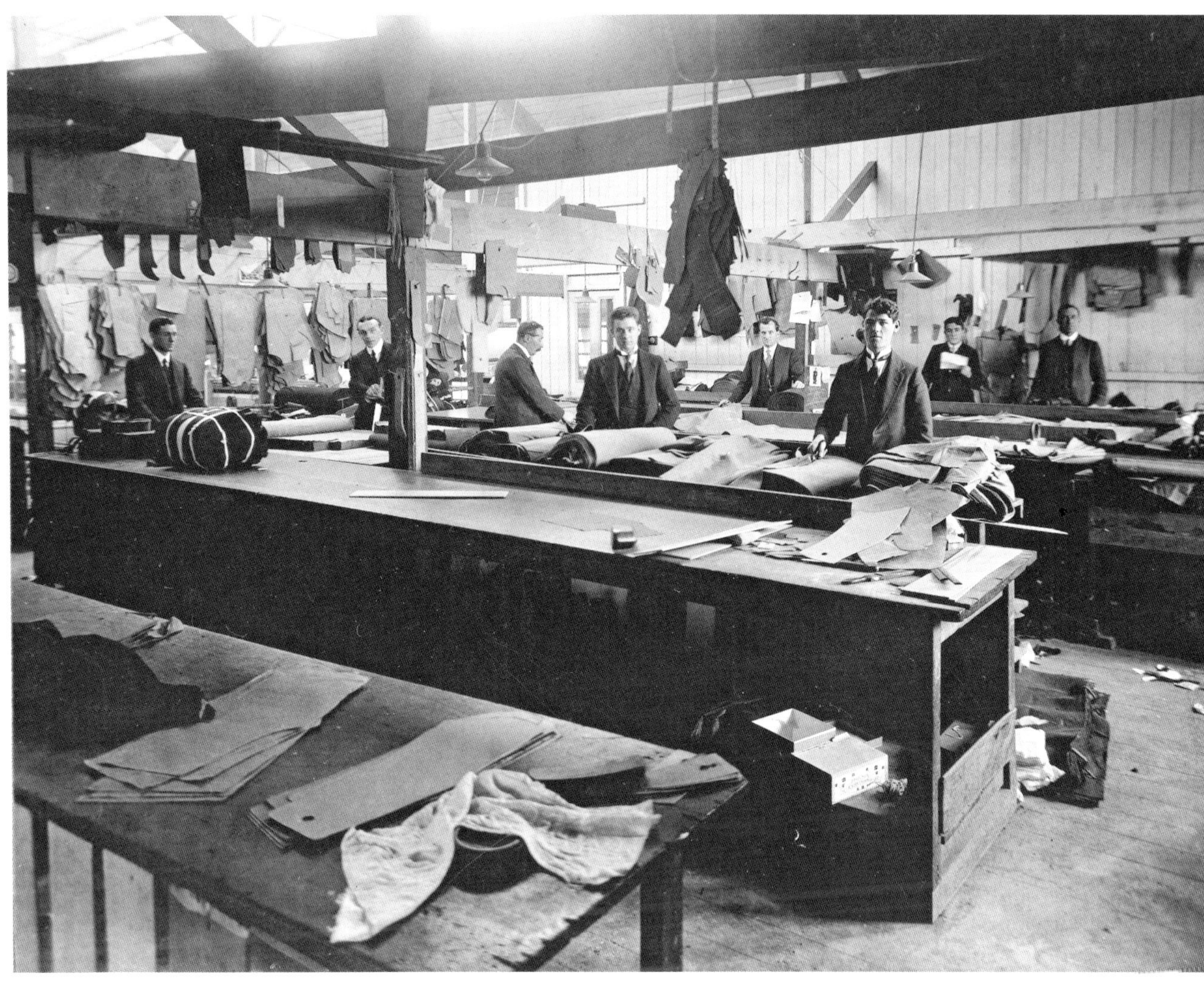

be engaged in quality control. The woman standing is the supervisor. Note that all the women workers are wearing protective aprons, each one different, indicating that the women provided (and probably made) their own. We are not sure when this photograph was taken or who took it but it may be another series by Armstrong & Greer, in which case it is probably 1899. According to Dr Jane Malthus, who was indispensable in interpreting these photographs, the clothes being worn confirm that date.

Hocken Collections, S13-278c.

ABOVE **The male cutters were the elite of the clothing industry,** whether working in factories or bespoke workshops. Here – in the New Zealand Clothing Company's top floor – the pattern-maker made patterns from manila cardboard; the layers-out laid out the cloth in as many layers as could be cut together, and the cutters then used scissors to cut out the pieces needed for manufacturing trousers, jackets and waistcoats. The company had the contract to manufacture uniforms for various organisations and firms, including the Territorials. Note how well dressed the men are, each wearing a three-piece suit, some with modern wing collars, others with starched high collars. When the cloth had been cut, it was sent downstairs depending on what garment it was needed for.

From 1873 until 1905, the New Zealand Clothing Company used steam to power its machinery, the coal-fired boilers being on the ground floor, but in 1905 they switched to gas, using a six-horsepower Otto gas engine.

Hocken Collections, S13-278g.

OPPOSITE **This photograph captures a distinct aspect of a worker's life around 1910–11.** Employers in Dunedin accepted a paternal responsibility for the wellbeing of their employees. Although the institutions of company welfare have been ignored by historians in this country, they need to be understood in the wider context of a concern for the moral wellbeing of their employees (most of whom were between 14 and 24 years old). Evangelical Protestants were also keen to reach out to the working classes in their places of work, as they were reputed never to attend church. Here we see a group of employees, all young, listening to two Baptist Evangelists from the United States, known as the Henry Potts Mission, who travelled the country in 1910 conducting an 'industrial mission'. These well-dressed employees of Ross & Glendining's Roslyn Mill were given time off by Glendining to attend the mission.

Hocken Collections, *Otago Witness*, 20 July 1910, p. 45; image S13-590a.

Catholic, Joseph Ward, became Seddon's Postmaster General and Minister of Railways, some evangelical Protestants suspected that his departments actually gave preference to Catholics. While there is no evidence of this, Catholics were by 1920 heavily over-represented in both the Railways and the Police. Working for the government was a smart idea, however, because it offered security, and many Irish preferred security to opportunity. Indeed, the only tradition subverting the value of work was the Irish-Catholic one exemplified by that notorious work-shy 'swagger', Edmond Slattery, known as 'The Shiner', a legend no less for his alcoholic benders than for his skill at avoiding work. John A. Lee, another boy from Dunedin, although he only ever visited the Flat to stay with an aunt, immortalised the Shiner in his wonderful *Shining with the Shiner* (1950).

In Britain the bonds of a common, persecuted religious denomination sometimes proved as effective as kin in underwriting business partnerships. The memory often allowed such ties to play the same role here.

IV

The gulf in status that separated skilled and unskilled, and manual from non-manual workers in Britain simply did not exist here. If you saw a contemporary photograph of a joiner or fitter in Britain, you could be reasonably sure their fathers had belonged to the same class and had even worked in the same trade; you would be quite wrong to make that assumption of those who appear in the photographs in this chapter. Boys rarely followed their fathers into an occupation and surprisingly often did not even end up in the same social class. Indeed, the freedom of choice was such that families quickly, in many cases, became cross-sections of their society. Although the processes matching persons with jobs are not well understood, even in more rigid societies the matching process is structured by preferences and bargaining. In this new world of the Flat, preferences and bargaining had much larger sway, although social networks and inertia were not unimportant. This is clearly underlined by the manner in which second-generation boys of Irish-Catholic parents abandoned the first generation's obsession with settling on the land and headed into skilled and white-collar jobs. It is also underlined by the speed with which young women abandoned domestic work and headed first for the new frontiers of freedom offered by factories, then department stores and finally offices.

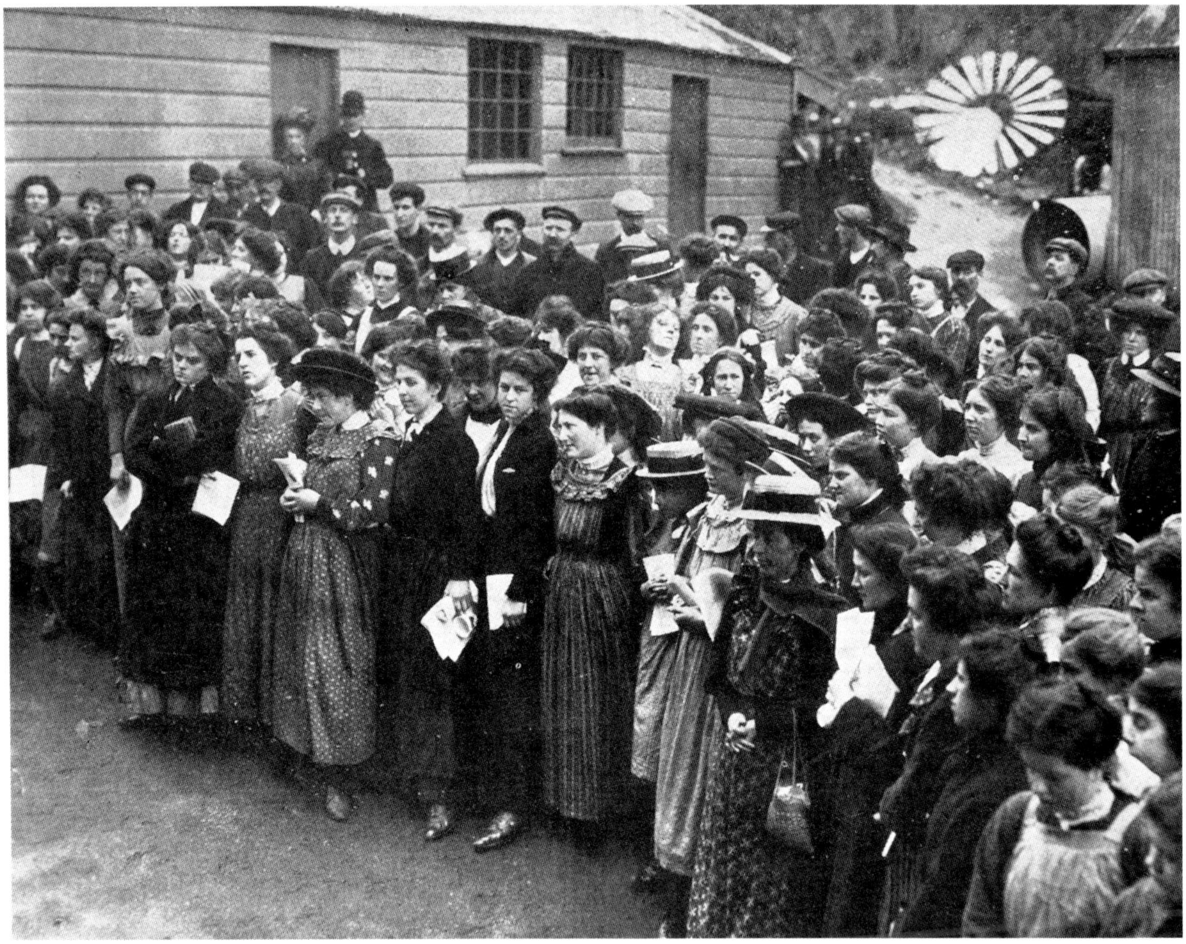

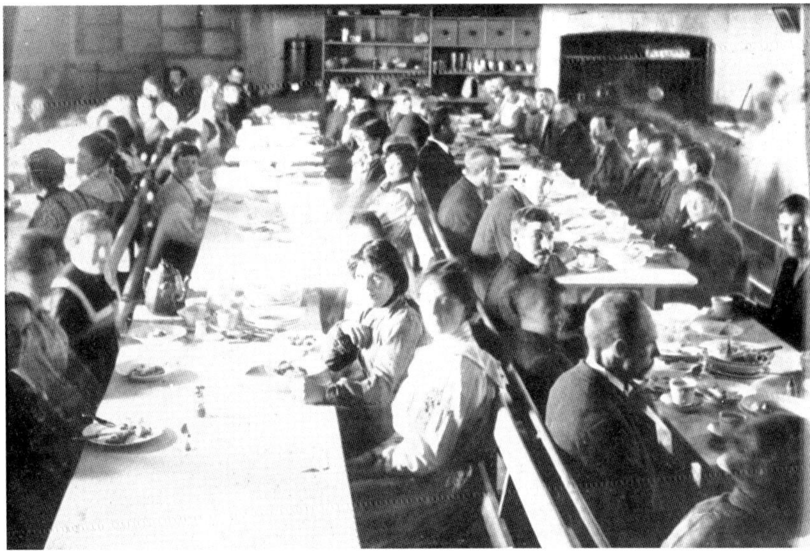

RIGHT **Workers in the dining room at Irvine & Stevenson, Preserved Provision Manufacturers, Moray Place, Dunedin, c. 1895.** The firm, which ran from 1864 until 1977, employed 17 or 18 women at this time. According to some of those interviewed during the Caversham Project, factories employed the 'lowest class of girl'. The mixed dining room was unusual at the turn of the century, when most shoemaking factories still provided separate entrances and dining facilities for women and men. A small number of photographs taken at Irvine & Stevenson, mostly by S. Collins (like this one), have survived. No photographs or archives relating to the Phoenix jam-making factory on McClaggan St have been found.

Hocken Collections, Irvine & Stevenson Ms-696-01379; image c/n E5992/30.

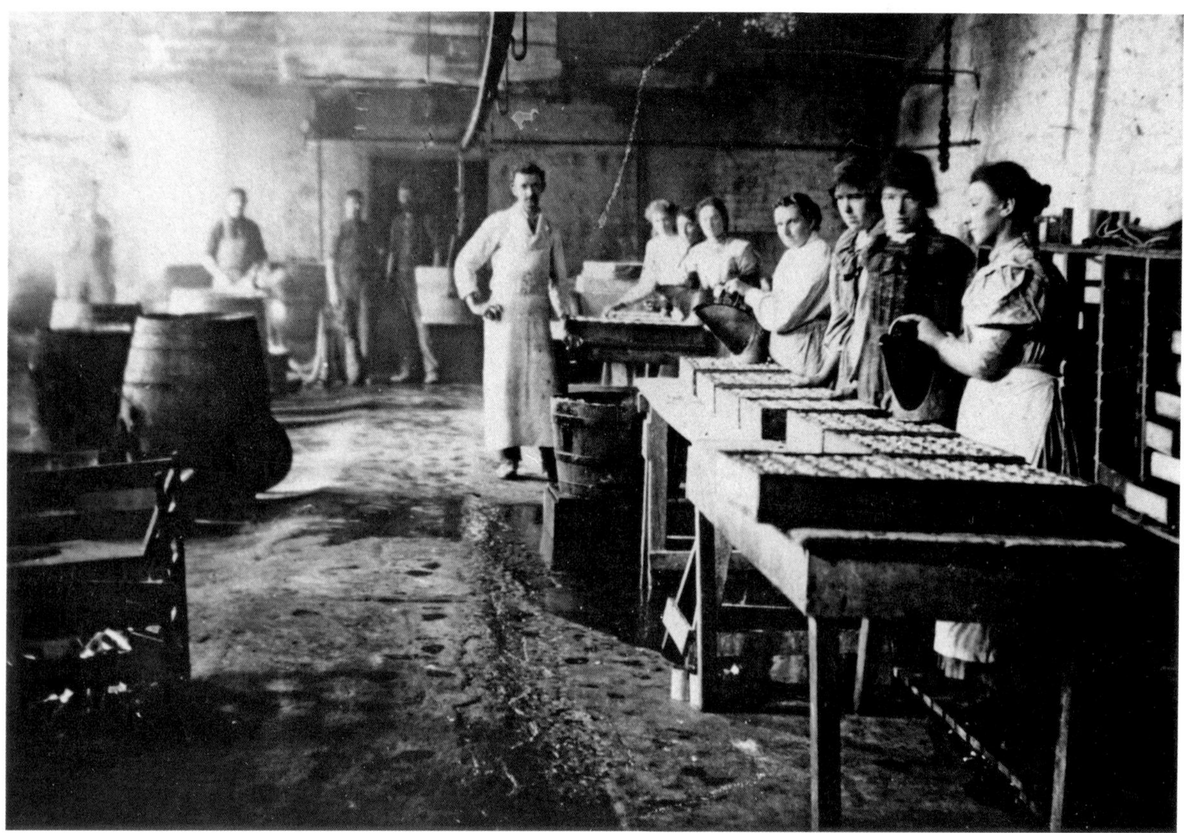

ABOVE **The St George jam room at Irvine & Stevenson's factory, c. 1895,** where the largely female workforce, supervised by James Irvine, are packing boxes with jars of jam. Irvine & Stevenson had two stores nearby, another in Wellington and exported to Australia, the Pacific Islands and England. Jam-making was a rapidly growing branch of their business, and they were the city's most successful jam-maker, although the firm made many other products as well. Thanks to a patent Irvine & Stevenson held for SO2 gas, which helped ensure the fruit did not deteriorate, they were able to ship fruit from their orchard in Motueka for jam-making in Dunedin. Until c. 1896, the jam was made and packed at the 'No. 2 factory' in Moray Place (bought from Peacock & Co in 1891). By 1897, all manufacturing was consolidated in a new purpose-built factory in Filleul Street. Photograph by S. Collins.

Hocken Collections, c/n E 6000/33.

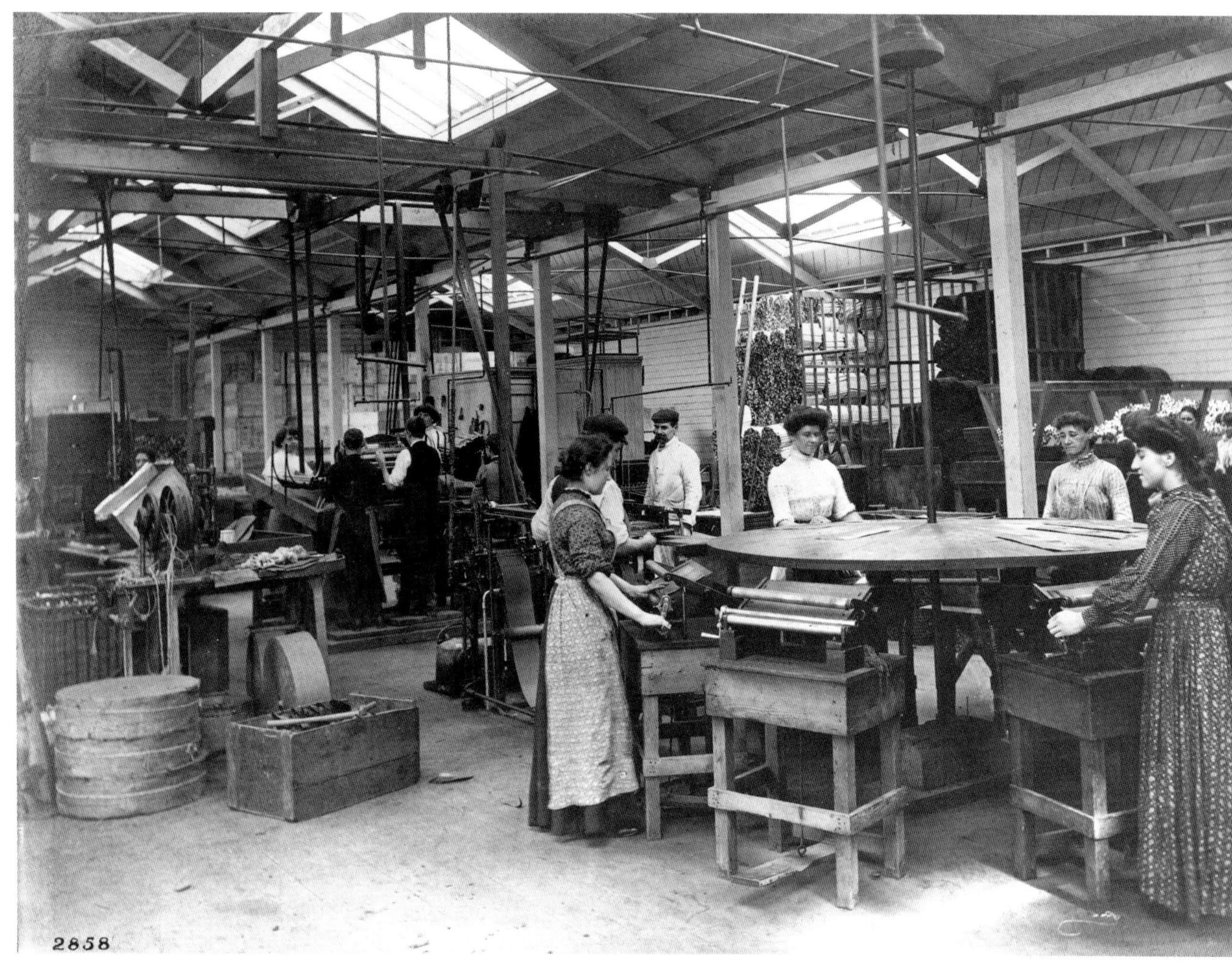

ABOVE **Most of the workers in Wax Vesta factories were women**, apart from the manager and two technical staff. Robert Rutherford and his son, also Robert – cousins of the grocers – opened the New Zealand Wax Vesta Company's first factory in 1894 in the old Immigration Barracks and later built a new factory on David Street, near Forbury Corner. Their partner, John Watson, was a fitter with a passion for machinery. No photographs taken inside the Dunedin factory in this period appear to have survived, but the industrial processes were identical to those of Bell's Wellington Wax Vesta Factory, shown here. This was the most labour-intensive part of the process: the making and filling of the boxes. The process of coating cotton thread with wax, rounding and cutting it and dipping it in phosphorous, was highly mechanised, with only a few women being needed to supervise the machines. Almost all the employees were young women, although as time went on a few retained their jobs even after they married and had children. The work was regular and the pay reasonable. Factory 'girls' had a poor reputation in general, and the 'matchy tarts' were no exception; they often slanged passersby as they stood outside during lunch breaks. The Rutherfords worked very hard to improve their female workers' sense of self worth by insisting on a dress code, providing a library, and later encouraging the formation of a marching team.

Alexander Turnbull Library, Wilkinson Sword Collection, Labour Department Album, F77173 1/2.

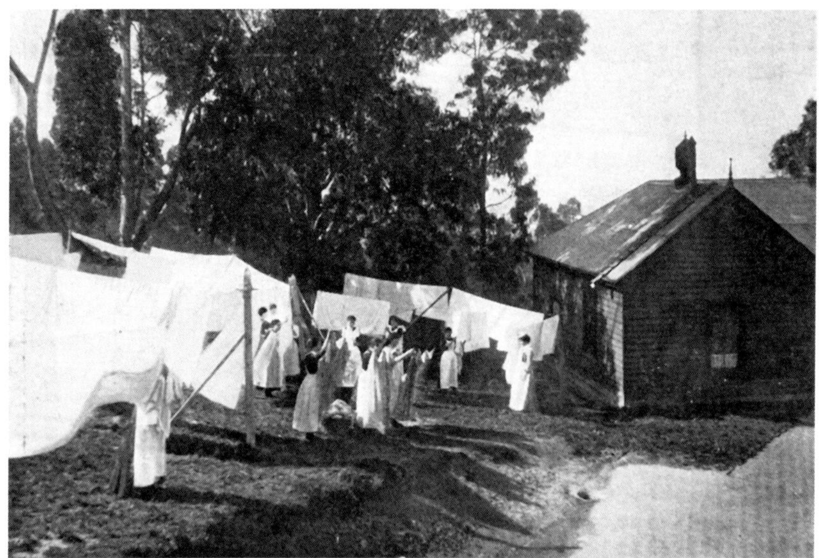

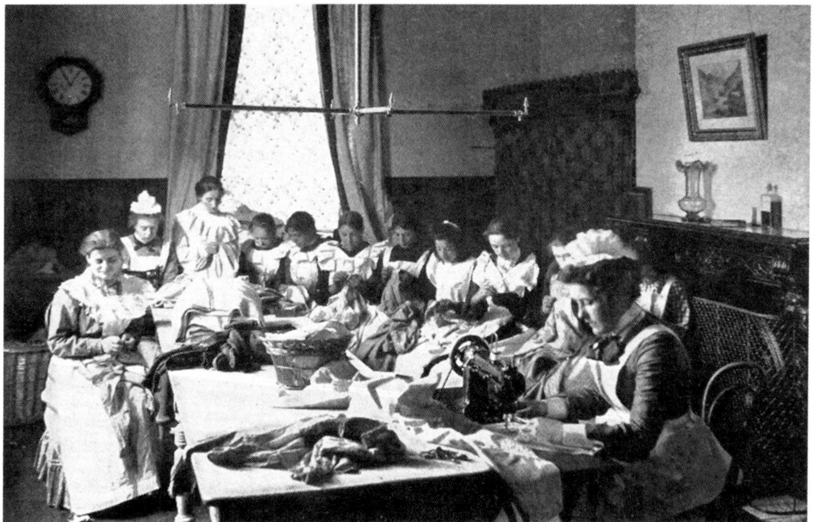

ABOVE **The Industrial School, Caversham.** After a scandal, the Department of Education took over the country's industrial schools. They promptly separated those children who were juvenile offenders from those who were there because they had no adult able or willing to look after them. The Caversham school became predominantly a home for orphans and neglected children. Such institutions aimed for self-sufficiency, the males learning to work in the gardens, and the females learning to do the washing and cleaning. At this time, boys could also learn the principal trades; girls were trained for housework. This was not necessarily to ensure a supply of domestic servants for the well-to-do (at this time, every woman needed help around the house when nursing a baby); rather, this training prepared them for running their own households.

Hocken Collections, *Otago Witness*, 28 September 1904, pp. 42–43

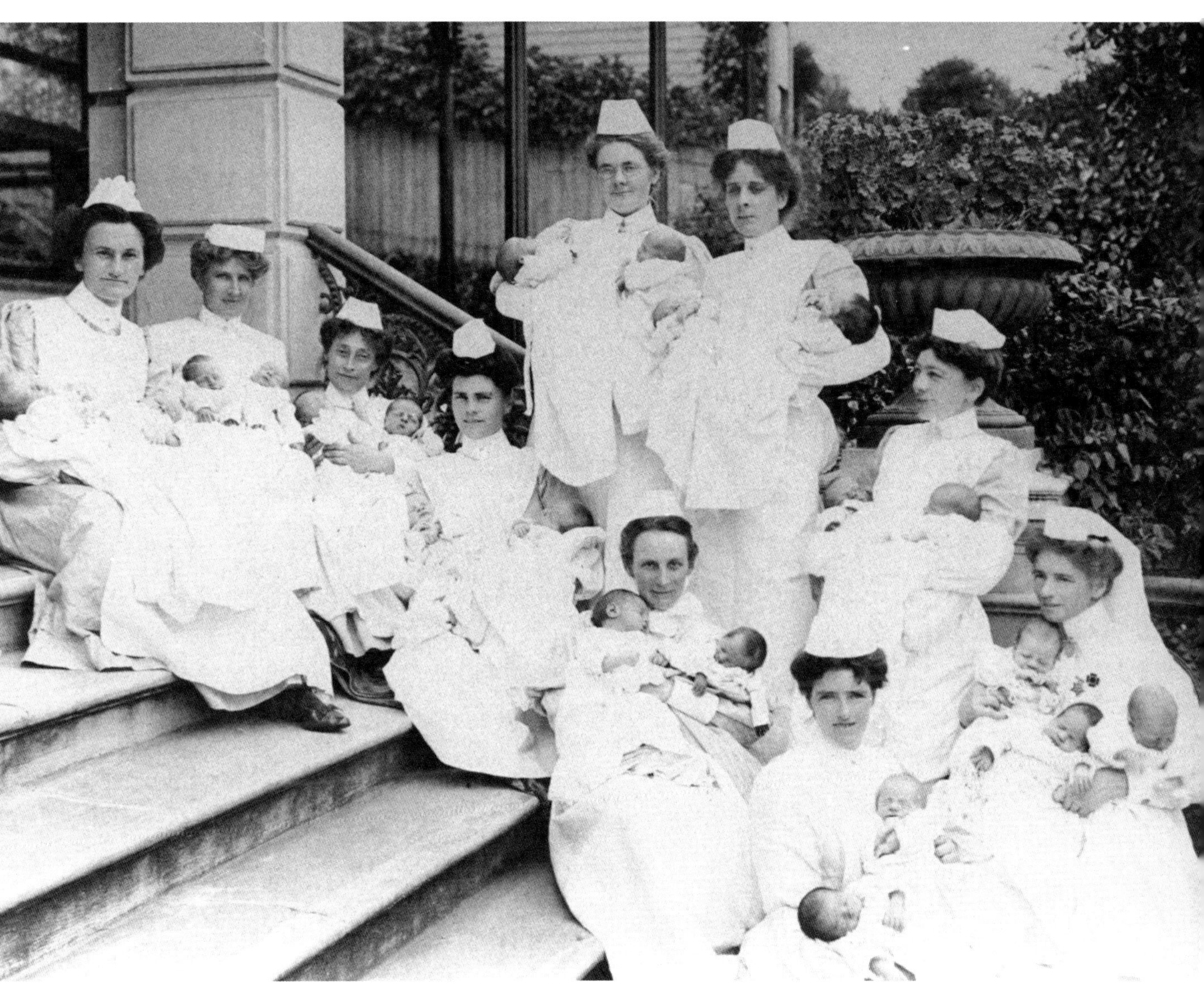

ABOVE **A group of St Helen's nurses, c. 1908–09,** with other women's babies, pose outside the St Helen's maternity hospital in Forth Street. The Seddon government had instituted this maternity system for the benefit of those who did not have the costs of birth covered by a private provider (most friendly societies introduced such benefits in this period). Like Dr Truby King's Karitane Nurses, who specialised in looking after any babies admitted to the Karitane Home, St Helen's nurses were not as skilled as general nurses; however, the pay was good, job security excellent, there were opportunities for promotion, and the work provided a rich social life and the self-respect that came from belonging to an acknowledged profession. Photograph by Guy Morris.

Hocken Collections, S03-202e.

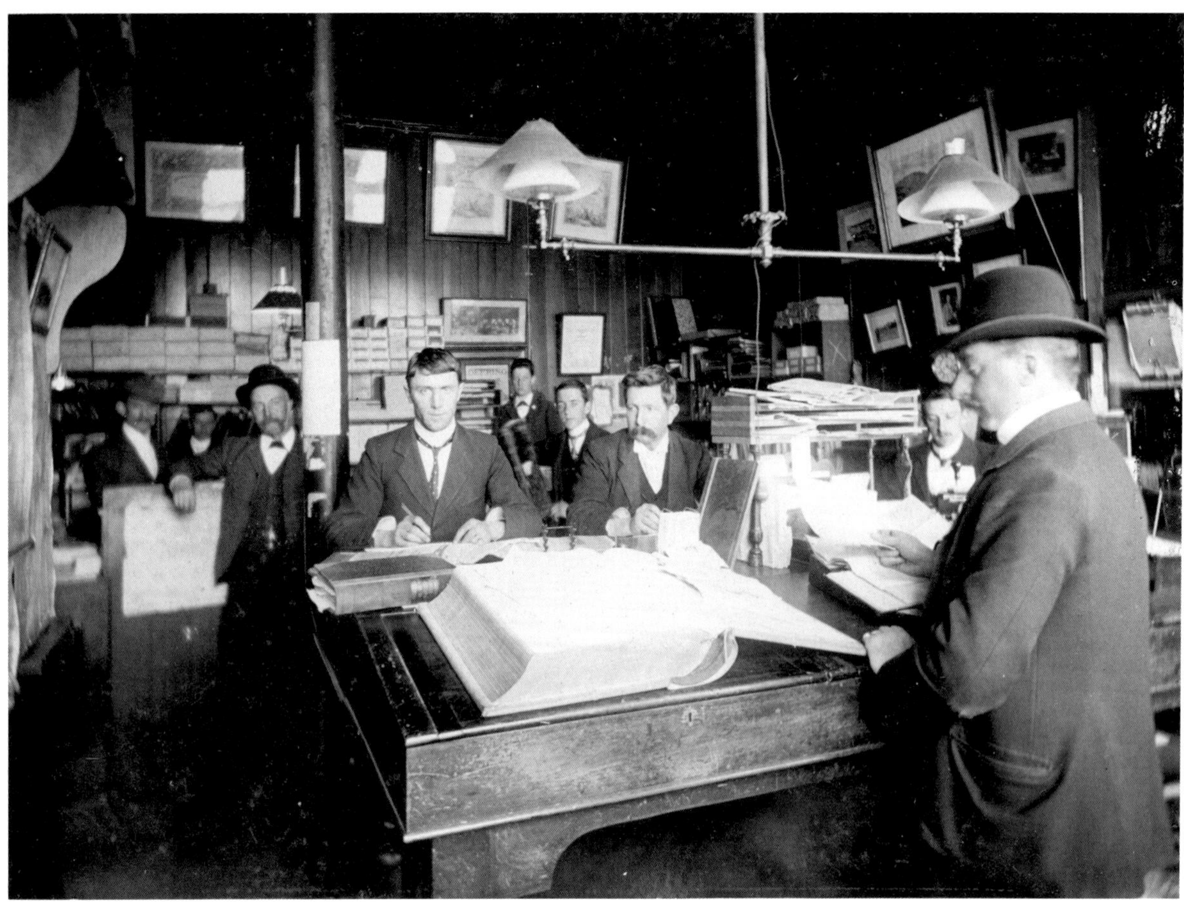

ABOVE **The head office of Reid & Gray, Princes Street South, c. 1900.** As manufacturers of machines for farmers, Reid & Gray had no rivals in the South Island. Their extensive workshops and offices were immediately to the east of Kensington. At the turn of the century, most clerks were men who had usually completed at least two years of secondary school. After schooling to the age of 14 became compulsory, the educational requirement increased. These men enjoyed considerable responsibility and excellent prospects for promotion to the position of company accountant or even manager. They pose with their ledgers for tracking the movement of all orders, goods, money and weekly wages. At this time, it was still unusual to find women employed in offices.

Hocken Collections, Reid & Gray Photograph Album, Ms-1165-063.

RIGHT **The Dunedin telephone exchange, 1920s/1930s.** When the first telephones were introduced in the 1870s, men staffed the exchanges. In the early twentieth century women increasingly challenged and, during World War I, overthrew their dominance in most aspects of the work. Like other government departments, the Post and Telegraph Department forced women to retire from the service when they married, but the policy proved unworkable in suburban and country post offices, where married women often served for decades, as happened in Caversham. The union, completely dominated by men, successfully excluded women from all work other than in the exchanges and typing pools.

World War I delayed plans to introduce automatic exchanges. Rotary automatic exchanges of the WE7A variety (a dial phone connected to an automatic exchange) were installed in the 1920s, Dunedin receiving its in 1930. During that decade, the number of subscribers rose from 72,000 to 157,000.

Hocken Collections, *Otago Witness* Collection; OW neg. 517, c/n E5663/30, 695-00900.

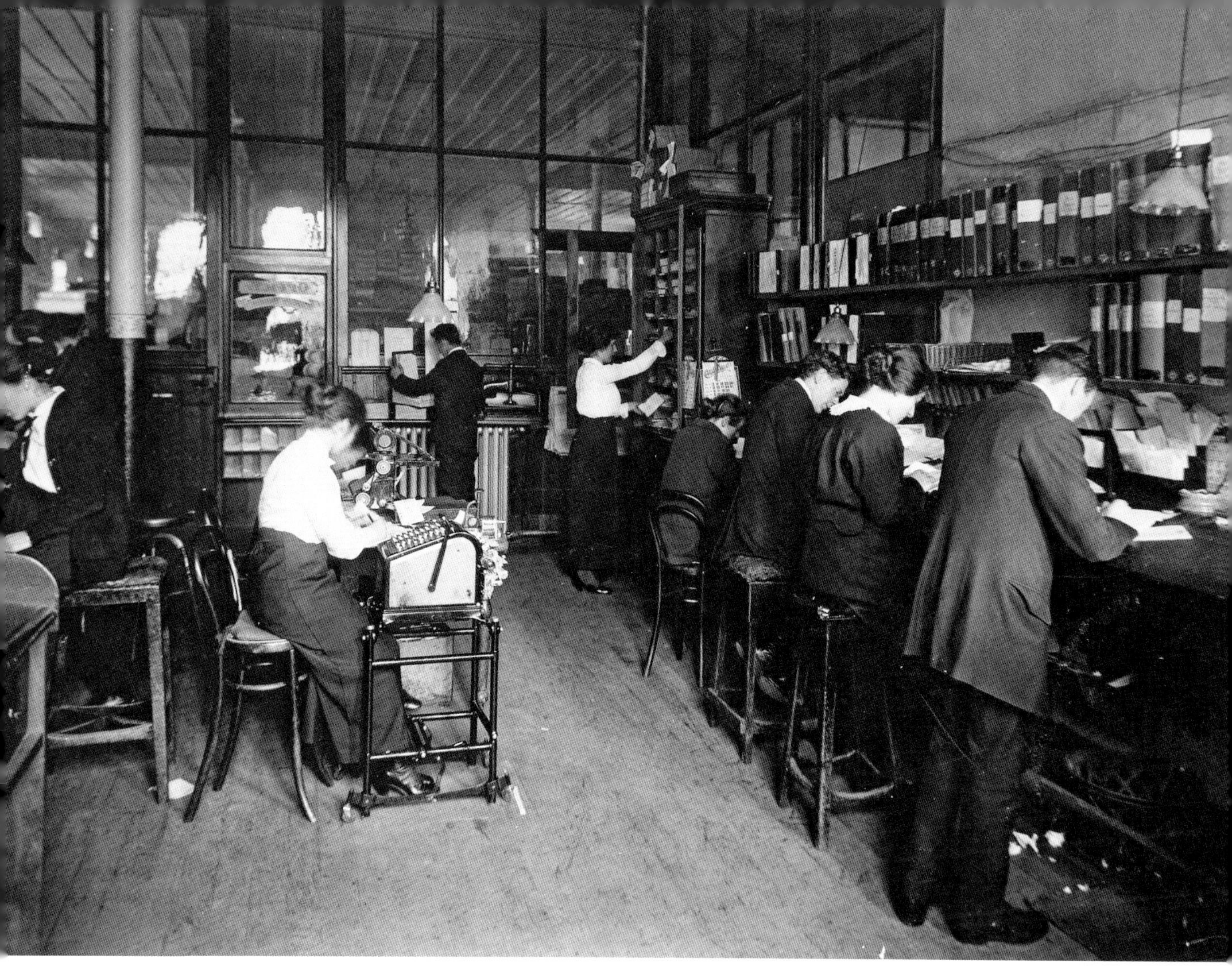

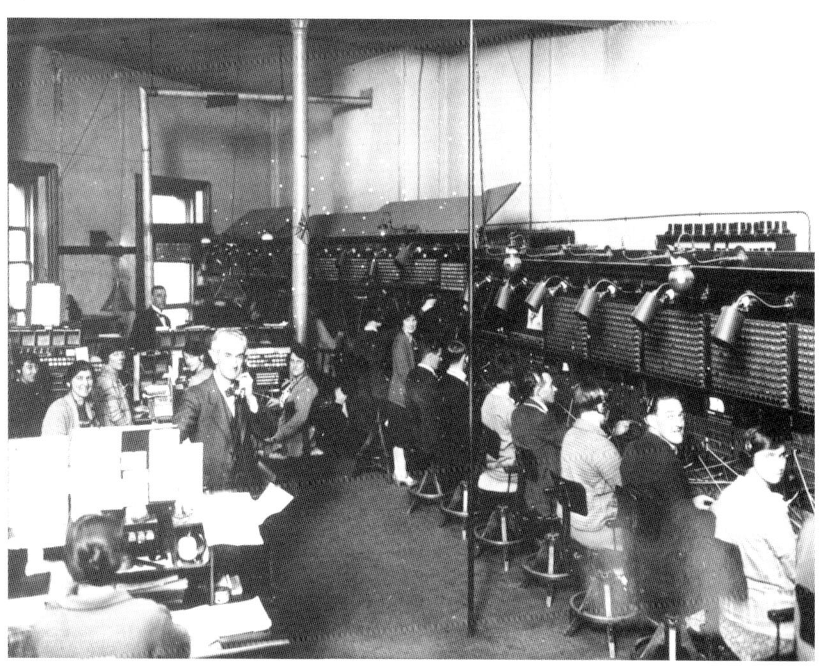

ABOVE **Few offices employed women as early as Bendix Hallenstein's New Zealand Clothing Company, c. 1912,** but over the next 30 years the office became a new domain for women workers. Whether operating the new-fangled typewriter, taking down dictation, filing or writing out invoices, women made themselves indispensable. Before long, they were also being sought as secretaries and receptionists. Many preferred such work and increasing numbers of young girls, especially those from skilled families, stayed on longer at school to acquire the necessary qualifications. Office work rewarded verbal aptitude, finger dexterity and clerical perception. Here we see a woman typing in an office which has roughly equal numbers of both sexes. The presence of women, most of whom would leave the office when they married (somewhere around their mid-twenties), underpinned male authority.

Hocken Collections, Hallensteins Ms-ARC-0041; image S13-279d.

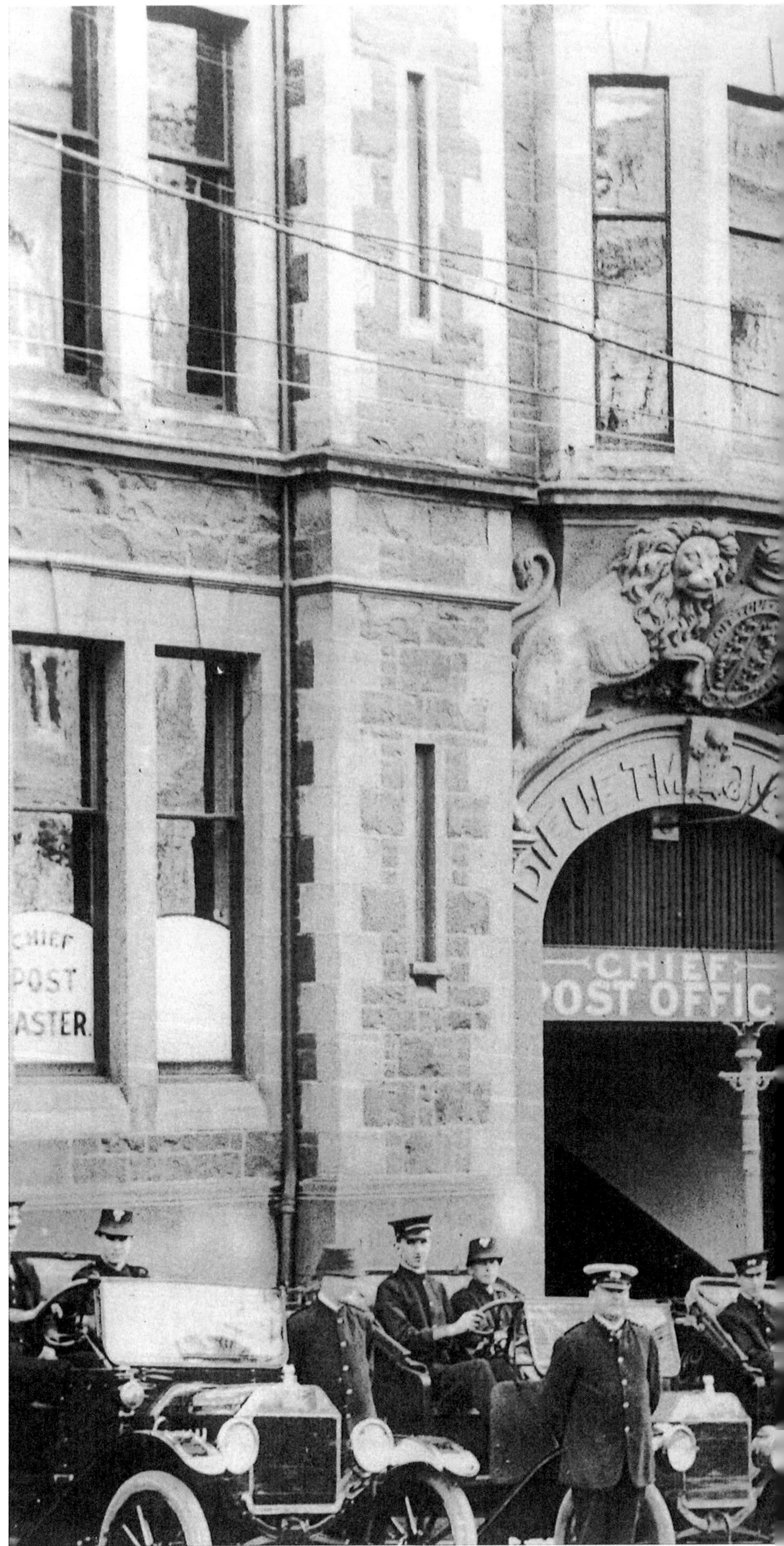

The uniformed delivery staff of the Dunedin Chief Post Office pose outside their new but temporary premises in the city's famed Garrison Hall, Dowling Street, in April 1921. The Post and Telegraph Department was second only to the Railways in the first half of the twentieth century, employing 8000 persons in 1909, or one fifth of all public servants. Employees of these two largest state-owned operations, comprising around 8 per cent of the total workforce, and substantially more in the main towns, enjoyed comprehensive classification schemes (which protected each trade from any threat) and promotion by seniority. Permanent employees also enjoyed generous superannuation schemes and a high level of job security. Their entitlements shaped the aspirations of the broader labour movement.

The evidence from Britain suggests that uniformed jobs appealed to the morally conservative and patriotic. They were often taken by the sons of skilled workers, who preferred security to risk and were likely to belong to other uniformed organisations, such as the Territorials. They were also much more likely to be active in the Labour Party and its predecessors. Their union was one of the largest in the country.

The photograph also reminds us that for deliveries within the larger towns the horse remained competitive with the truck until the 1920s, although the car had triumphed over the horse for all but the largest parcels. Eventually stables disappeared, except at race tracks, and grazing paddocks became in-filled with suburban homes or parks. The demand for oats and hay crashed. Drivers and bullockies disappeared, to be replaced by motor-vehicle drivers.

I am indebted to William Cockerill of the firm OCTA, now the owners of Garrison Hall, for this photograph, which according to the annotation on the back of the original was donated by Neil Wales.

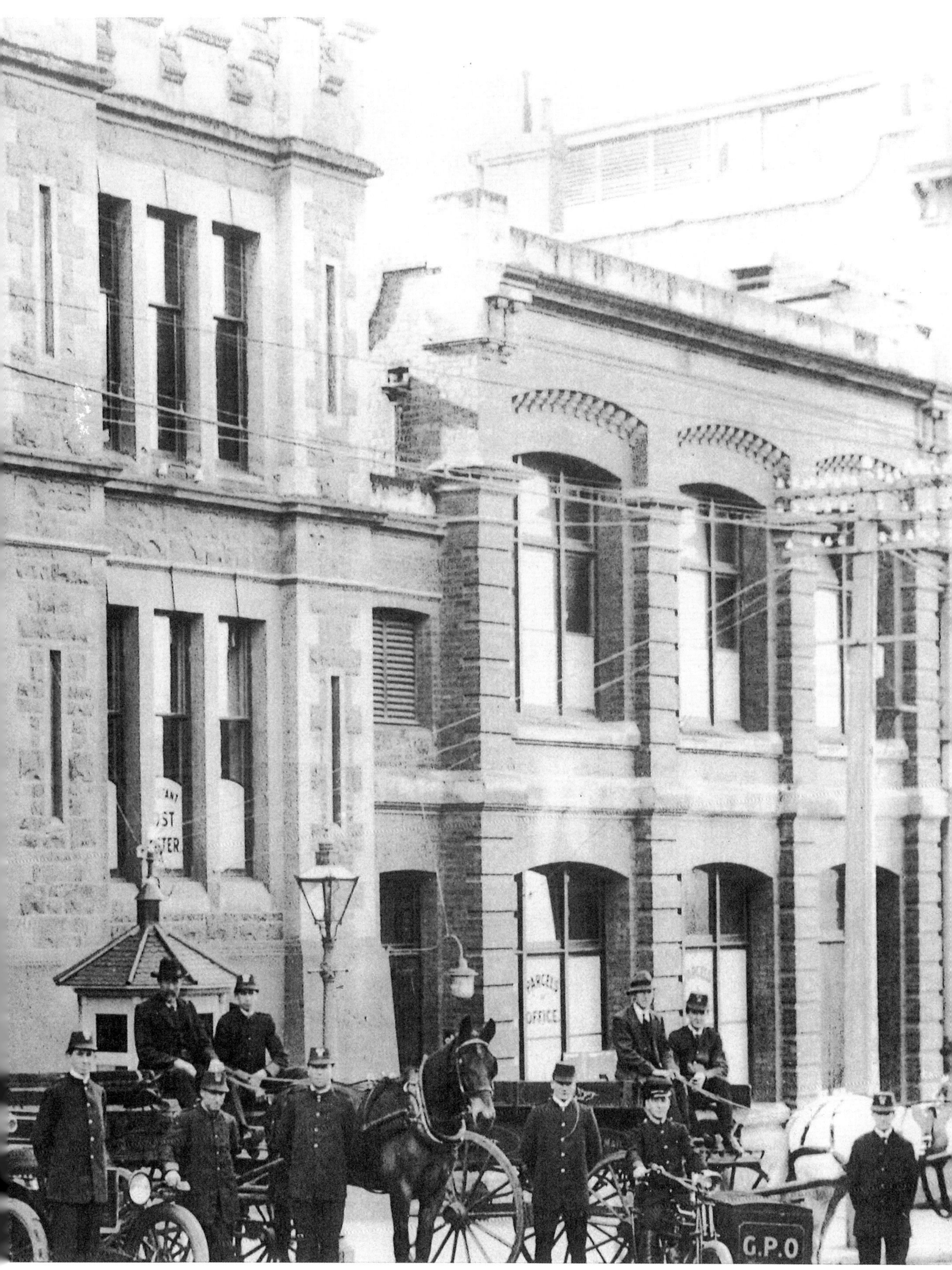
PARCELS OFFICE
G.P.O

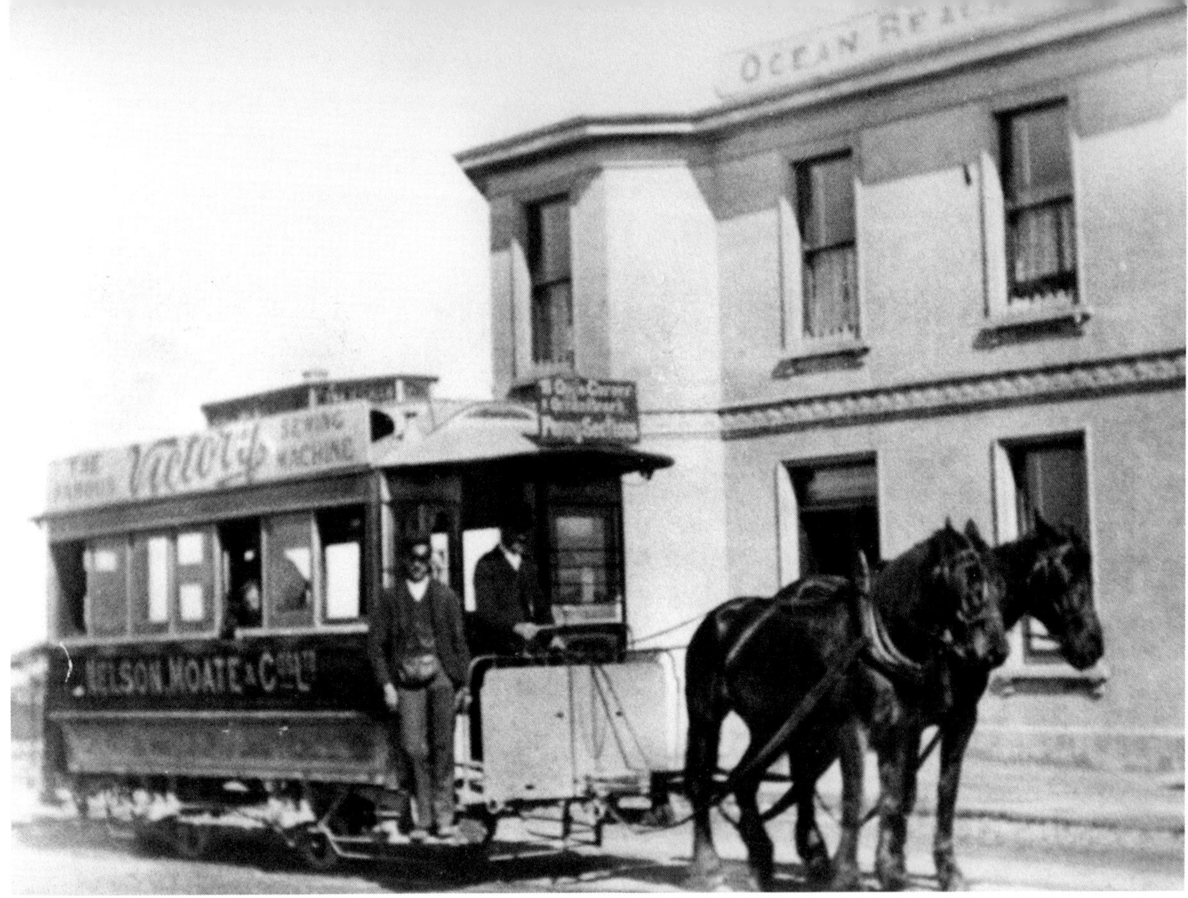

ABOVE **A horse tram waits outside the Ocean Beach Hotel at the St Kilda terminus,** some time between 1881 and 1905 (the era of the horse tram). Tramlines had already halved the cost of the journey and made life tough for omnibuses. When replaced by electric trams in 1905, the horses and their drivers became redundant (as we would say). Trams were amongst the first to abandon the horse, but plenty of opportunities remained for drivers until World War I. Work with horses was no longer a smart choice for a school-leaver, however.

The horse tram began the process of creating a single urban market, a process greatly accelerated by electric trams and cheap fares. Suburban businesses came under pressure as a result. In Caversham, the ratio of grocers, hairdressers, tailors and dressmakers per 1000 persons began to drop quite sharply.

DCC Archives.

OVERLEAF
Shopping. Gordon Burt took this photograph of a Wellington department store at some time in the 1920s, the precise year being unknown. Photographs of women shopping in a department store are rare.

In South Dunedin, Wolfenden & Russell opened their grocery and drapery stores in 1910 and expanded it to become the Flat's largest department store. An inner-city department store, the Drapery Supply Association, established a sizeable branch in Hillside Road, Kensington, in the 1890s. By 1900, there were 13 department stores in Dunedin.

In the nineteenth-century West European-American world, women became shoppers and shopping became a defining characteristic of modernity; department stores emerged to cater to the demand. Thanks to enterprising entrepreneurs like Dunedin's Bendix Hallenstein, department stores appeared here almost as quickly as they did in England and France and these businesses continued to benchmark against overseas stores. Because women played such an active part in shopping, many of the departments within the stores were entirely staffed and managed by women; even in departments where men were employed, as in this one, they had to be most attentive to women's wishes.

In the 1900s department stores, like most shops, were open from 7.30 am until 8 pm. On Fridays, they stayed open until 11 pm and on Saturdays, midnight. Although the store closed on Sunday, somebody had to feed and groom the horses that pulled the delivery carts. Shop hours had been a major political issue since the Lib–Lab government enacted the first Shop and Shop Assistants' Act in 1894. Subsequent extensions in 1901 and 1905 generated considerable controversy. Until recently, this photograph was said to be of Kirkcaldie's, but the curators are no longer sure of that.

Alexander Turnbull Library, F117814-1/2.

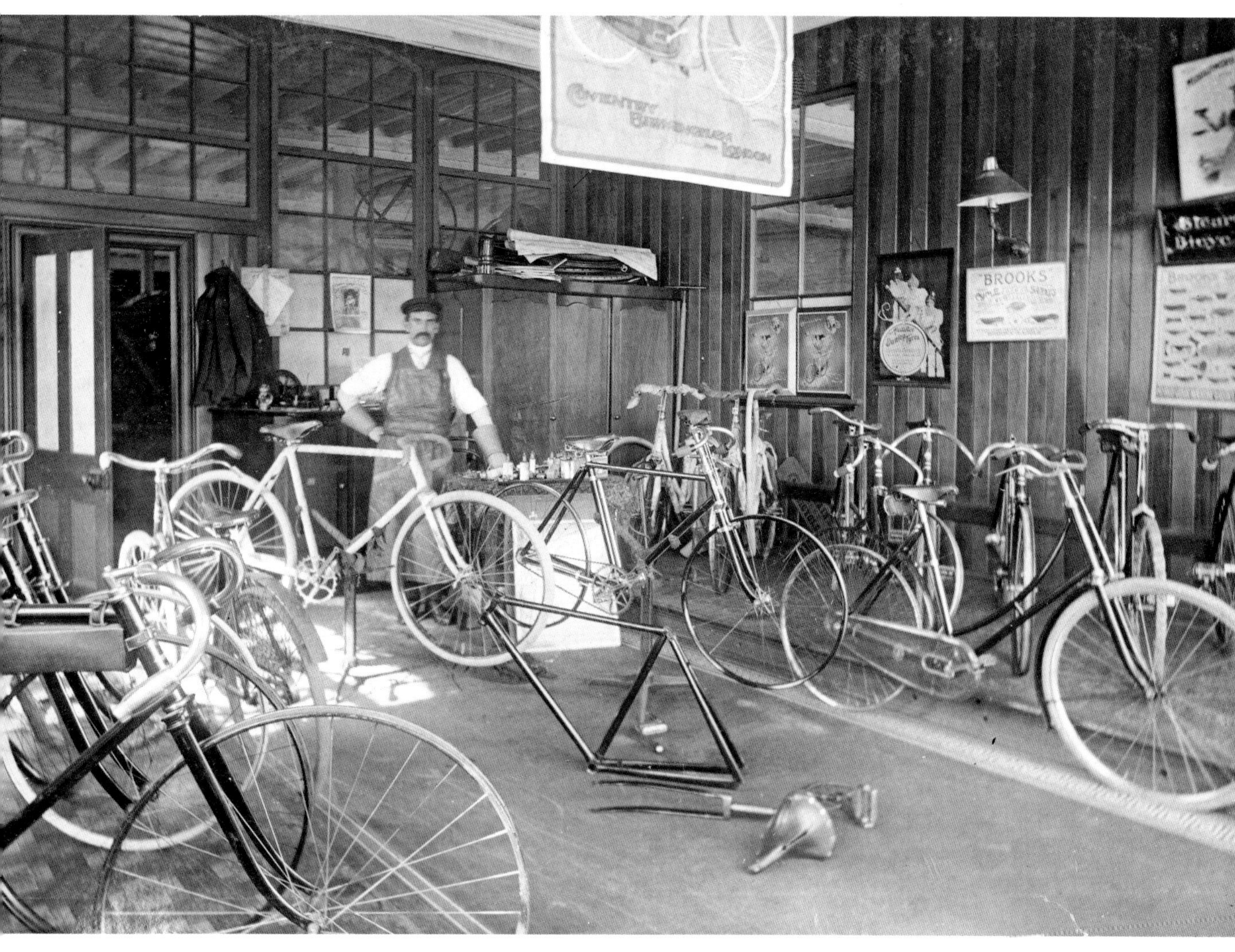

ABOVE **The cycle salesman and his showroom.** When the Long Depression ended in 1895, Reid & Gray obtained the agency for two makes of bicycle and opened a showroom within their factory complex on Crawford Street. Here we see A. Thompson, one of the new breed of white-collar workers, standing in the showroom. The firm had the agency for the Rudge-Whitworth Bicycle, a British make, and the Yellow-Fellow, a 'Yankee' bike.

Hocken Collections, Reid & Gray Photograph Album, Ms-1165-063; image S12-528c.

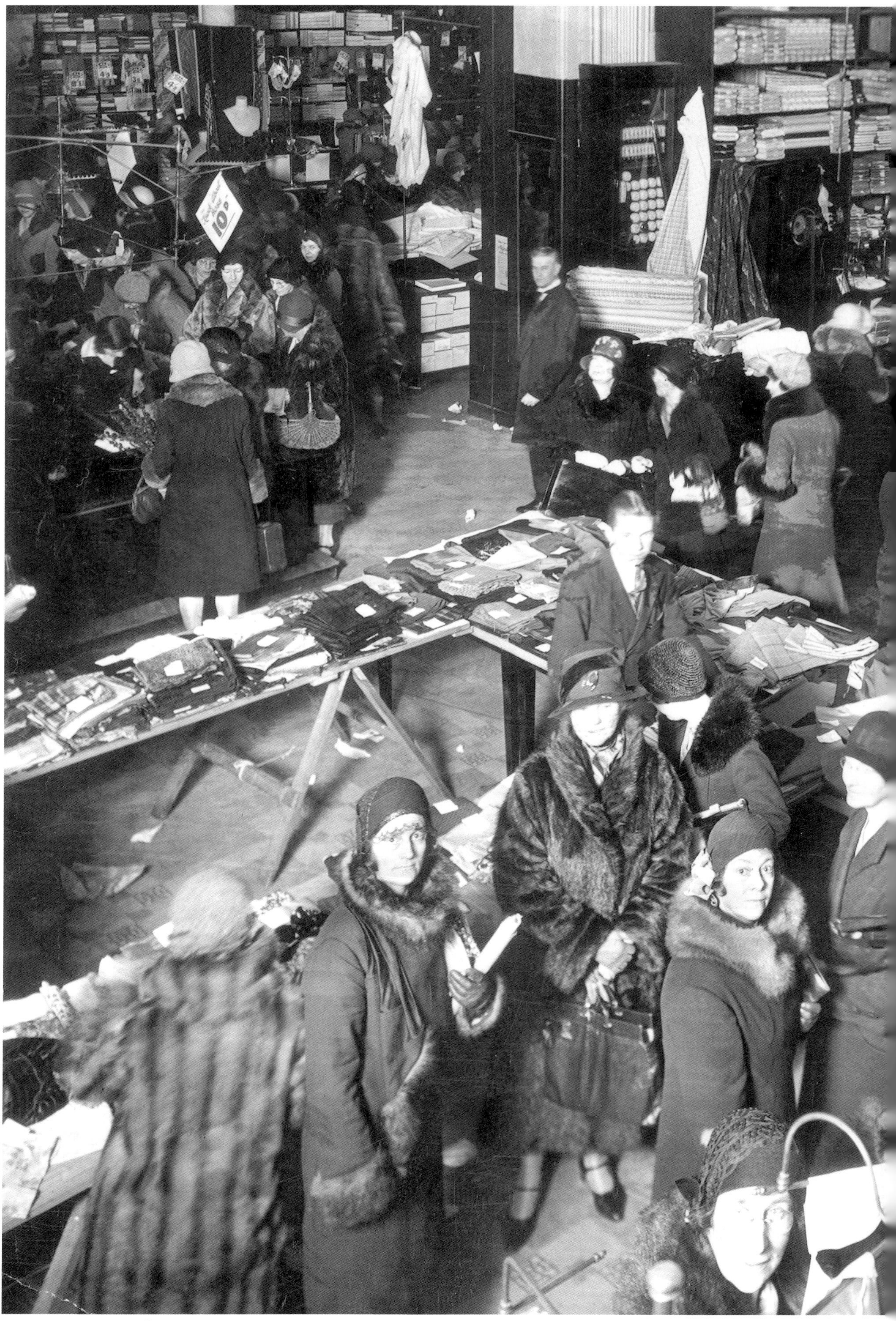
10d

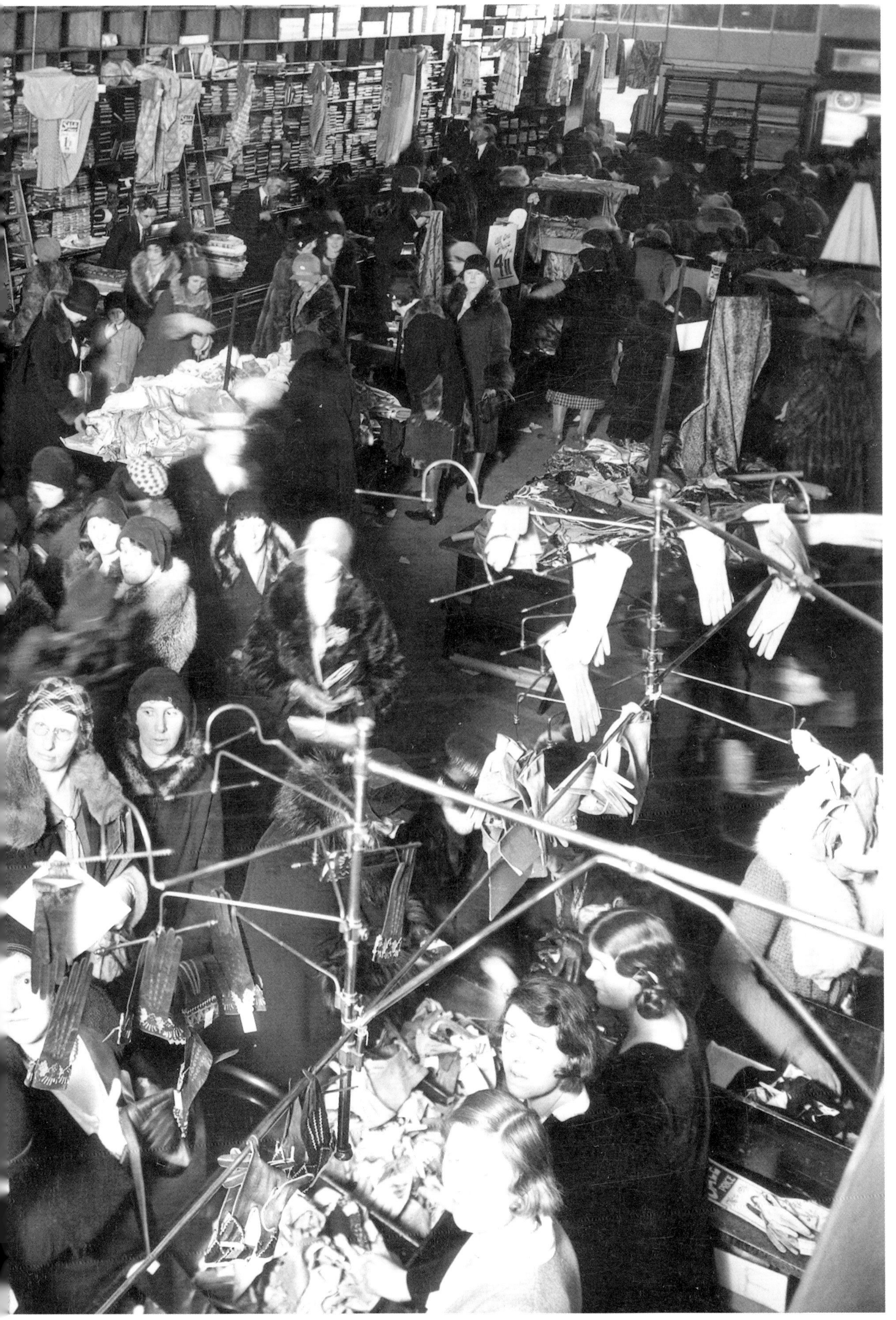

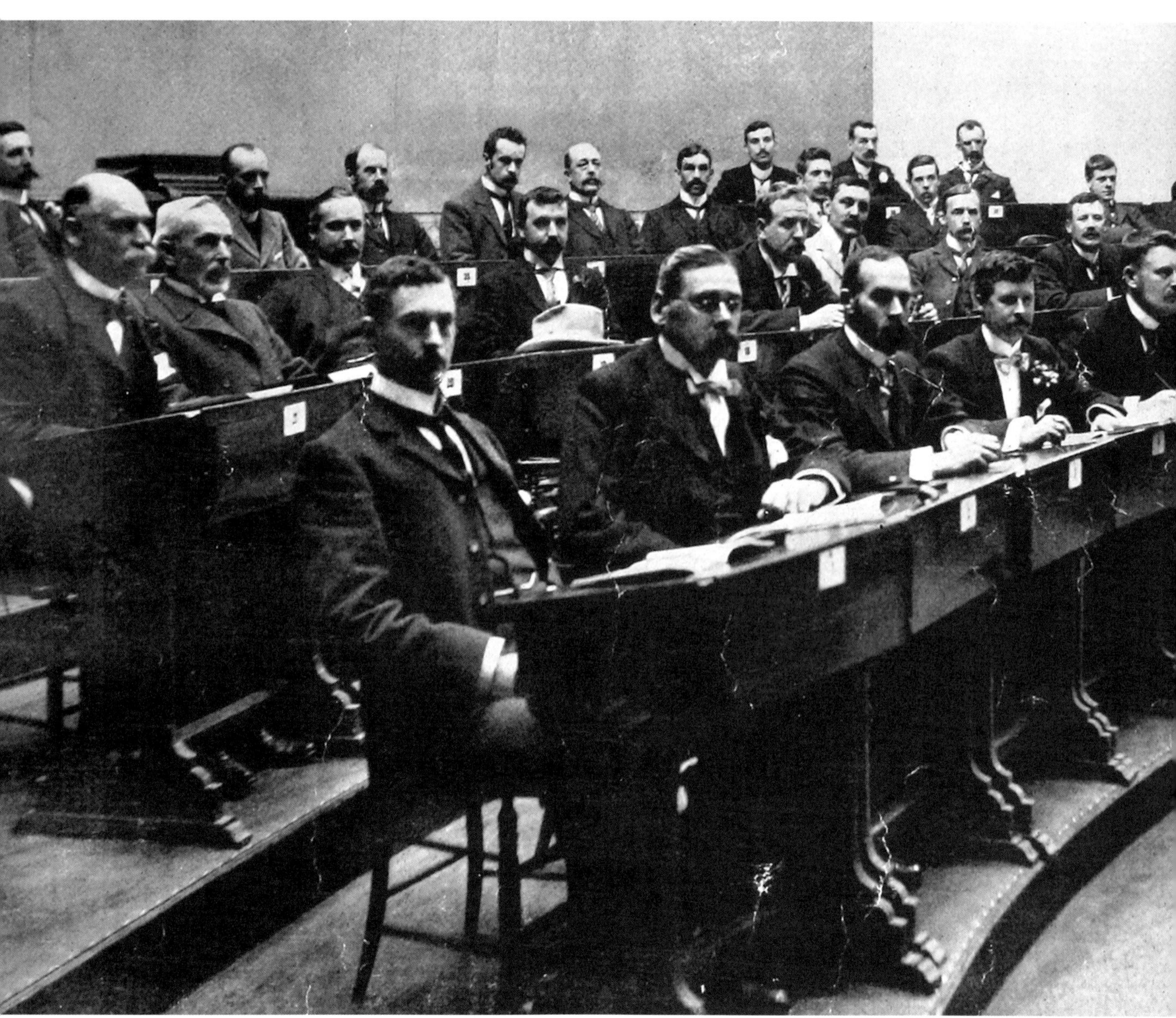

ABOVE **The Dunedin Stock Exchange in 1900,** at the height of the gold-dredging boom in Central Otago which saw the number of dredges in Otago, thanks to the contagion of rich strikes, soar from around 30 in 1896 to at least 171 in 1899. A fever of speculation swept the province. The Exchange boomed (the Dunedin Exchange was the oldest of three exchanges in the city at the height of the dredging boom). Most of the members were valuers, accountants and civil engineers; several were recruited from the banks and insurance companies. There were also several speculators. All of these professions could be entered by virtue of on-the-job training. Most of the men in this photograph had at least three years of secondary schooling, 15 of them at Otago Boys' High School, and belonged to one of the city's two clubs for gentlemen. There were few requirements for becoming a broker but the fee for admission to the Dunedin Exchange was £250 a year, more than twice the annual wage of most tradesmen. Guy Morris, an employee of the *Otago Witness*, took this and several other pictures of the Exchange in 1900.

Hocken Collections, *Otago Witness*, 19 September 1900, p. 47.

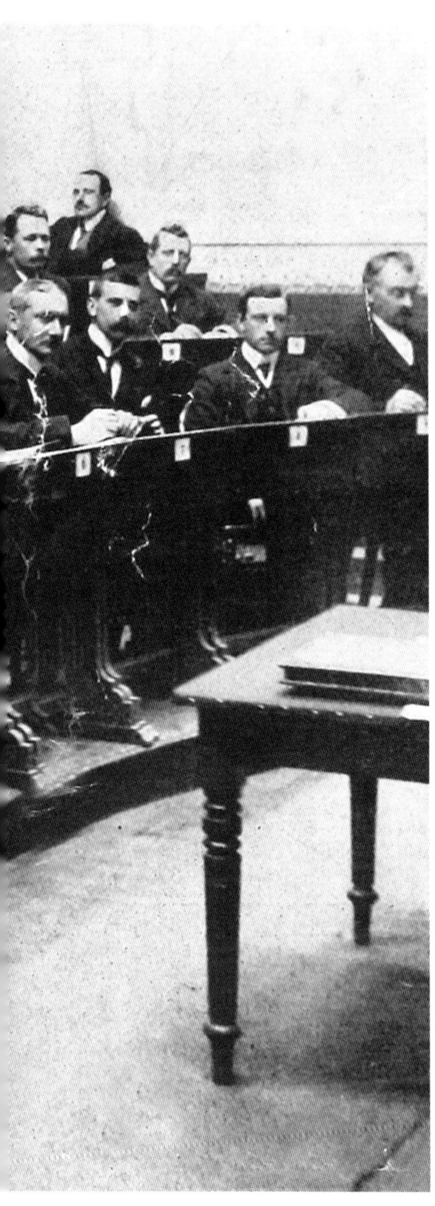

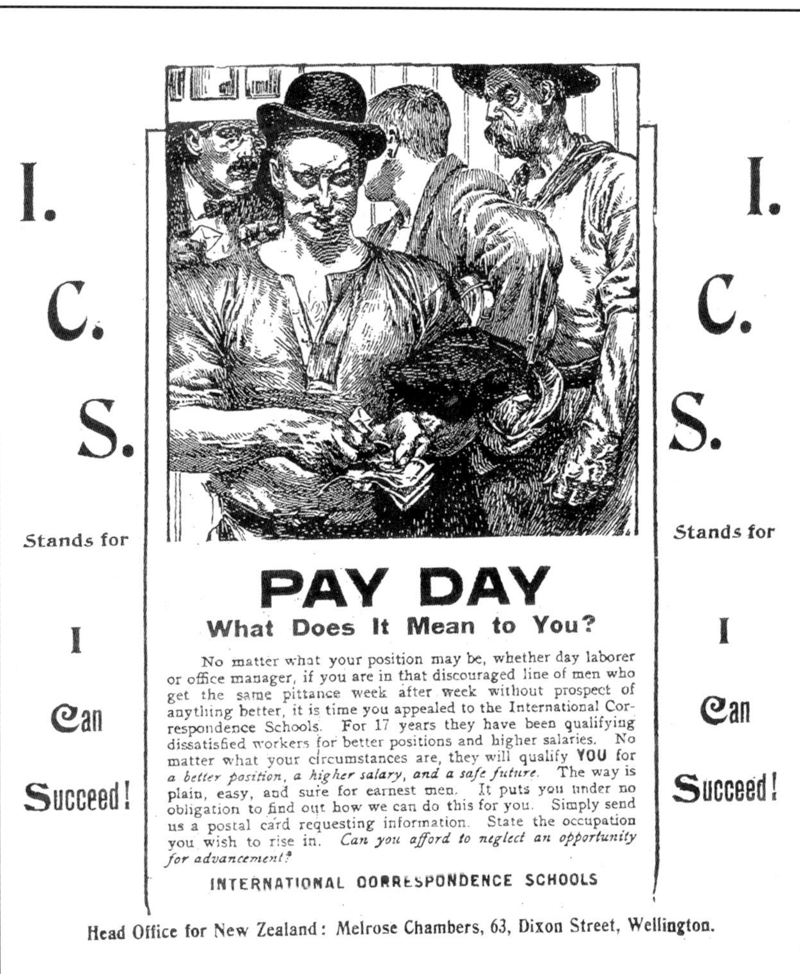

ABOVE *Pay Day: What Does it Mean to You?* **Such advertisements were unusual but not unknown.** The system of free-place technical schools enabled most young people to acquire the necessary skills before entering the workforce, but the International Correspondence School appealed to those who became more ambitious later. It is not known how successful this firm was, but it had no local branch. Apart from Rossbotham's, which concentrated on teaching typing and shorthand from the mid-1890s, as did a small handful of private teachers, nobody else offered post-school training.

CHAPTER 4

A Less Unequal Society?

ALL SOCIETIES ARE UNEQUAL, but some are more so than others. Despite a deep desire among some on the left to believe that England's class structure was transplanted to the antipodes, in the past New Zealand was less unequal than most other capitalist democracies. According to the customary statistical measures, New Zealand became less and less unequal between 1890 and 1940.

Which is not to say that everyone had the same income or wealth. Nobody thought the Levellers' ideal sensible or feasible. It was accepted that men should be rewarded for talent, enterprise and industry and that virtue ought to be rewarded, but almost every immigrant who spoke on the subject insisted that they did not want to replicate the extremes of wealth and poverty that existed in Britain. Nor did they. By 1900, few urban households boasted more than one live-in servant, usually a cook or maid, although some of the wealthiest businessmen often had a coachman and a gardener to manage an extensive suburban estate, as well as three or four domestic staff. Incomes ranged, give or take. William Dawson, the brewer and founding partner at Speight's, the city's most successful brewery by 1900, received £7500 a year from his shares in the brewery, together with a salary of £1200 a year. His origins were humble and his schooling limited. Caversham's brewers, Briggs and Cowie, made nothing like this, and in tough years probably made little more than their skilled men, around £220 a year.

We do not know how Dawson dealt with success. Perhaps he arrived at work at around 9 or 10 am, took as long as he liked for lunch, possibly at the Dunedin Club, and knocked off when he chose. Until very recently, however, he had put in the same hours as everybody else and always had done. The stokers at

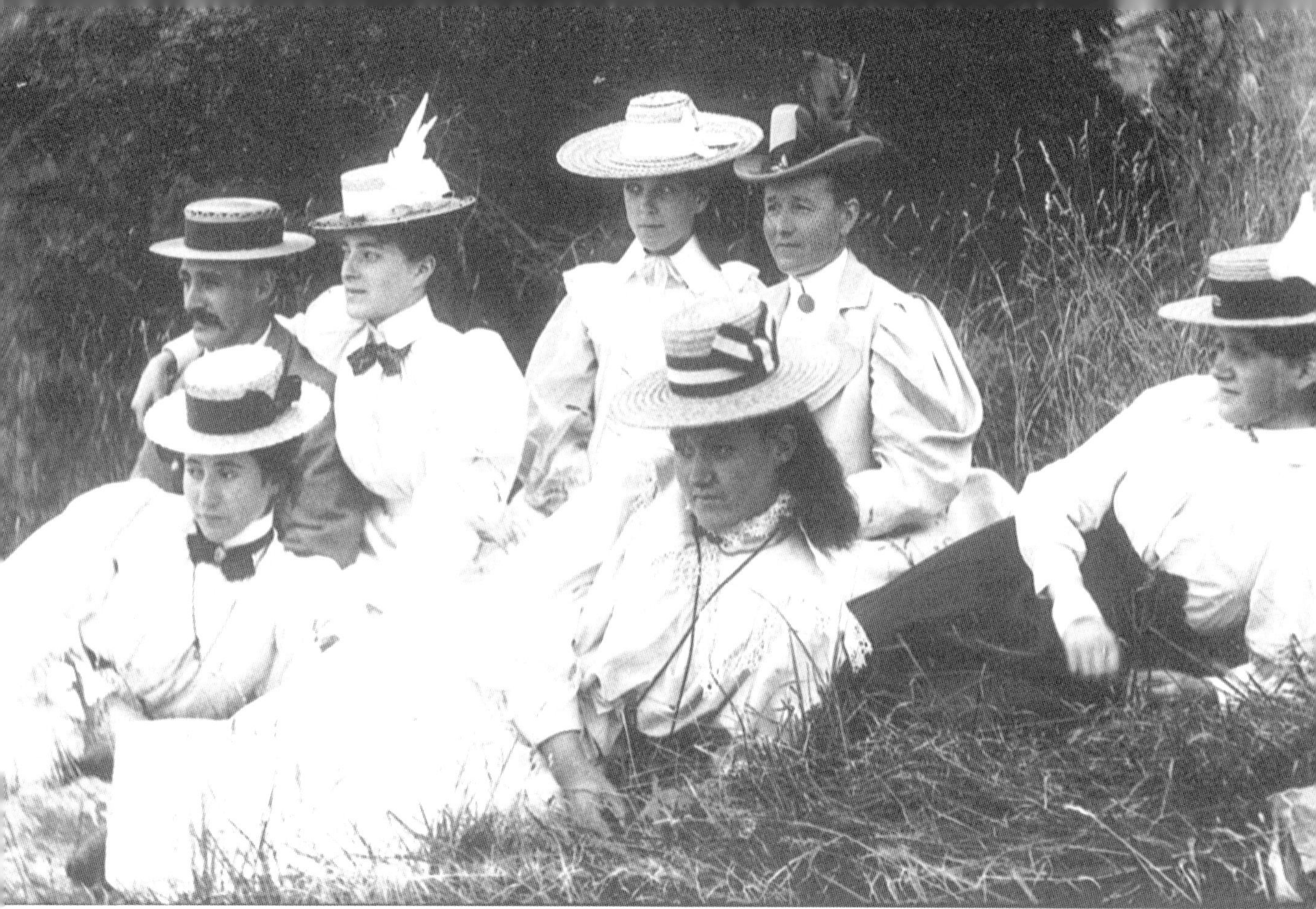

the gasworks, to move to the other extreme, worked seven-hour shifts in summer and 12-hour shifts in winter, the employer determining exactly how long. They also worked seven days a week. In winter, they worked an 84-hour week. The work was hard, filthy and dangerous. The day and night shifts switched each fortnight, invariably on a Sunday, when the day shift actually put in a 24-hour shift. In 1900, the pay was one shilling an hour for labourers. If you could survive the hours and conditions, you took home £4/4/– a week. This was comparable to what many skilled men got for a 44-hour week. Over the next 20 years, that difference was steadily reduced.

Many of the labourers who worked at the gasworks took off in the summer, however, to work on the wharves or at a freezing works, or to head up-country to join a shearing gang. The pastoral economy boomed in spring and summer; the gasworks boomed in winter. Even before overtime became normal, as it did in the next decade, the so-called unskilled could work long hours and earn big money. This was especially true of young men in their physical prime. In some respects, the unskilled also had more freedom to come and go and learn about opportunities in the wider economy.

Housing most visibly reflected the extent of inequality. Edward Cargill's Italianate mansion, 'The Cliffs', like Sir William Barron's two-storeyed wooden mansion, 'The Willows', represented success in this colony. By English, Scottish, Australian or American standards, however, neither

A group of young women employed in Richard Hudson's biscuit factory, which was a block north of the Central Railway Station and no more than a 10 minutes' tram or train ride from Cargill's Corner, enjoy themselves in all their finery at the company's annual picnic in 1899. Almost all firms held an annual picnic. The photograph confirms what almost every contemporary observer remarked on, that high rates of pay allowed the working classes to dress as well as their mistresses and masters. These young women probably had but one 'going-out outfit', however. Thanks to the ubiquity of the new-fangled sewing machine, and 'off-the-shelf' pattern books, they may even have made their clothes themselves.

Hocken Collections, A.N.L. Clark Collection, glass neg. ANC 114 & c/n E2498/20.

OPPOSITE, TOP **City Corporation Tramway Employees, 1914.** To many young working-class lads, the appeal of uniformed jobs was strong; like the railways, the tramways modelled itself to some extent upon the army. In publicly owned systems, as Dunedin's became in 1901, these men enjoyed secure jobs, guaranteed minimum hours per week, and career ladders that allowed them to advance to the highest-paid jobs, as motormen, and even to become inspectors. Commitment to the organisation, especially among uniformed staff, often provided another source of cross-class loyalty, and helped sustain a pre-industrial 'middling class' of skilled and semi-skilled, foremen, clerks, inspectors and managers.

Hocken Collections, Bathgate photograph, c/n E 3125/17.

OPPOSITE **City Corporation Tramway Officials, c. 1914.** Again like the railways, the tramways needed a back office of clerks, engineers and managers to ensure the efficient operation of the system and the business, as well as navvies to look after the tracks, and skilled motormen and conductors to run the trams. Whatever their origins – and many were recruited from working-class homes – these men almost invariably joined 'associations' rather than trade unions. These associations modelled themselves on the oldest and best organised professions, medicine and law.

With the exception of nurses and teachers, women in the white-collar sector had no organisations to represent them.

Toitū Otago Settlers Museum.

house was especially large nor grand (although you would have been lucky to find a better architect / engineer than Petre). As can be seen from some of the photographs in this book, in the 1860s and 1870s most labourers lived in a two-room cottage with a lean-to at the back, in which the fire or range would be found. In 1900 most of the rich, like everyone else, had to use a privy and pay a nightsoil man to empty it. What marked out the colony for labourers, and even skilled men, was the fact that a cottage was usually occupied by only one family and it would also boast its own section of land. Compared to living conditions in Dublin, Glasgow, Manchester, or even Wigan, this was luxury.

The worst cottages, of course, were older ones in the inner city and tended to be occupied by men who could not earn a decent wage because of age, ill health or injury, or even a combination of the three. Many of these men increasingly used alcohol as an antidote to the injuries and indignities of class, poverty and age.

Around 1900, servants were ceasing to mark off the wealthy from the comfortably off. Neither Cargill nor Barron employed more than a housemaid and a cook. Indeed, the colonial wife was usually not much older than her maid and cook, often worked alongside them, and like her husband never disdained physical labour. The Theomins' grand mansion, Olveston, was Dunedin's largest and most expensive when completed in 1905, and while it had 35 rooms it never employed more than five live-in servants (the butler, coachman and gardener lived elsewhere). When the local Member of the House of Representatives, T.K. Sidey, who greatly extended the family home at Corstorphine so that it eclipsed 'The Willows', wanted to hold a garden party, he and his wife hired young lasses from the township to assist. These lasses relished the opportunity. In England, of course, gentlemen needed several servants, and families of higher status employed many more, as well as gardeners and coachmen who lived under the same roof.

By 1900, the best-off had fewer children than their inferiors and could give them more privileges, such as a bedroom of their own, private tutoring (although that was becoming unfashionable), and even perhaps a private school for their daughters. The urban middle and working classes promptly followed suit. So, too, did the Catholics. At the other end of colonial society were those lasses who found themselves pregnant to a man who evaded his responsibilities. Their children got a much poorer start to life and were much more likely to die or reach puberty damaged by some illness or other. Locals were well aware of the dangers to young lasses of letting a man go 'too far' or 'have his way' because of the Salvation Army's home for 'fallen girls' on Lindsay Street. Another hazard of life was that the untimely death of a parent could see a child confined to the Industrial School, just above Caversham township, which also made for a poorer start in life.

The unskilled and the large employers, plus some of the more successful professionals, marked the poles of colonial society; in between, by contrast,

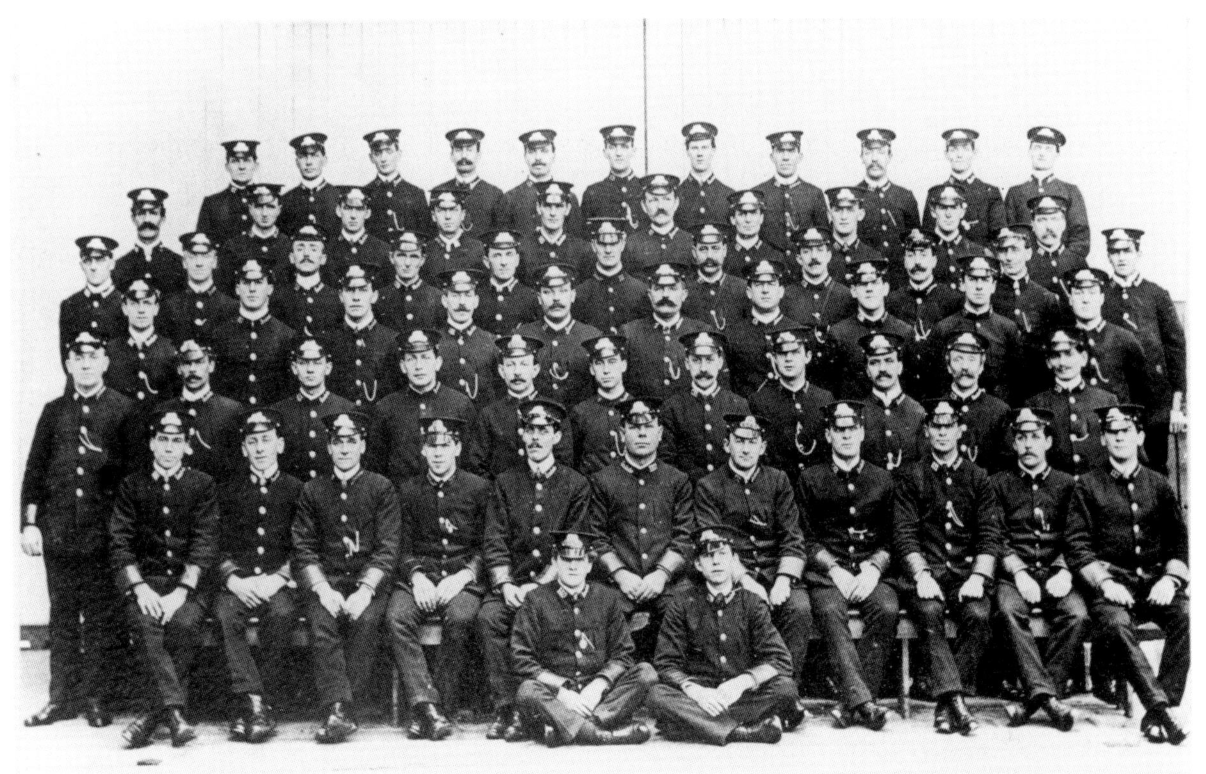

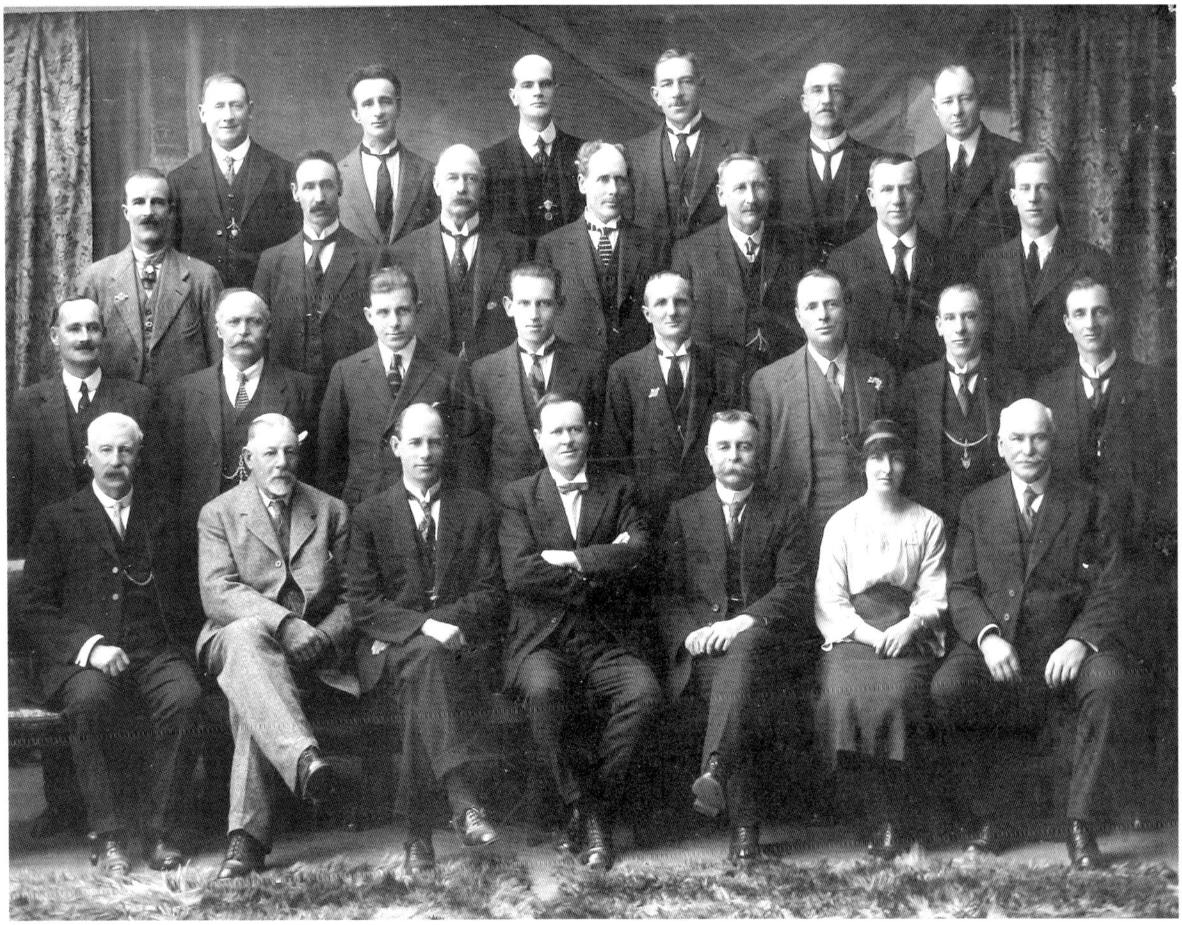

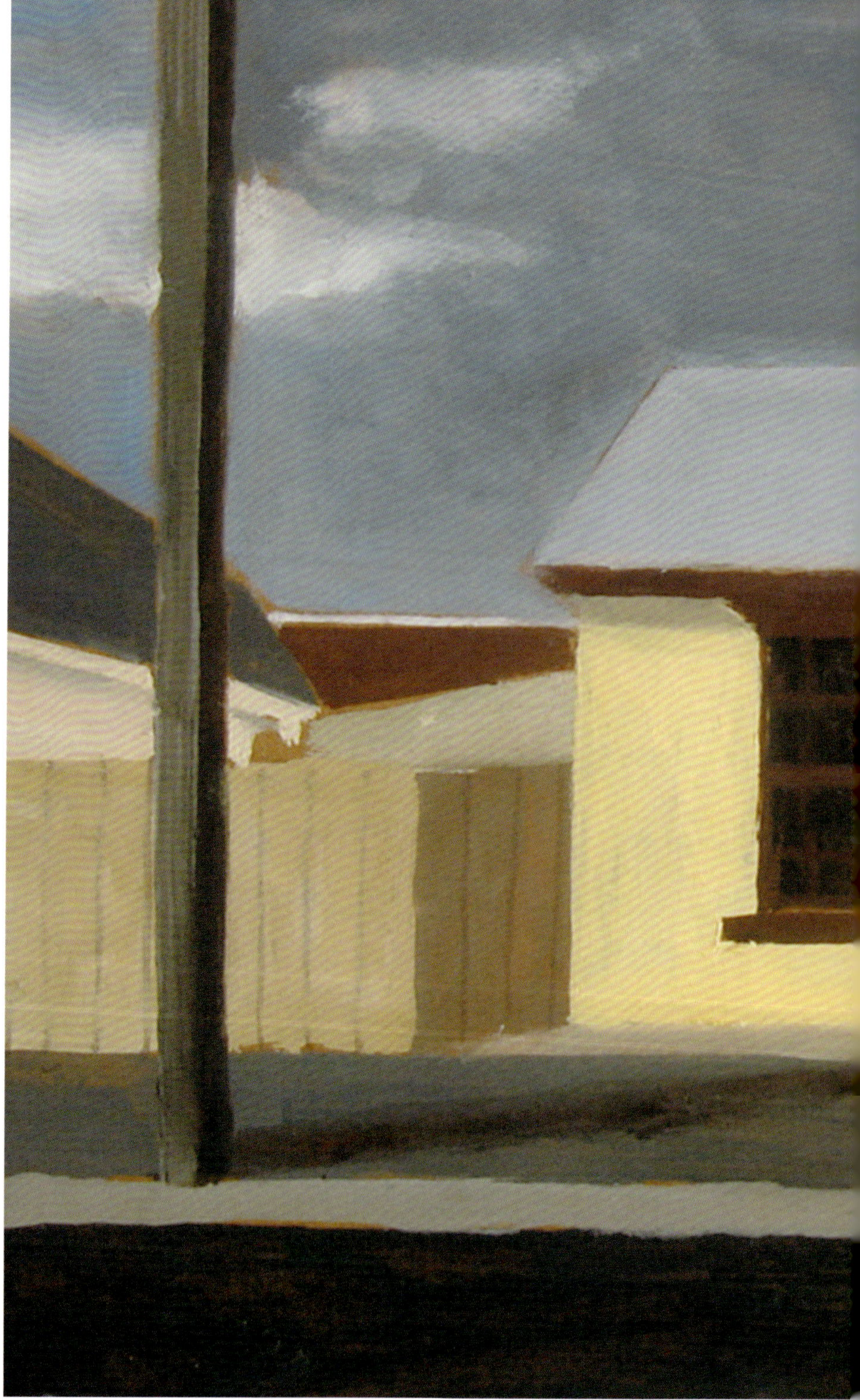

Patrick Hartigan, *Dunedin House 2*, 2010, oil on board, 208 × 328 mm, courtesy Patrick Hartigan and Brett McDowell Gallery, Dunedin. Hartigan, an Australian, fell in love with the Fingall Street precinct, which includes the Greek Orthodox church. He recognised at once that these houses represented something distinct about New Zealand. Nowhere else, including Australia, is the least valuable working-class housing essentially a modest variant of the houses owned by much wealthier people of much higher social status – a stand-alone single unit, secure on its own section, small, easy to heat and cheap to maintain.

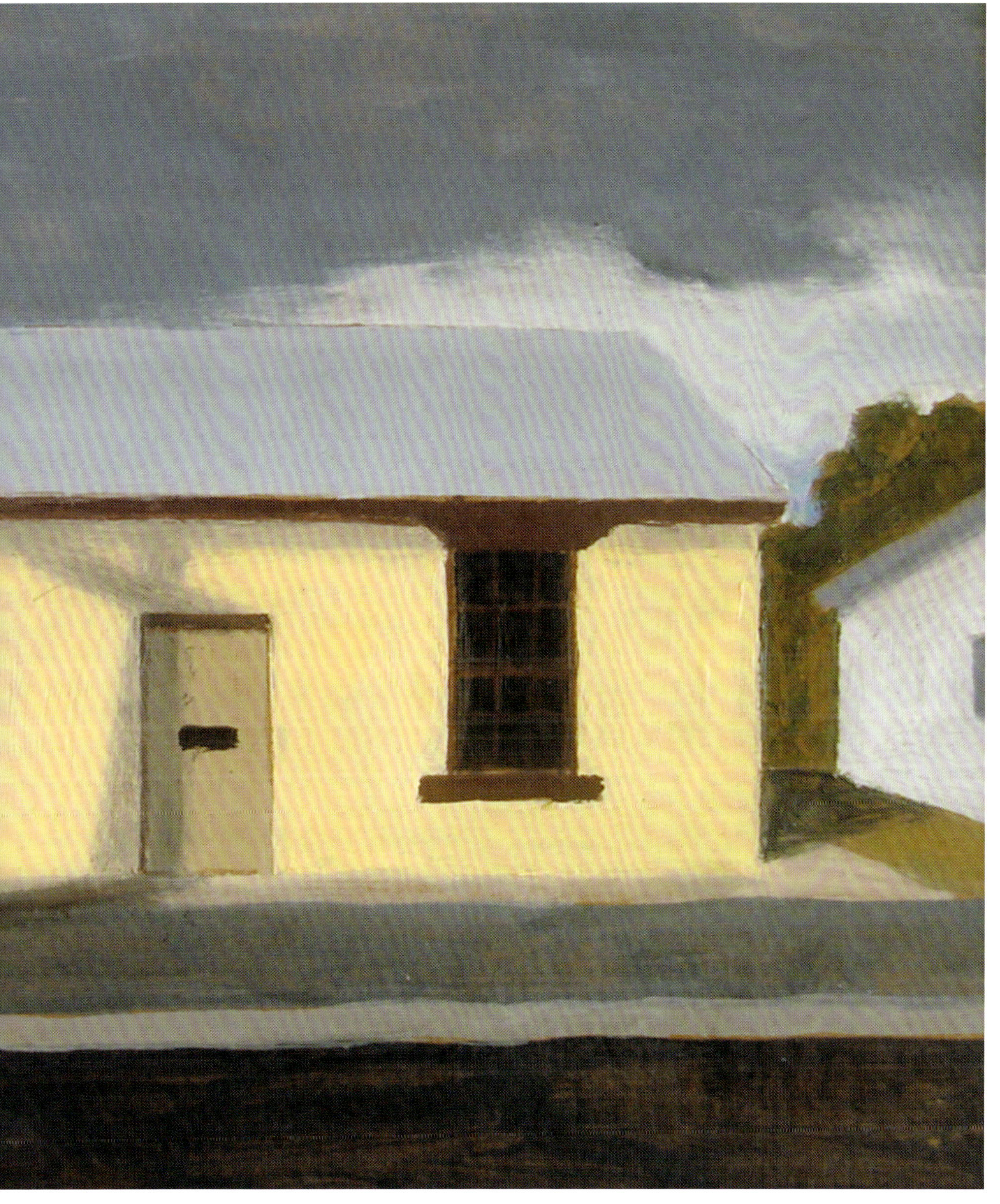

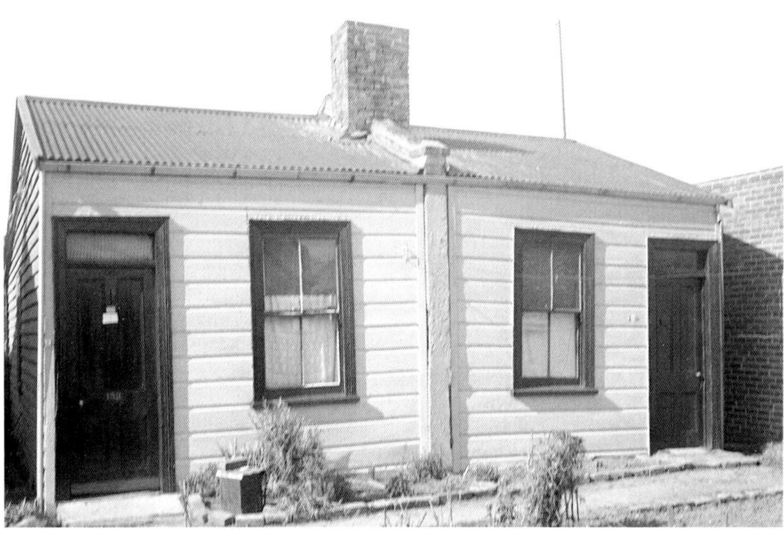

ABOVE **Cottage at 159 Castle Street, Dunedin, c. 1938.** Bruce Elwyn Orchiston, a Wellington architect, took many photographs of Dunedin's inner-city 'slums' in the late 1930s, when 'slum' clearance was a political issue. Cargill's Castle may have had few rivals in the city, but the meanest cottages and lanes of southern Dunedin were superior to those of the inner city, not least because they were built much later. By the 1930s, the joists and floorboards of many inner-city cottages were rotting due to inadequate drainage, wet rot being more common then than dry rot or infestations of borer. Given the majority of the cottages had been built in the 1860s, most had no water source and the fire was primitive; many never saw the sun. As a rule, on the Flat as in the city, the worst cottages were inhabited by old people, women as well as men, who lived alone, or by men who could no longer earn a regular wage because of injury or ill health. Dramatic improvements in housing, not least in the amenities available (such as internal flush toilets and gas or electricity), marooned many old people.

Alexander Turnbull Library, Bruce Orchiston Collection, F-86816 1/4.

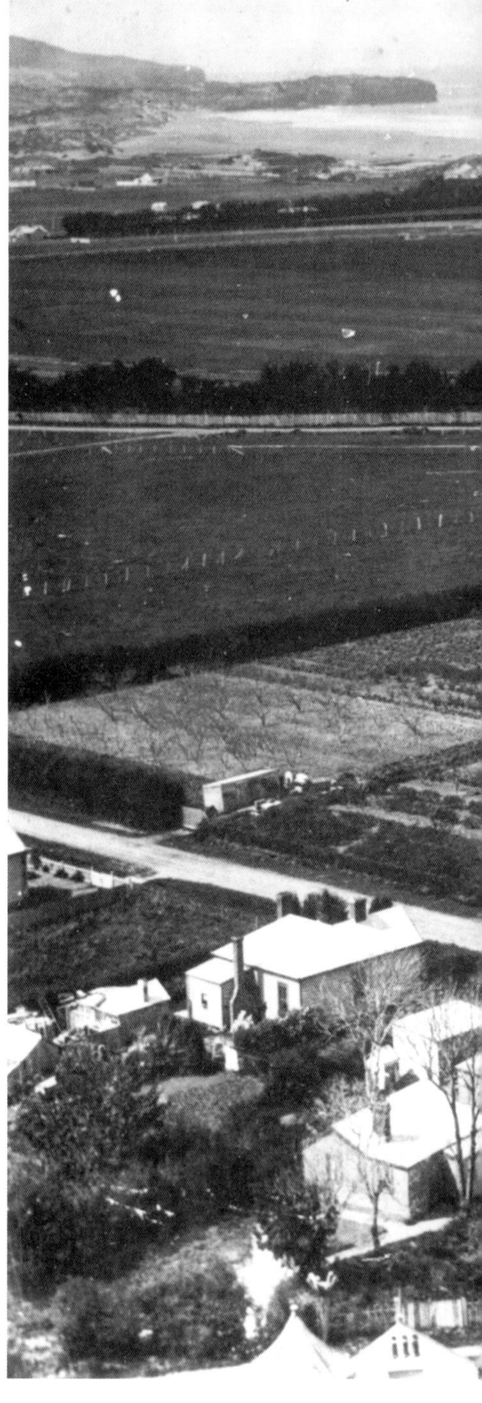

ABOVE, RIGHT **This photograph captures the range of housing on the St Clair flat.** It was taken in the 1880s by Ebenezer Ings, son of the prominent Baptist and nurseryman, from somewhere near Shiels' quarry. At the intersection of Bay View and Forbury roads is Dr Milan Coughtrey's substantial villa, which he had turned into a private hospital and sanitorium. On the far side of Bay View Road is the home of William and Elizabeth Ings and their nine children. A slaughterhouse sits to the right, and to the left, facing Bay View Road, is a cottage in which their son, Eben, lived.

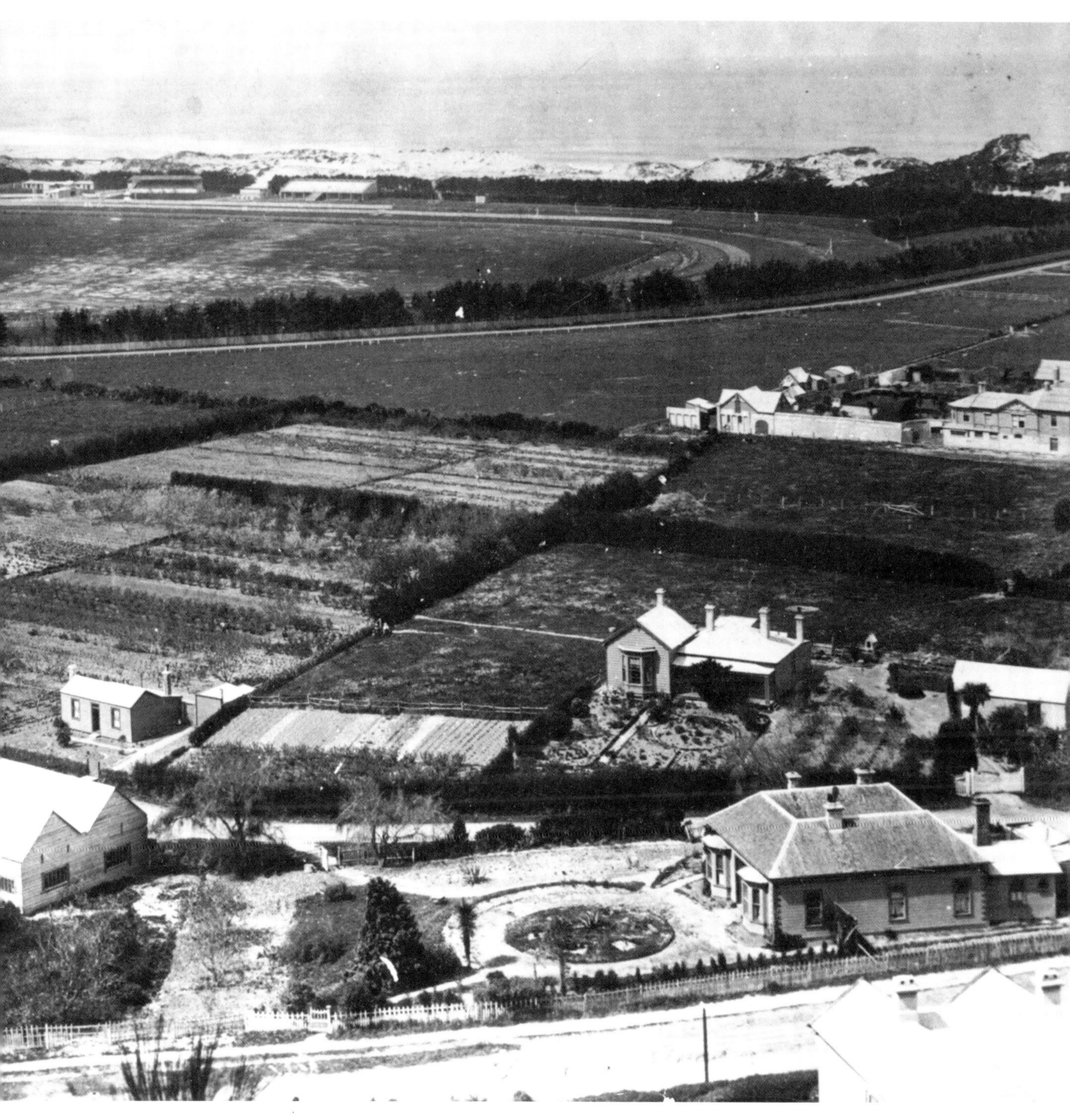

William and his brother John cleared and drained the swamp, digging out ditches and laying dried manuka in them to create the first drains. Shortly after buying the land, he sold a sizeable section to the Forbury Park [Racing] Club for the race course that lies beyond the nursery. By 1902, he had reclaimed some 35 acres from the swamp and planted 24 in pasture and six in fruit trees. He also ran a commercial garden, supplementing his income by subdividing his land. He made substantial benefactions to the Baptist Church. In 1900, Eben bought the orchard and garden from his father for £988 and started his own market garden. As we have seen earlier, and Barbara Newton has shown in her history of St Clair, there were several nurseries and market gardens in this area.

Scanned from G.N. Stedman, 'The South Dunedin Flat', MA thesis, Otago University, 1966, following p. 145.

was largely undifferentiated. Inter-generational mobility, marital mobility, and even worklife mobility underline the extent of the mix. Streets, neighbourhoods and suburbs were mixed, as were school catchments, and every voluntary organisation, including churches, recruited from across the social structure. In broad terms, colonial society was much more free and equal than the 'Old Country' and everyone was expected to pull their weight.

I

The Labour Party was one of the most left-wing parties in the world in the post-war period. Its agenda for revolution would be severely modified, however, before the Depression of the 1930s helped it to victory. Because political historians have focused almost exclusively on Labour's rise, in the process demonising Reform, they have not noticed the speed with which the great reforms of the Liberal–Labour government of 1891 to 1912 not only came to be accepted by almost everyone but became a matter of national pride, part of our sense of our nation as exceptional.

Although it has often been remarked how quickly and completely New Zealanders move on from their debates and confrontations to achieve a viable consensus, there have been few attempts at explaining this (apart from *An Accidental Utopia?*, that is). Undoubtedly, the relative homogeneity and smallness of the country's population are part of the story. So is our commitment to sending our children to the state primary schools, as well as our national preoccupation with fairness. What has less often been remarked is the impact of high levels of geographical movement and social mobility. In particular, high levels of marital and inter-generational mobility provided the coping stone for the political consensus that made the contentious Lib–Lab reforms of the 1890s part of a new political consensus by the time Massey and Reform finally won office in 1912. The rise and leftwards move of Labour only deepened that consensus. A generation later, even Labour's astonishing achievements of the 1930s had become the subject of consensus and a matter of national pride!

It is well nigh impossible to document the remarkable levels of geographical movement with photographs, as no photographers found moving house of interest. Even social mobility is hard to illustrate. Worklife mobility cannot be illustrated, nor photographed, although every photograph of a large or small employer in the industries and businesses that dominated the Flat captures somebody who once worked as a journeyman and probably served an apprenticeship. We can show photographs of couples who have married across a social faultline, although that can only be made clear in the caption. We can also show photographs of special wedding anniversaries, especially golden weddings, but again it requires a caption to explain the extent to which the lads and lasses chose their own jobs and careers, regardless of what their parents had done.

OPPOSITE **Sir William Barron had this house overlooking Forbury Corner built for his family in 1876.** He named it 'The Willows'. Sir William (1837–1916) came to New Zealand during the gold rushes, made his pile and invested in runholding before retiring to Dunedin, where he became a developer and company director. He also represented Caversham in the House of Representatives from 1879 until 1890. His views were Liberal, especially on educational matters. He and his wife had 11 children. His eldest daughter taught piano and violin, and later woodwork and other handicrafts, while two younger ones ran a millinery workshop and store in Caversham. Some boys remembered, many years later, retrieving tennis balls for the Barron girls in return for an occasional penny. This photograph by Robert W. Rutherford is one of the few of his to survive and probably dates from c. 1880.

Hocken Collections, 77.1258.

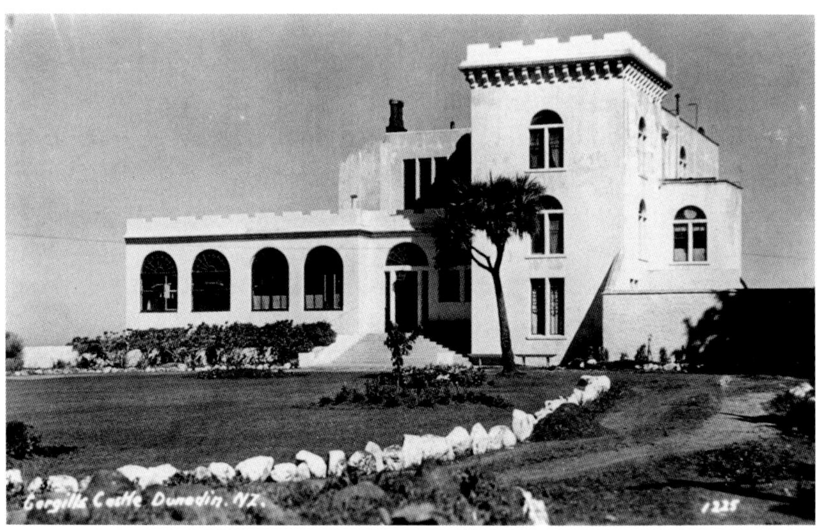

LEFT **'The Cliffs'**. Edward Bowes Cargill, Captain William Cargill's eighth child and seventh son, alone of the second generation, achieved a comparable status to his father. He had considerable experience of commerce in East Asia and Australia before settling in Dunedin in 1859, where he became one of Otago's leading businessmen, with interests in runs, banks, coalmining and refrigeration. In the early 1870s, he commissioned a recently returned young architect and engineer, Frank W. Petre, to design his great house, 'The Cliffs', overlooking the Flat and St Clair. He had fixed views on architecture, insisting that the living rooms face south to avoid the sun and that the toilet be in the foyer, close to the living and dining rooms, on the grounds that such functions were natural. In 1876, the family moved into their new home.

DCC Archives.

In New Zealand, mansions, like fortunes, were invariably modest compared with mansions in Britain or even other New World societies, for our wealthy would not have ranked in Britain, the United States nor even Australia. Yet residences dramatised more than any other visible marker the extent of inequality. Here, E.B. Cargill's 'Castle' and Sir William Barron's house represent the country's mansions.

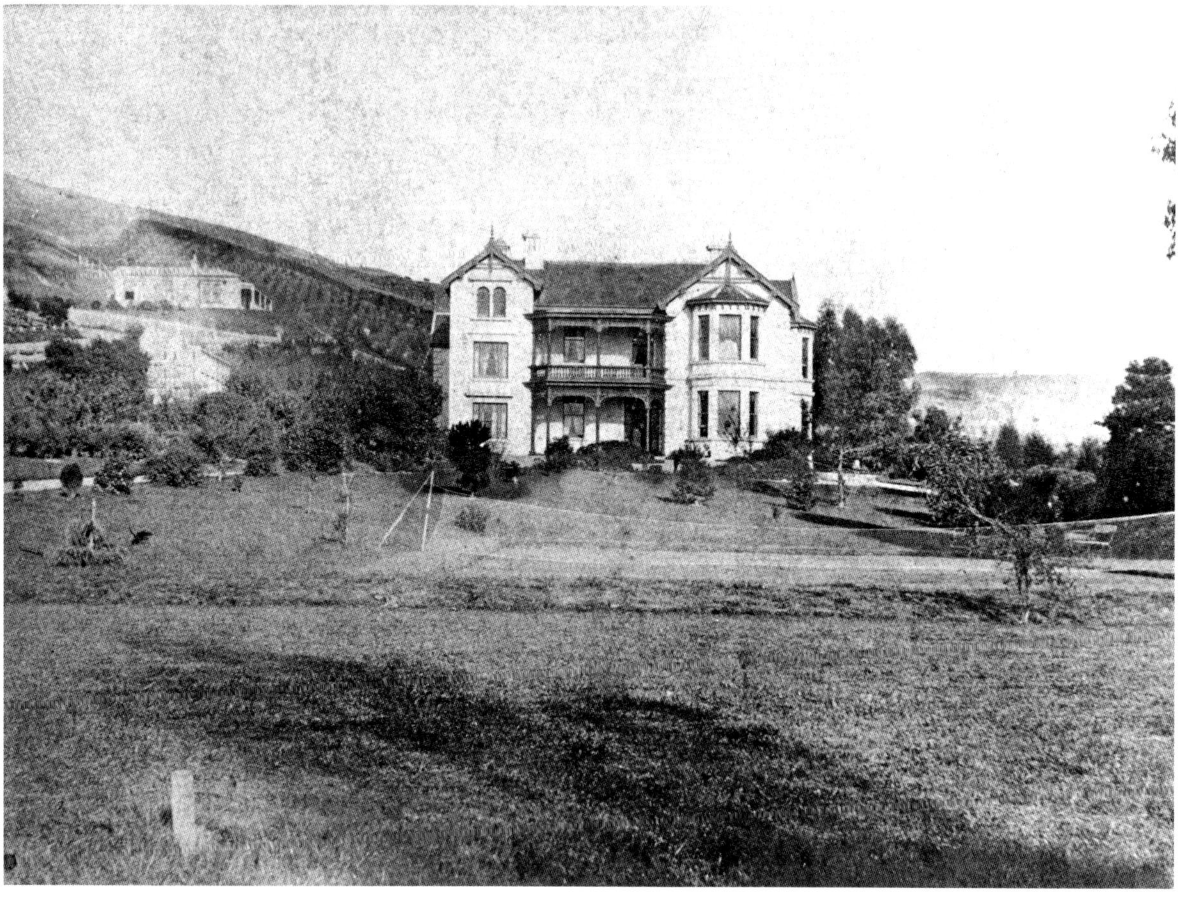

II

The Flat was the most ethnically and religiously diverse area in Dunedin and one of the most diverse in New Zealand. The dominance of the Presbyterian Free Church – the 'Wee Frees' – within the original city of Dunedin, seems to have made the Flat especially attractive to those who resented 'Wee Free' claims to pre-eminence. The English and Irish were more numerous than elsewhere in Dunedin. So, too, were English non-conformists and Irish Catholics and Protestants from Ulster. In the 1900s, the English-born were roughly as numerous as Scots in southern Dunedin and Irish Catholics comprised around 20 per cent of the population of South Dunedin and St Kilda. The mixing of three of Britain's four distinct peoples, who mainly lived apart in Britain, provided the coping stone for a new society. By New Zealand standards, the Flat also included sizeable ethnic-racial minorities: Chinese, Jews and even a handful of Kai Tahu. Between 1908 and 1910, Greek Orthodox from modern Lebanon settled on the Flat.

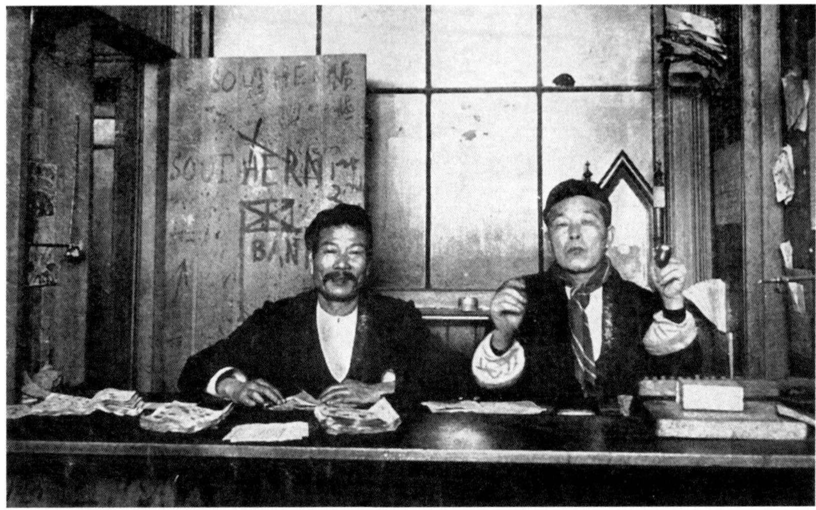

RIGHT **Although sojourners, by the 1900s Chinese men were widely accepted in southern Dunedin** and usually took part in local parades and festivals. Their exotic dress and customs, no less than their spectacular fireworks, enhanced many occasions. Several established local businesses, converted to Christianity, took European wives, and even became British subjects (although they lost this right in 1908). In 1904 the *Otago Witness* published four pictures, three of a large wooden boarding house that had been condemned under the Public Health Act 1902, and this one. Along with other Chinese gambling games, Pakapoo had been made illegal in 1881: the start of a sustained period of government discrimination and persecution. The photograph of the 'Pak-a-Poo' lottery shop's interior must have been designed to show how the Chinese ignored European laws and conventions. The police, it seems, turned a blind eye unless forced to notice, which may have been the intention of this photograph. Pakapoo was not especially popular with Europeans.

Hocken Collections, *Otago Witness*, 21 September 1904, p. 45.

OPPOSITE **Clara Calder's school leaving certificate 1882,** showing that she had completed Standard IV to the satisfaction of her teachers and the local school committee. Although we cannot identify all of these children, a surprising proportion of them were the children of successful local businessmen. Back row (one child is unnamed): George Cochrane, son of a local brewer; Dick Harrison; W. French; Alf Briggs, son of another brewer (and a cousin of George Cochrane); and Flora Fawlkes. Middle row: Edwin Briggs, Alf's twin; Dick Grimmett, son of a local bricklayer and builder, and brother of Rowland, who migrated to Sydney and became a builder while his son, Clarence, became the world's most famous leg-spin bowler; Arthur Anderson, son of the postmaster; and Walter Peddington. Front row: Clara Calder, whose father owned a quarry and ran a market garden; Edith Pearce, daughter of the Town Clerk; Jessie Rutherford, daughter of grocer Robert and his wife Jesse; Lot Allan; and Joanna Barron, one of Sir William's daughters. The Briggs twins, Cochrane, and Jessie Rutherford all proceeded to high school.

Toitū Otago Settlers Museum.

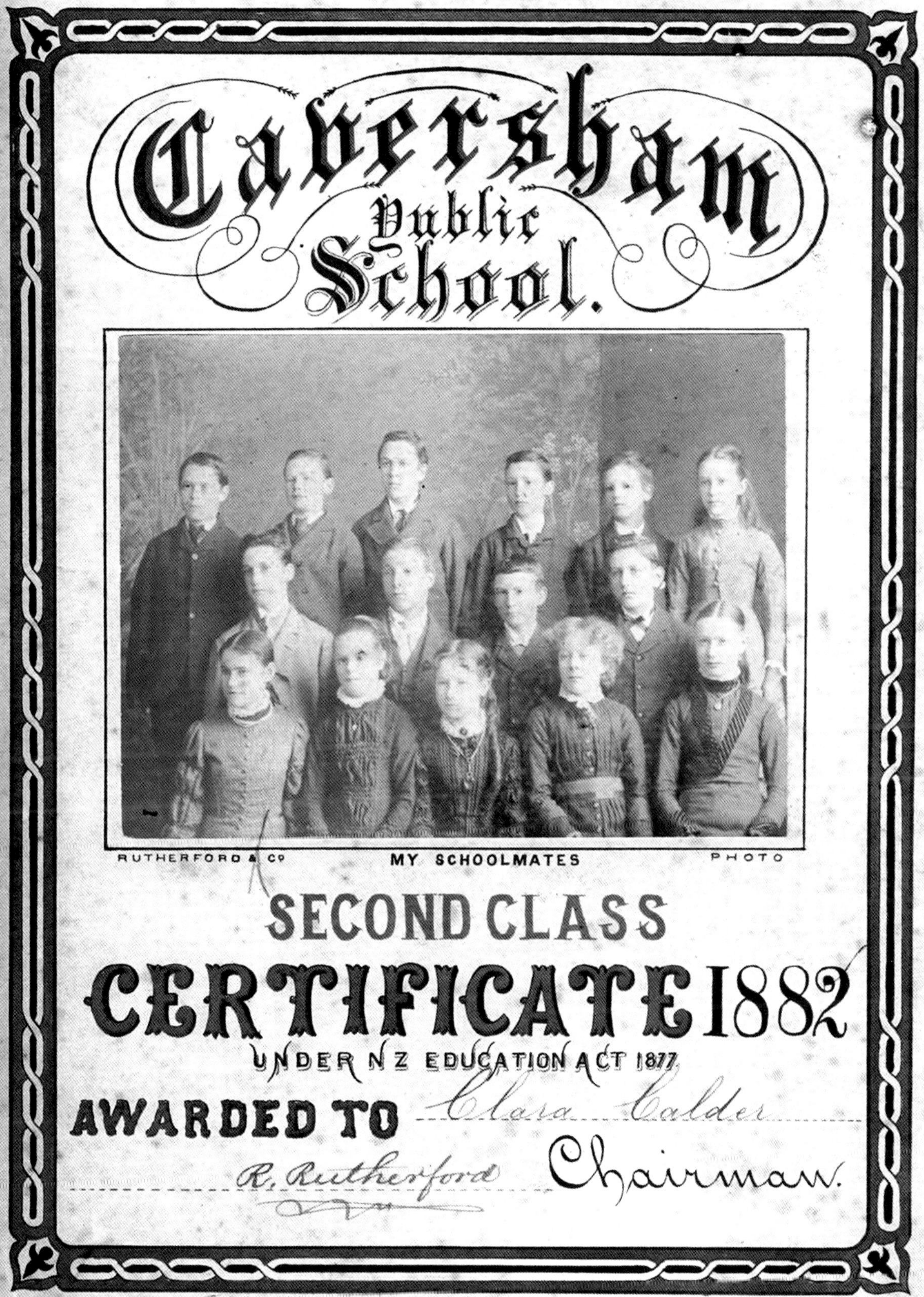
Caversham Public School.
RUTHERFORD & Co
MY SCHOOLMATES
PHOTO
SECOND CLASS
CERTIFICATE 1882
UNDER N Z EDUCATION ACT 1877
AWARDED TO Clara Calder
R. Rutherford Chairman.
COULLS, CULLING & Co, PRINCES ST., NORTH

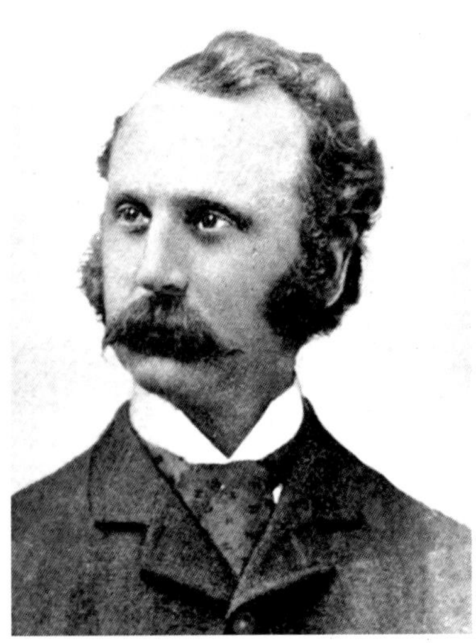

ABOVE **Frank Petre, architect of 'The Cliffs'.** While overseeing construction of E.B. Cargill's new house, 'The Cliffs', an innovative work in concrete, Frank fell in love with Margaret Cargill, despite the fact that he was a Catholic and she a Free Church Presbyterian. Frank's parents, who had settled in the Hutt Valley in 1840, where he had been born, were dead; Margaret's parents, by contrast, were alive and strongly opposed. Her father attempted to bribe her with a grand piano – music being her passion – then sent her to Europe to study. When it became apparent that Margaret was determined, her parents gave way. Bishop Patrick Moran, a close friend of Frank's, married them at 'The Cliffs' in 1881. In those days, nobody took photographs of newly weds. While two carte-de-visites of Frank survive, no photograph of Margaret is in a public collection.

Toitū Otago Settlers Museum.

When Margaret Cargill accepted Frank Petre's proposal of marriage in 1880 and legitimised mixing, Petre, born in the Hutt Valley but descended from one of England's oldest Catholic aristocratic families, had but recently returned from England, where he had gone to study architecture and engineering. Margaret's father, Edward Bowes Cargill, retained Petre to design a new house to affirm the family's wealth and standing. And so it came to pass that Frank laid siege to the heart of Edward's oldest daughter and the grand-daughter of Otago's Presbyterian Moses. Edward, most unhappy, tried to distract Margaret by sending her to Europe to study music. It did not work. He finally accepted the couple's wishes and gave his daughter away at the new Cargill family home, 'The Cliffs'. On the Flat, where the Catholic church had a formidable presence and its own schools, many married across this boundary even when the non-Catholic, unlike Margaret Cargill, refused to convert. Indeed, another of Edward's daughters married an Italian Catholic aristocrat, while yet another lived for most of her life with an English Catholic woman. They founded, and for many years ran, the tearooms at the bottom of the Spanish Steps in Rome. There is no photograph of any of these daughters in any Dunedin archive.

Occupations and social classes were also mixed. This mixing process began as you disembarked in the country, if you were male. One's background and skills undoubtedly influenced where one might look for work, and what one might hope to do, but the key to survival, let alone getting on, as one guide book after another affirmed, was a willingness to be flexible in using one's knowledge and skills. Nor were those born here confined by what their fathers had done. The sons of the unskilled moved freely into skilled and white-collar jobs. Many became self employed or small masters. Most owned or rented enough land to grow sufficient food to achieve partial independence from the labour market. The sons of the skilled and white-collar men also headed off across the full range of choice, including unskilled jobs. True, the sons of professionals and skilled men were more likely than not to enter another profession or trade, but they rarely stayed in their father's occupation. Significant numbers also kicked over the traces.

In any decade, and especially the 1900s, a considerable proportion of men moved back and forth between wage labour and self employment. Some combined the two, the proportion depending on the state of the labour market. In the colony, social strata, like classes, were like a tram at lunchtime: always full, but always full of different people. The rate of turnover was the key. Even larger employers in the 1900s had remarkable rates of turnover. Indeed, if one looks at large employers in 1910 and asks what they were doing in 1900, it transpires that 31 per cent had been small employers or self employed, over 12 per cent had been skilled workers, and 6 per cent had been unskilled. Around half had not been large employers.

If one looks at marriages, one finds an equally remarkable mixing process. Astonishingly by the standards of England or Scotland, around 32 per cent of upper-middle class women and 32 per cent of upper-middle class men

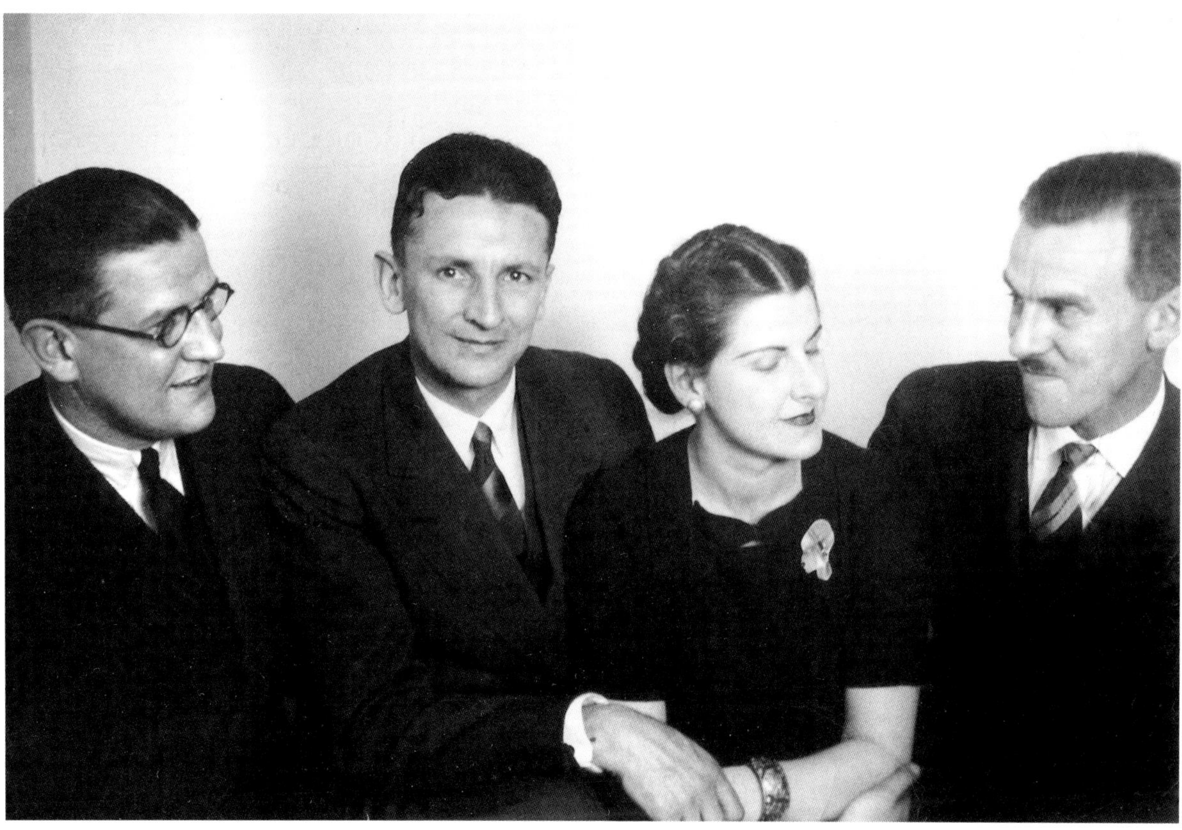

ABOVE **The four children of John and Mabel McIndoe** illustrate the complex ways in which kin complicated inheritance. When John unexpectedly died in 1916, the printing firm was barely viable. His widow, Mabel Hill, the well-known artist and daughter of Wellington's famous hatter, found herself with four children younger than 18 years of age, her daughter, also Mabel, being only seven. The eldest, also John, was sent to Sydney to learn the trade, including the business side (a manager being appointed to run the business in his absence); Archibald, b. 1900, wanted to become a doctor, and his mother reluctantly agreed, as the family's financial future was uncertain. Kenneth, b. 1904, and Mabel, b. 1909, were still at school. Archie (far left) won a scholarship, which saw him through his studies, and later established himself as one of England's most significant plastic surgeons. John (second from left) completed his training and installed one of the country's first colour printing machines, just in time to take advantage of the Dunedin & South Seas Exhibition of 1925, thus ensuring the firm's success. Mabel and Ken proceeded to Otago University and postgraduate study; Ken became a plant geneticist in the United States, and Mabel a botanist in Scotland.

I am grateful to John McIndoe for the photograph and to the late Reg Graham for having copied it.

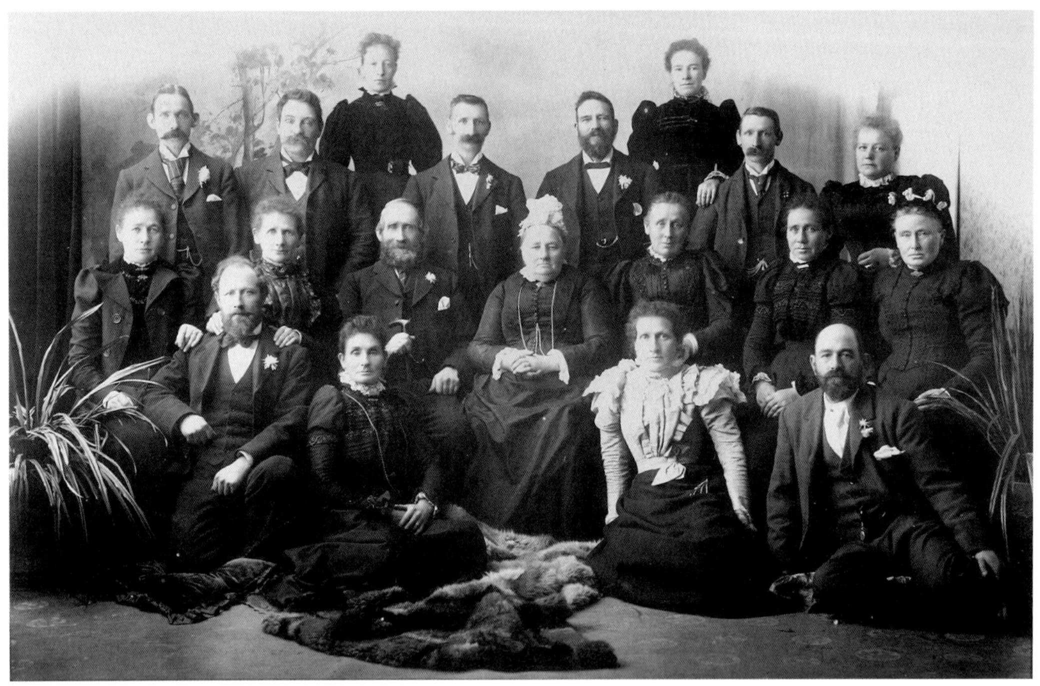

ABOVE **William Blackwood (b. 1824–25) and his wife Martha** (née McIndoe), both from Kilmarnock in Ayrshire, arrived in Otago in 1859, and can represent the pre-1860 immigrants. This photograph was taken on the occasion of their golden wedding. They farmed in Caversham and when the township's population grew he opened a bakery and store, somewhere around 1870. He died in 1900 and his wife then ran the store and bakery until she died in 1904. The youngest of their three sons (and eight children), William, who had served his apprenticeship as a tea-blender, then took over the store. The other two sons served apprenticeships, one as a tinsmith, who moved to Lawrence, and the other as a fitter at the Hillside Workshops, who listed himself as a mechanical engineer in *Stone's*. By 1906 two daughters had married blacksmiths, one married the manager of the Tuapeka Gold Sluicing Co, another a farmer, and the youngest a flaxmill hand. At the time of Martha's death, there were 43 grandchildren and nine great grandchildren, scattered even more widely across the occupational structure.

Toitū Otago Settlers Museum.

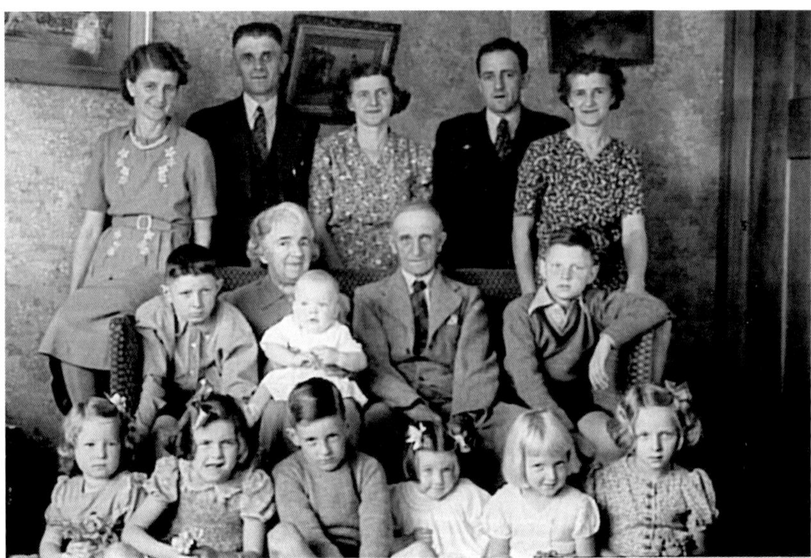

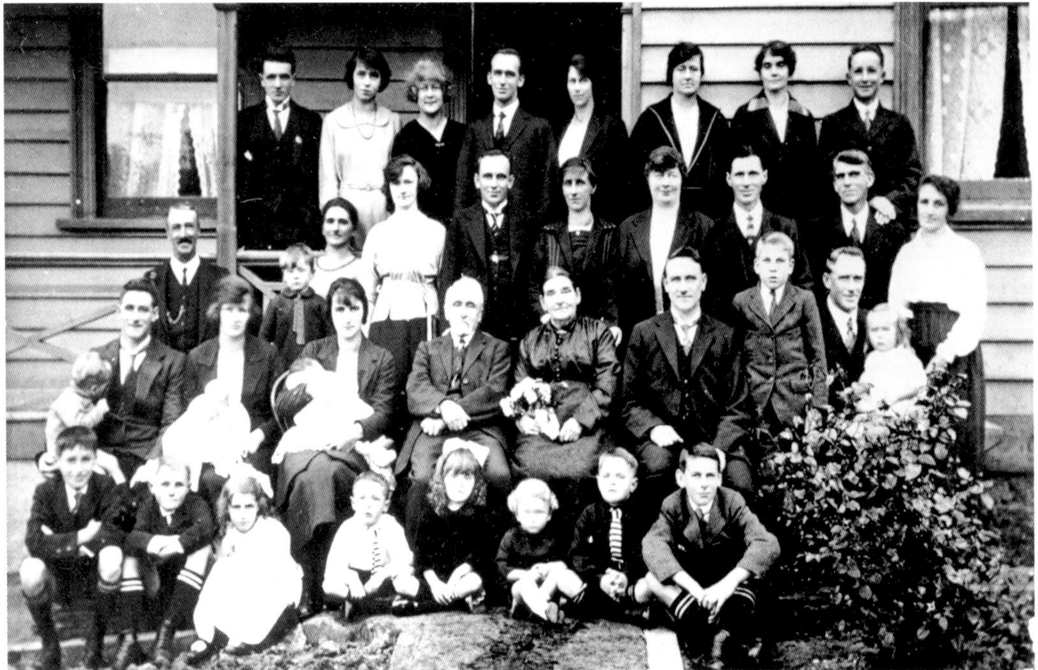

ABOVE **John and Sarah Hudson's golden wedding.** A Londoner, John Hudson (b. 1846) worked as a sailor before he married Sarah, the daughter of a shoemaker, in 1873. They migrated to Dunedin in 1875 with their first child, and settled in Sydney Street, Caversham. He worked as a brewer. They had six sons and four daughters. All but one son entered a trade, the sixth supervising explosives after serving in World War I. All but one daughter married tradesmen (the exception marrying a lorry driver). One son-in-law, Fred Jones, a boot clicker, was very active in the labour movement and converted Walter Hudson, one of the sons, to socialism and Labour. Both eventually became Labour MPs. All the males in the third generation entered the public sector, most of them in white-collar or semi-professional jobs, in search of job security and superannuation. By contrast, all but one of the females took white-collar jobs in the private sector. The third generation were all Labour supporters, and most of them became activists in either their unions or the Labour Party.

Courtesy Noel Hudson. Hocken Collections, Caversham Project, Ms-2690-196.

LEFT **Rowland and Emma Grimmett's golden wedding.** Rowland was a bricklayer who became a builder, and was the son of a bricklayer and stonemason. His wife, Emma Harris, was the daughter of a staircase builder, one of the most highly skilled trades. Her three brothers became an auctioneer, an antique dealer and a fur dealer. Their five children, standing in the back row, included three daughters – Myra (Thompson), Rita (Jory), and Louise (Wellman) – who were set up as hairdressers by their father and ran a salon until they left on marrying. The skills they learned in running their business became essential to the success of their own families. The two sons, Arthur and Bert, left the building trade. Arthur actually served an apprenticeship as a wood turner, and later became a taxi driver; Bert attended Otago Boys' High School, then worked as a clerk and storeman, before serving in the Pacific during World War II. He later worked for various firms before becoming manager of Kelvinator House. The next generation scattered across the occupational structure, two becoming professionals, another a butcher who later became a warehouseman and then a union official, and the rest entering white-collar occupations.

I am indebted to Noel Jory and Ross Grimmett for this information and to Ross Grimmett for the photograph.

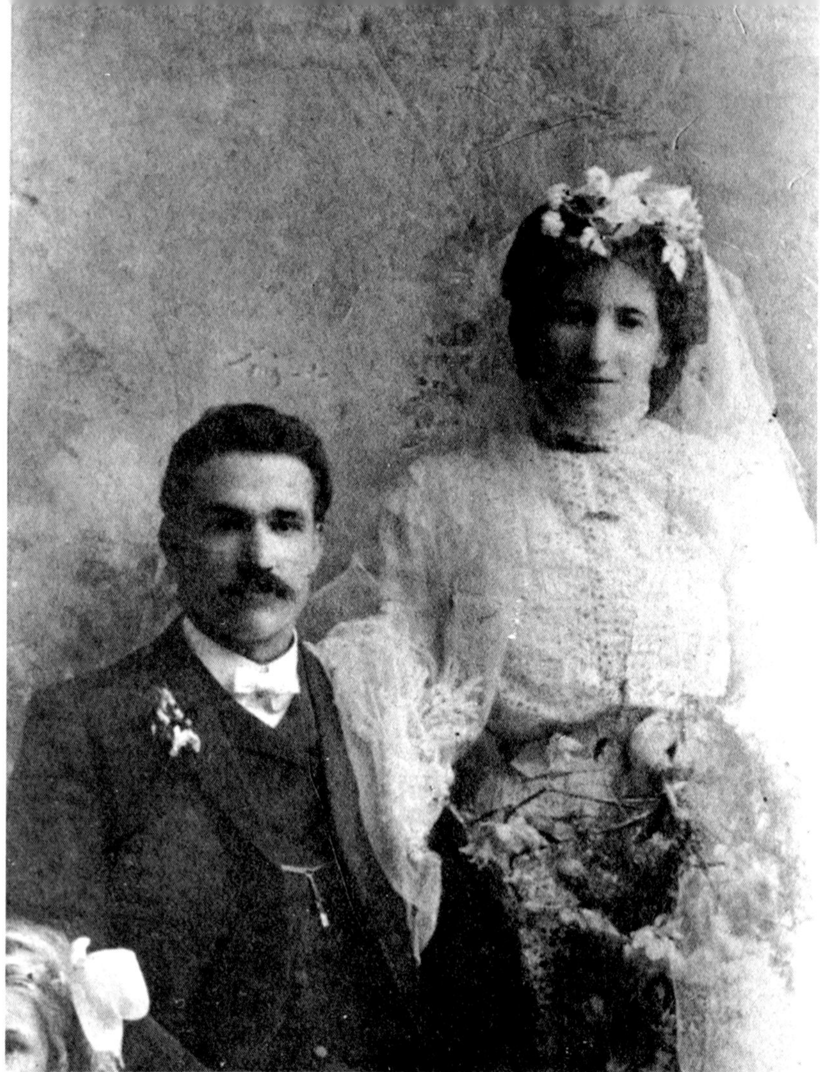

RIGHT **Jack MacManus and his bride** (née Gore) on their wedding day. MacManus was born in Ireland, while his parents returned from Australia to visit family, and he and his three brothers all became shearers, like their father. They also became active members of the Australian Workers' Union, a general-purpose rural union at the forefront of championing White Australia and moderate socialism. In 1906, MacManus joined the rush to New Zealand and soon married and became a union organiser for shearers here. He later became organiser for the local Labourers' Union and a branch of the revolutionary Socialist Party, served in World War I, and worked tirelessly to establish the Labour Party in local body and national politics. Like many Irish, he dreamed of becoming a farmer, and ended up scrabbling a bare subsistence from a small farm on the Chain Hills, his 12 children providing a source of free labour. Four of his daughters worked in either the Mosgiel or the Roslyn woollen mills before marrying, three sons entered the public service, two joined the City Council's tramway service, and two entered the building trades.

Hocken Collections, Pearl Preston, 'Pearl Preston and the MacManus Family', Caversham Project, Ms-2690-196.

chose a spouse from the manual working class. Even more startling, 20 and 22 per cent respectively chose a spouse from among the unskilled. It was much the same with every class, although the working class, being the largest, had enough men and women choosing working-class spouses to make class-based political action possible. The smaller classes, by contrast, were more quickly re-made through marriage, intergenerational and worklife mobility. At the breakfast, dinner and tea table, families worked out a consensus on how to maintain social harmony.

The streets, neighbourhoods and suburbs were also diverse in their occupational and social-class mixes. Although the proportions varied, and varied considerably, each neighbourhood and suburb contained a mix of people from every occupation and social class, all the main religious denominations, and the three principal nationalities. All but the shortest streets also contained a mix. At least one unskilled man and his family lived on every street in upper-middle class St Clair, just as in grimy Kensington each street boasted at least one major businessman or professional. In this period, most employers, whether large or small, lived close by their businesses, or even upstairs. In Ings Avenue, one of St Clair's best streets, the Hegartys, Irish Catholics, lived alongside the Ings, English Baptists, and

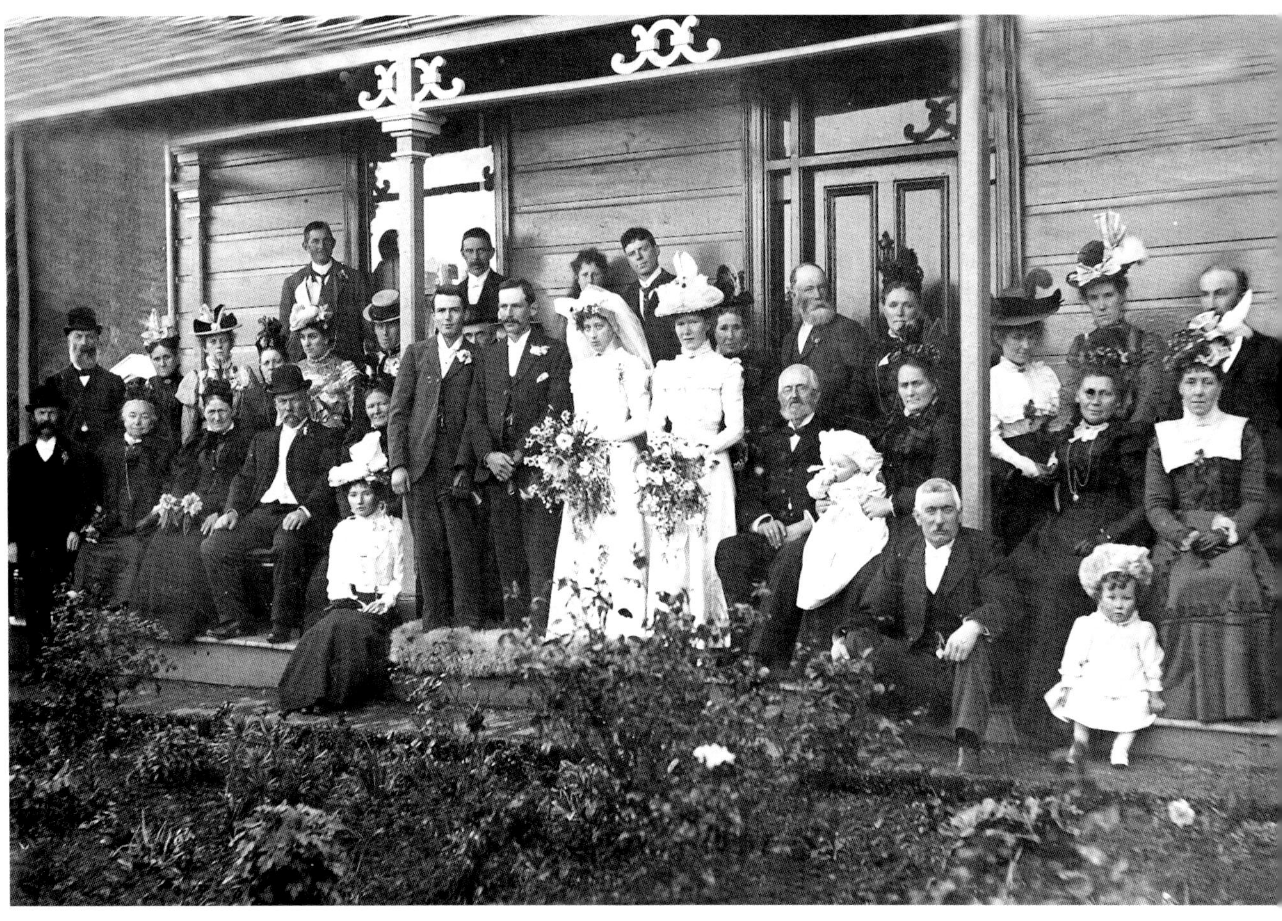

the Wolfendens, Irish Brethren. Some of Ings' gardeners and the Shiels' quarrymen and brickies lived close by, usually in a cottage. On first arriving on the Flat, both the Ings and Hegartys had lived in even more primitive cottages.

In learning to master their urban social structure, New Zealanders to varying degrees became experts in negotiating cross-class differences. They differed on the extent to which the state ought to help the weak and vulnerable, but they did not differ much. They also differed about what was fair, but agreed on the importance of being fair. To a lesser extent, they became experts at negotiating differences of religion, nationality and ethnicity. It was not all plain sailing all the time, but there can be no doubt that they built a New World that negated the organising principles of the Old Country. Some wanted to go further, and did, building voluntary organisations and political parties to advance their new nation towards greater equality and justice. Twice in less than 50 years, political parties considered revolutionary elsewhere drove through legislative changes that quickly came to be accepted by all. As Sam Lister remarked in 1890. 'We will be a different people. We certainly shall not be English – we shall not be Irish, we shall not be Scotch. Neither shall we be called British.'

ABOVE **John and Janet Rutherford** (née Thompson), photographed on the verandah of their new home in Peter Street, on the occasion of their wedding in 1900. The newly weds are flanked by John's older brother, Robert (the grocer), and his wife, Jesse. To the left are the Rutherford grandparents, Robert (1855–1942) and Jeanne (1855–1926). They, together with Robert Sr's brother, founded the grocery shop. Janet's father was a nurse at Riverton Hospital. Their sons all ended up working in the family business, except for Robert T., who did an apprenticeship as a boilermaker at Hillside. He married Agnes Stewart, daughter of a Scottish ploughman who became a driver for the City Council; their son, also Robert J. (known as Jack), became a teacher. Jack's memoir, 'Some Rutherford memories' (2003), has been deposited in the Hocken Collections.

I am indebted to Jack Rutherford for lending me the photograph for use in this book, and to the late Reg Graham for copying it in 2007.

CHAPTER 5

The Labour Movement and the Labour Party

OPPOSITE **The first New Zealand Trades Union Congress, 1885.** Robert Rutherford, the photographer, is second from the left in the middle row and a bearded Charles J. Thorn, President of the Otago Trades Council, is sixth. Both men were masters. The Trades and Labour councils of the 1880s were not militant but tried to act as boards of conciliation or arbitration. Thorn, a native of Essex who had worked for many years in London, was a founding member of the Dunedin branch of the Amalgamated Society of Carpenters and Joiners. In 1881, he took the initiative in establishing the Otago Trades and Labour

ALTHOUGH THE DEGREE OF INEQUALITY was small compared to England or Scotland, not to mention Ireland, the experience of the Long Depression (1878–95) in Otago convinced a growing number of people that inequality might become unacceptable. Fear of being displaced or undersold by the Chinese, unskilled whites, boys or even women made many artisans and mechanics uneasy. The mechanisation of the bootmaking and clothing industries intensified their fears. Many skilled men formed unions, as had become the practice in Britain, but during the Depression these did not prove very effective in protecting wages or conditions. Even the great imperial unions, the Amalgamated Society of Engineers and the Amalgamated Society of Carpenters and Joiners, both strong locally, proved ineffectual. (In Scotland and England, Parliament had accepted trade unions as legitimate methods for working men to protect their interests and in 1878, thanks to the Member for Caversham, Robert Stout, New Zealand followed suit.)

As the Long Depression settled on the South Island, men were thrown out of work or forced to work short-time. Many worked on in the hope of being paid when conditions improved. Poverty, drunkenness and disorder became more visible. Men increasingly fretted about how best to solve these problems. Small numbers of women also joined in the discussion. Some unionists became politically active. They formed a Trades and Labour Council, a parliament of unionists, which spelt out a programme of reform and helped to elect several radicals and liberals to the House of Representatives in the 1881 elections. These radical liberals proved ineffectual. In 1885 Charles Thorn, Caversham's joiner and undertaker,

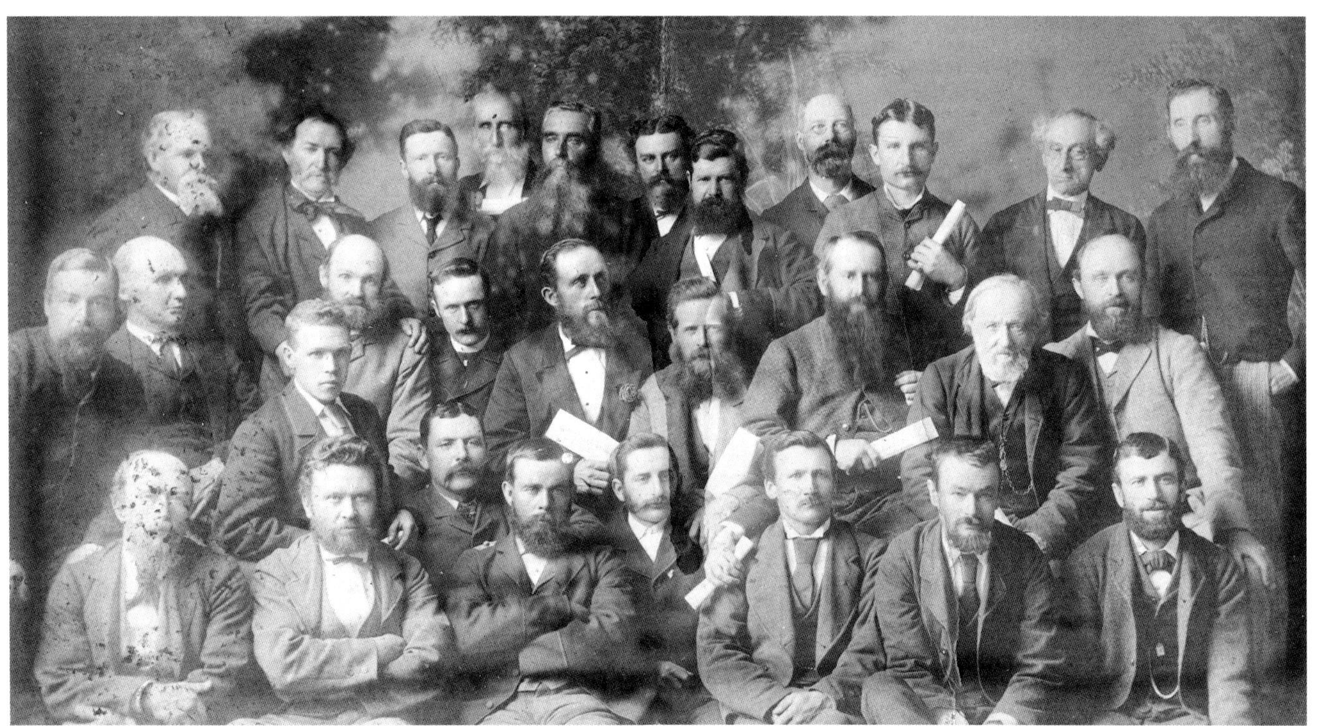

convened a national conference of Trades and Labour Councils. Most working men remained apathetic, as were working-class women, who at the time were still not eligible to vote. In 1887, Sam Lister began publishing his trenchant and witty radical weekly, the *Otago Workman*, a potent antidote to apathy. In 1889, New Zealand introduced universal male franchise. The local Trades and Labour Council collapsed.

In 1888–90, politically active masters and journeymen formed a Protection league (to demand tariff protection for local industries), mobilised in large numbers into unions and revived the Trades and Labour Council; they then formed a Labour Party to contest the 1890 election. Caversham's Bob Slater, a presser by trade and a Baptist by conviction, played a key role as secretary. In the 1880s, the seamen also formed a union. John A. Millar, ship's captain and son of a major-general, led the seamen to a famous victory in 1887–88 and ignited a passion for union. He organised the seamen and watersiders into a Maritime Council, based in Dunedin, and won victory after victory. Organised labour throughout the Empire was on the march. Even London's despised 'dockers', not to mention the 'matchy girls' in Bryant & May's London factory, struck and won. In Dunedin, around 10,000 men had joined a union by mid-1890: a quarter of the city's population. Unionisation became the foundation for political mobilisation.

Ironically, even as the revived Trades and Labour Council prepared to nominate their own ticket, the seamen and watersiders at Port Chalmers, over-estimating their strength, came out on strike in support of their brothers in Sydney and Brisbane. Through the Trades and Labour Council, the non-maritime unions fully supported the strikers. The Bootmakers' Union and

Council to maintain the eight-hour day and 'improve the conditions of the working classes'. The Otago Council, unlike most of the northern ones, was strongly committed to ensuring that Labour was represented in Parliament. Thorn's Otago Council convened the first national conference, the Trades Union Congress, which met in Dunedin's Oddfellows Hall in January 1885. The fundamental purpose was to organise all workers, including women, in order to advance Labour's legitimate interests in the legislature.

Hocken Collections, Otago ASC&J Ms.

OPPOSITE **Samuel and Helen Lister** (née Miller) with their first born, probably in Edinburgh where they married in 1862. Helen was a dressmaker and Sam a lithographic printer. Three years later, they embarked for New Zealand with two children. They had eight more after settling here, two dying. All three sons entered their father's trade.

The Listers settled in South Dunedin. Lister worked as an engraver and a printer, often on his own account but also in several short-lived partnerships. After his eldest son died tragically in 1875, Lister took solace in strong drink, was censured by the Kirk, and left it.

Sam Lister's main claim to fame was his success in establishing the *Otago Workman* in 1887. His keen espousal of Labour's rights, his caustic denunciations and excoriations of the 'unco guid', whether clergy or capitalists, and his witty commentaries on the events of the day saw his *Workman*'s circulation rival that of the *Otago Daily Times*. The traditions of artisan radicalism, republicanism and anti-clericalism had found their tempestuous local champion. Lister and his *Workman* played a critical role in engineering the great working-class mobilisation which climaxed in the Maritime Strike of 1890. He championed not only trade unionism but a Labour Party, and saw his dream come true when a local Labour Party swept the city and captured the seats of both Caversham and Peninsula, the latter then including all of South Dunedin and the lower reaches of Caversham.

The *Workman* vigorously championed the Lib–Lab government, but Lister's strong hostility to both Prohibition and Women's Suffrage saw him become increasingly cantankerous. Life at home – which he reported on for his adoring public – must have been stressful for Helen, whom he always addressed formally. She supported both causes and signed the 1893 petition. He calmed down after 1893 and strongly supported the Lib–Lab governments until he sold the *Workman* to the Trades Council in 1901. He died in 1913.

Hocken Collections, Caversham Project, Ms-2690-196; image S03-208.

the recently formed Tailoresses' Union were particularly generous. Mass meetings and rallies became the order of the day. Discipline and close stewarding ensured that no drunks or roughs threatened the hard-won respectability of the unions. Some unions even stopped meeting in pubs. The unions also organised a spectacular Labour Day parade, which became an annual feature locally until Millar (who had entered Parliament in 1893) pushed through the Labour Day Act in 1898 and made it a national holiday.

Throughout the Western world the 'Labour Issue', like the 'Woman Question', aroused fierce debate. In Dunedin the Reverend Rutherford Waddell, whose inner-city parish included Kensington, helped make sizeable sections of the city's middle classes sympathetic to the working classes by denouncing 'sweating' in the clothing industry, one of the blots of the Old World that he had found here. He served on a royal commission that concluded that trade unions would help ensure that such Old World diseases would be kept at bay. In 1889, a group of prominent men promptly organised the country's first union for women, the Dunedin Tailoresses. Much of middle-class Dunedin now accepted the need for unions. So did most manufacturers. The local Trades and Labour Council bought a daily newspaper, *The Globe*, and even the *Evening Star* (often known as the 'Twinkler') swung into a mood of impatient reformism. Such works as Edward Bellamy's *Looking Backwards* (1888), a socialist evangel, raced through several local editions.

Although thrashed on the industrial front, the unions' fledgling Labour Party swept Dunedin and picked up Peninsula and Suburbs (the last two seats included the Flat). The Labour Members, led by David Pinkerton, one of Sargood, Son & Ewen's finishers, achieved so much influence in the new Liberal government that it became widely known as the Liberal–Labour government. In the 1893 elections the Workingmen's Political Committee (WPC), a union-based organisation that allowed radical political groups such as the Women's Franchise League to affiliate, retained all the seats won by the Labour Party in 1890 and helped Millar to win Port Chalmers. The local Members dominated the newly established Labour Bills Committee, which played a crucial role in drafting the laws that made New Zealand the world's first wage workers' welfare state (in which all able-bodied men had a job and a wage sufficient to support a family).

The key measure was the Industrial Conciliation and Arbitration Act (1894), long demanded by the local union leaders, which allowed seven persons to form a union that their employers had to recognise and reach agreement with, or the Court would simply issue a legally binding award. In 1897–98, the local seamen tentatively explored the new law's potential. The shipping companies had blacklisted all unionists following their rout in 1890, and 'Big Bill' Belcher, the secretary, tentatively proceeded before the Arbitration Court. The Union Steam Ship Company refused to negotiate, until it became clear that the Court would hold them in contempt. One union after another now rushed before the Court. In 1897–1900, unions began rapidly to grow in number and size. The Trades and Labour Council's Bob Slater had been elected the workers' assessor on the Court. He played a key role in converting a succession of judges to the views of the skilled craftsmen. Most masters agreed.

During the lean years between the 1890 rout and the first sitting of the Arbitration Court in 1896, Dunedin's labour-radical movement remained surprisingly buoyant. Lister's *Otago Workman* now preached the importance of unions, state socialism and the Liberal–Labour Party. On the Flat, James McIndoe, who saw himself as an antipodean Cincinnatus, formed a branch of the Knights of Labour. During the 1890s, the Caversham Knights flourished. An American organisation, the Knights promoted the social uplift of working men and women regardless of race or religion, rejected socialism, and demanded the eight-hour day. The Caversham branch had a long list of further reforms, most of which the Liberal–Labour government had enacted by the end of the decade. They excluded from membership all unproductive parasites: doctors, lawyers, sharebrokers, bankers and liquor manufacturers. Nor did the woman's movement disappear. After a massive public campaign, strongly supported on the Flat, women won the vote in 1893. The attempt to capture women for the radical–Liberal cause failed, but the Tailoresses survived the decade in better shape than most unions. Various women's organisations formed the National Council of Women in 1896.

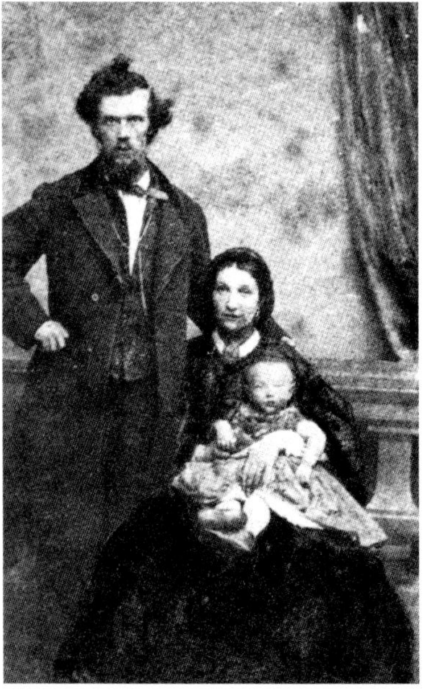

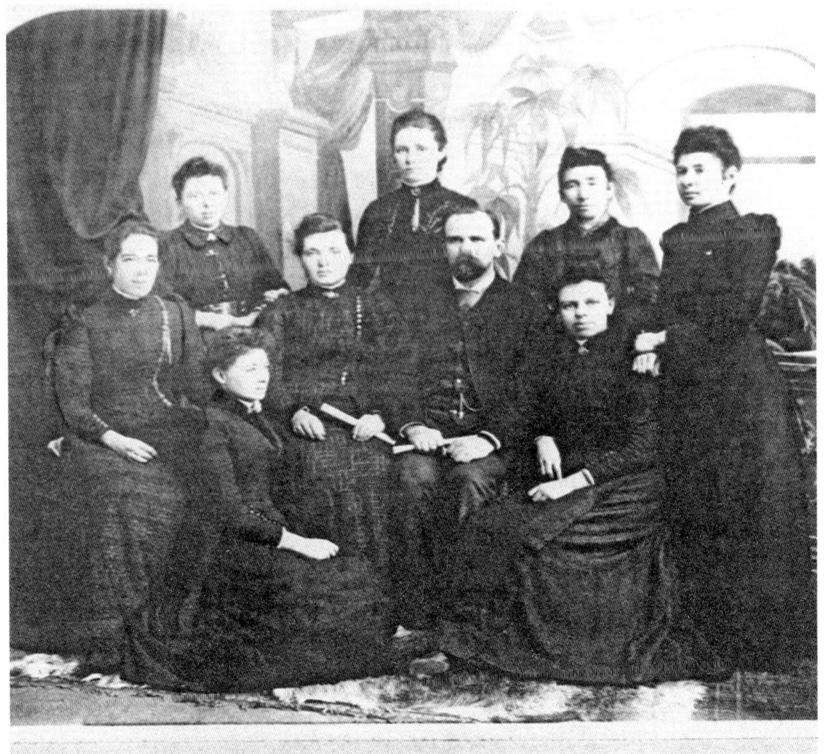

LEFT **The First New Zealand Federated Tailoresses' Conference, Christchurch, 1891.** After the Sweating Commission concluded that where unions exist 'the condition of the operatives has improved, wages do not sink below a living minimum, and the hours of work are not excessive', several of the city's leading gentlemen took the initiative in enrolling several hundred young women into the country's first union for women, the Dunedin Tailoresses' Union. The president and secretary were men but an articulate young seamstress, Harriet Morison, was elected vice-president. Under threat of strike action, local employers reduced hours and raised wages. In 1891, Morison, a Bible Christian from the Flat, became the union's first full-time secretary, forcing an aspiring male to withdraw. She dedicated herself to organising tailoresses throughout the country. She also mobilised her tailoresses behind the Women's Suffrage League, which she helped found and lead, and was active in the Women's Christian Temperance Union. Photograph by Eden George.

Hocken Collections, S10-062a.

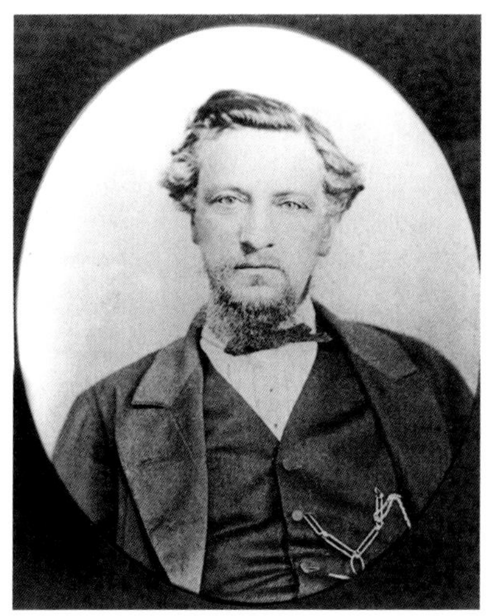

RIGHT **James McIndoe (1824–1905),** from a family of pedlars and shop-keepers, became a merchant and town councillor in Rothesay, Scotland, and married Elizabeth Gillies in 1852. After the death of two children in 1859, they decided to migrate to Otago with their remaining three children. He became a merchant and auctioneer but also took up land in Caversham – his house still stands at the corner of Macandrew and Forbury Roads – and combined small-scale farming with local politics and journalism. He fiercely supported the eight-hour day and Provincialism. He also took a deep interest in the history of Otago. His *Sketch of Otago* (1878) remains useful.

During the 1880s, as the Long Depression settled on the south for the best part of 20 years, he read widely on social issues and became more radical. Early in the 1890s, he founded a local branch of the Knights of Labour, a radical labour organisation which had originated in the United States, and tried to organise workers regardless of skill, race or sex. In Caversham, as in New Zealand, the Knights became more of an educational society and a political lobby that supported various radical measures. The 22-point programme included that famous goal of making 'industrial and moral worth, not wealth, the true standard of individual and national greatness'. As with most radicals of the time, the Knights worked hard for the land and labour reforms that the Liberals are still famous for.

I am grateful to John McIndoe for lending this photograph.

I

The role of the unskilled in the mobilisation of 1888–90 is unclear, although they certainly provided most of the votes. As working men exploited the possibilities of the arbitration system, a second wave of unionisation occurred. Dunedin's skilled unions led the way, rapidly reversing their humiliation following the defeat of the Maritime Strike. In 1900, Dunedin was the country's most unionised town but over the next 10 years the initiative moved north, especially to Wellington and Auckland. Although Dunedin was no longer the decisive force in the national labour movement, the presence of its Labour MHRs within the Lib–Lab government ensured its ongoing influence in maintaining the forms of protection central to the wage workers' welfare state. Even when T.K. Sidey won the Caversham by-election in 1901, defeating the WPC's candidate, Labour lost little ground because Sidey worked closely with the radical pro-Labour group. Sidey, one of the best educated Members, was also a gentle man and a gentleman. His ladies' committee was especially effective.

Thanks to a succession of strikes in 1906–08, the unskilled became an influential force within both the union movement and the political Labour movement. Spurred on by these developments, and the arrival in town of 'Big Jack' MacManus, a veteran of the Australian Workers' Union and a committed socialist, Dunedin's labourers began to organise themselves. So, too, did Otago's watersiders and shearers. Even the coal miners at Walton Park, across the hill from Caversham, revived their union. Led by the coal miners of the West Coast, the unskilled demanded industrial unionism. The more militant saw industrial unions as the engine for achieving revolutionary socialism. They thought craft unions were doomed to go the way of the horse and buggy. Many craft unionists agreed.

In this ferment, more and more working men attacked the Liberal–Labour Party as a sham. 'King Dick' Seddon, the Lib–Lab Prime Minister since 1893 and a fitter by trade, had little difficulty managing his critics.

THE CAVERSHAM BYE-ELECTION : CANDIDATES FOR THE VACANT SEAT.

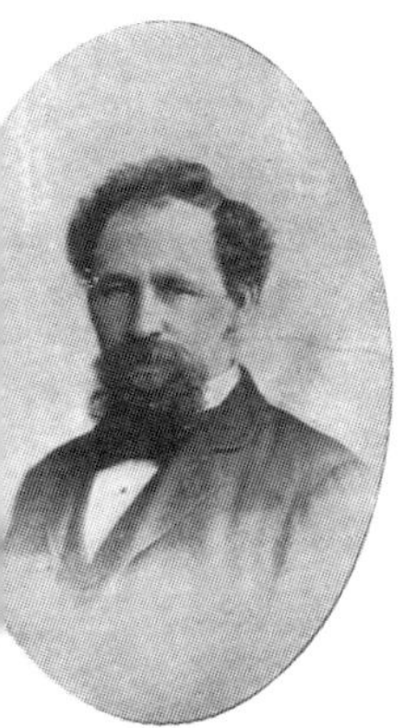

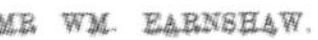

MR WM. EARNSHAW.

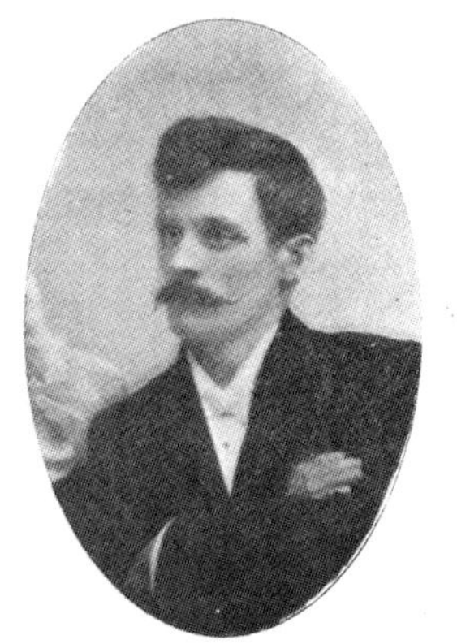

MR P. HALLY.

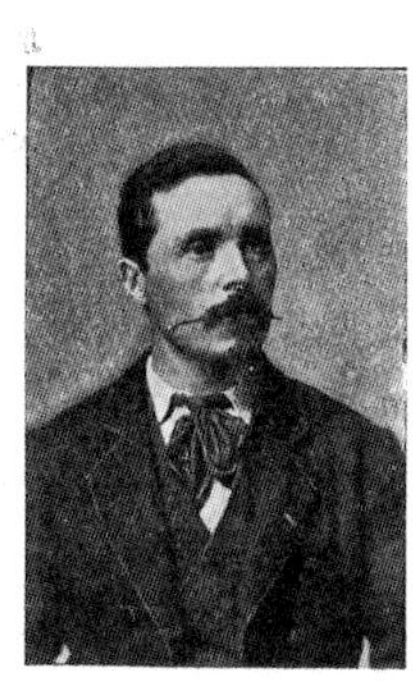

MR W. H. WARREN.

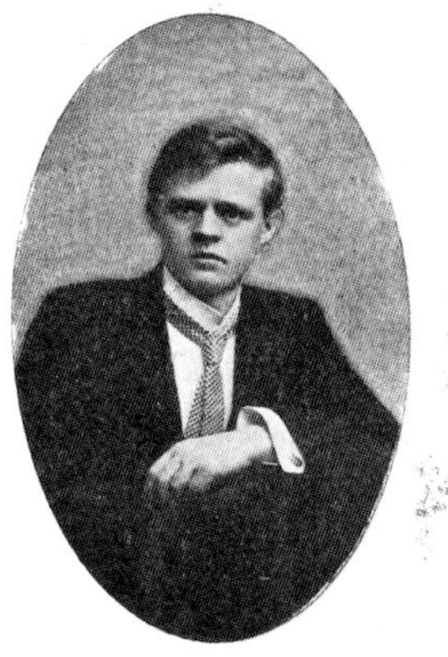

MR W. H. BEDFORD.

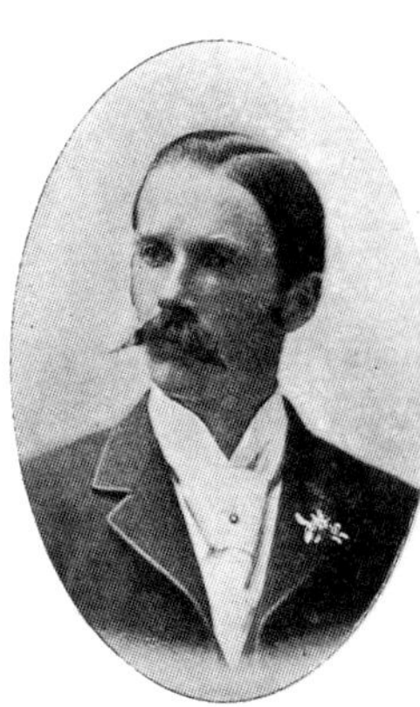

MR T. K. SIDEY.

1901 Caversham by-election. When Arthur Morrison, MHR for Caversham since 1893, died, five men threw their hats into the ring. Morrison had been the approved candidate of the Workingmen's Political Committee and that organisation nominated a Catholic master bootmaker, Pat Hally, who had been president of the union during the Maritime Strike and had also served as secretary in the early 1890s. The Liberal–Labour Federation backed him as well. Bill Earnshaw, once the Labour MHR for Peninsula, also stood and claimed to represent Labour, attacking Hally's Catholicism and demanding that all Catholics be expelled from the civil service. A sworn enemy of Seddon, Earnshaw believed alcohol the worst enemy of the working man and the root of larrikinism, destitution and poverty. Another independent Labour candidate, a carpenter, also stood on an anti-Catholic platform. The local solicitor, Thomas K. Sidey, son of John Sidey, who made his pile during the gold rushes and now farmed on Kew Rise, stood on behalf of the Liberal Party. New Zealand-born, T.K. had served on several local organisations, and for three years as Caversham's mayor. He was well known and liked. In a nasty campaign he personified good manners, civility and tolerance. Sidey narrowly defeated Earnshaw in 1901, but increased his majority in the elections of 1902 and 1905. In these years, he voted with the radical Liberals. He represented the Flat until 1928, when he was appointed to the Legislative Council.

Hocken Collections, *Otago Witness*, 1901; 'Picnic Group', S14-076 – Mr Sidey's Ladies' Commitee Picnic Group, December 1908.

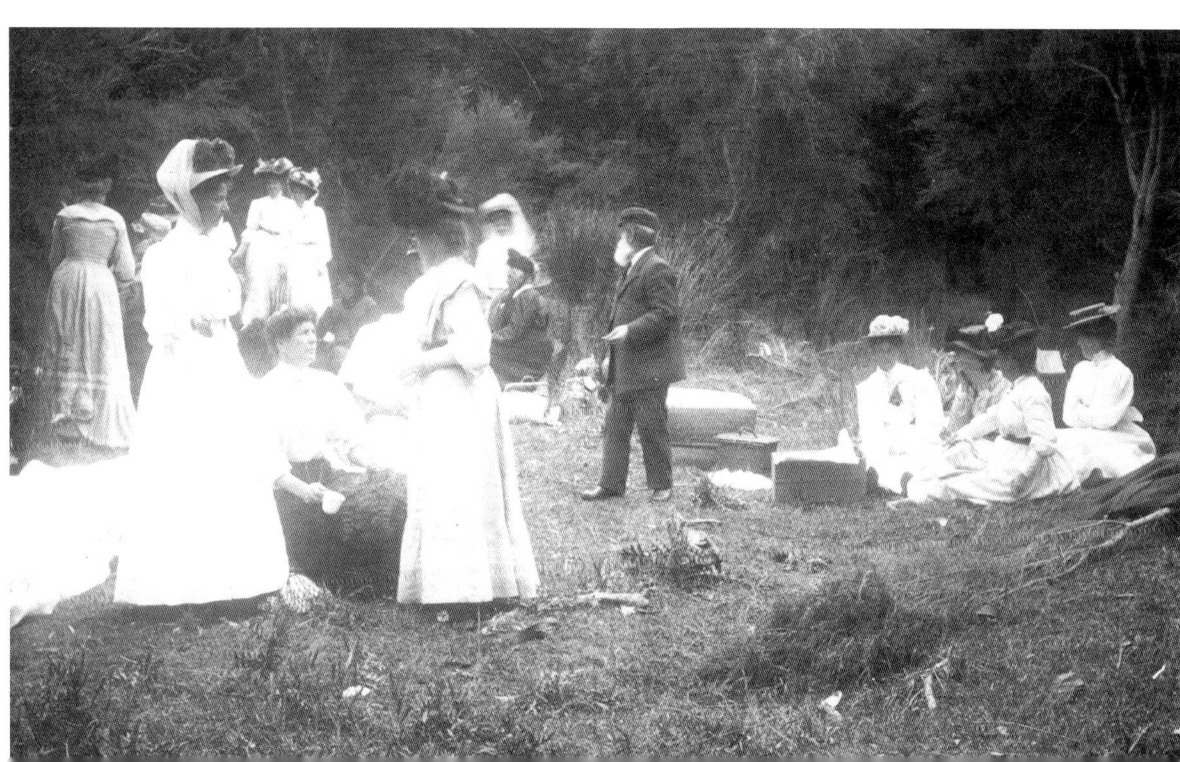

Once he died in 1906, however, no liberal could contain the threat. As a new and militant national organisation, the 'Red' Federation of Labour, marched from success to success, working men poured into unions. In the main towns, the demand for an independent Labour Party grew. Dunedin's unions had long been accustomed to exercise considerable political influence, thanks to their Labour MHRs, and even the unskilled locally proved resistant to the appeal of anti-political ideas. As early as 1908, a majority of the voters in Kensington and eastern South Dunedin preferred independent labour or socialist candidates. In 1911, some 70 per cent did so. When shifted from the Dunedin South to Dunedin Central electorate following the 1908 election, and no longer able to vote for Sidey, just over 50 per cent of voters in Caversham township preferred a socialist.

The mobilisation and radicalisation of the unskilled provided the basis for the defection from the liberals to an independent Labour Party, which kept reinventing itself while moving left across the period 1905–22. That mobilisation and radicalisation reflected the exclusion of the unskilled from the late-Victorian and Edwardian feast. Where men (but not women) of every other class enjoyed unprecedented opportunities in the decade 1901–11, the unskilled did not. Their situation worsened sharply in 1911–14, a period of heightened industrial conflict. The link between the unionisation of the unskilled and voting Socialist–Labour seems clear. The contagion was such that even the young women working for Ross & Glendining struck. And late in spring 1913 a second maritime strike began. Before long, the Dunedin–Port Chalmers branches of the watersiders and seamen had struck. When Massey called for volunteers to help the police to maintain order, clerks from the banks, insurance companies and warehouses, not to mention university students, rushed to assist. Hundreds of farmers enlisted. They camped at Tahuna Park.

These upheavals split the skilled. Some wanted to remain a distinct ginger group within the Liberals; others wanted a wider class-based unity on the national stage; and some threw their support to Reform, which preached 'A Square Deal' for 'sane Labour' (echoing Teddy Roosevelt's Progressive Party in the United States). The aggressive tactics of the unskilled, especially in the coal mines and the northern cities, deeply divided skilled workers not only in Dunedin but throughout the country. In Dunedin, unionised skilled men pursued their old strategy of local autonomy, but when they began to move away from the Liberals most of the non-unionised artisans did not follow and a sizeable number of skilled union men baulked. One suspects that working men hostile to unionism and Labour began to escape from Kensington and eastern South Dunedin, dominated by the unskilled with their enthusiasm for Socialism and their hostility to Prohibition, and headed for the suburban frontiers of Kew, St Clair and St Kilda.

It was not simply that the skilled split. As the unskilled unionised and moved left, most railway workers, now blessed with generous superannuation, distanced themselves. So did their union. Most of the men in handicraft trades too small to sustain any union also backed away from

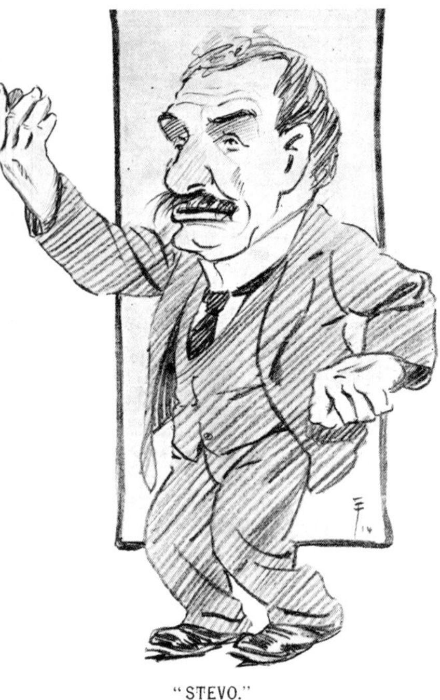

"STEVO."

ABOVE **'Stevo' Boreham, an immigrant from Australia back in the 1870s, 'gun' shearer and a key figure in founding the Shearers' Union in Otago.** After 30 years on the road, he and his family settled on the Flat in 1906, where he soon became prominent in organising the General Labourers. A born stump-orator, famed for his humour and his soapbox orations, he represented various unskilled unions before the Arbitration Court and demanded a living wage as a right. He was also the Flat's deadliest enemy of Prohibition, and never feared to speak his mind. 'I say if a man likes to take the risk and drink water it's no business of ours to prevent him taking his life.' He was killed in 1925 when hit by a car after leaving the pub late at night. Thousands attended his funeral, including the mayors of St Kilda and Dunedin.

Hocken Collections, *The Sketcher*, no. 4, Dec. 1914.

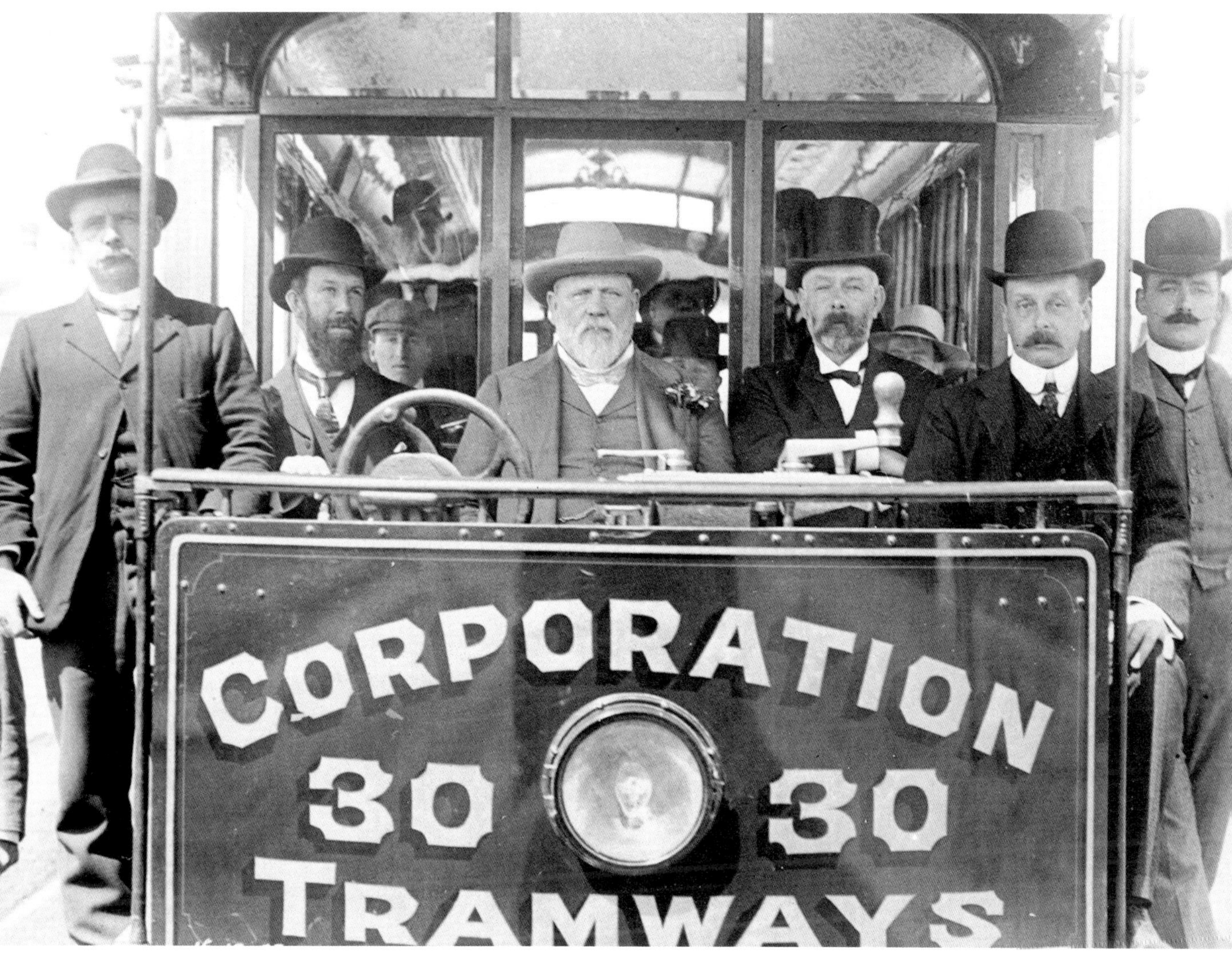

ABOVE **'The Premier, while in Dunedin [to open the electric tramway system], takes a jaunt on the electric cars',** flanked by various local notables. In 1898, Premier Dick Seddon had formed a Liberal–Labour Federation to institutionalise working-class and trade union loyalty. Seddon's Federation and the Workers Political Committee dominated Dunedin in these years. Seddon, who still holds the record as the longest-serving prime minister, held the country in thrall. A handful of radical Liberals, mostly Methodists and Prohibitionists from Christchurch, constituted the effective opposition, but when they mounted a challenge in the 1905 elections Seddon routed them. His broad humanitarianism, no less than his pride in New Zealand, made him a formidable leader. His political skills were already legendary and he was in the process of becoming a legend in his own lifetime.

James Arnold, the Member of the House of Representatives for City of Dunedin Central since 1899, is to Seddon's left. (He later represented Dunedin South, 1905–08, and Dunedin Central, 1908–11.) Arnold, a boot clicker, first won for clickers the right to join a union, then successfully led the union in one of the first cases before the Arbitration Court. In 1899, the Workers' Political Committee chose him as one of its three candidates for the City of Dunedin, the local branch of the Liberal–Labour Federation endorsed the choice, and he won. His ability to grasp complex issues, understand the legal process and argue his corner saw him appointed chairman of the House Labour Committee in 1903, when the other Lib–Lab MHR from Dunedin, John A. Millar, became chairman of committees. The other men represent the city and the contractor.

Hocken Collections, *Otago Witness*, 6 January 1904, p. 44; image from the Guy Morris Collection, P99-068, c/n 6681/71.

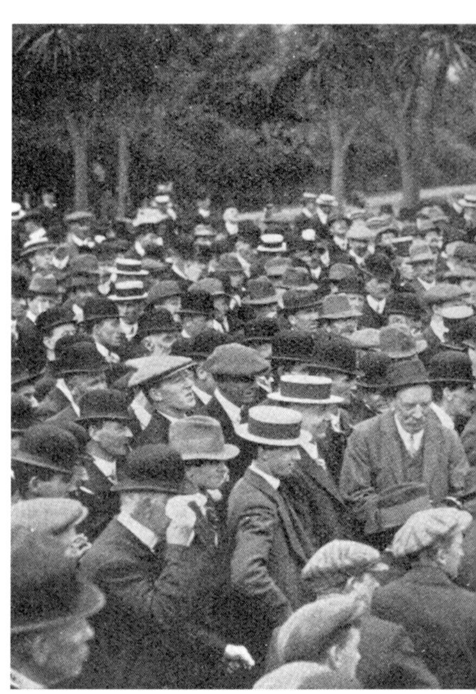

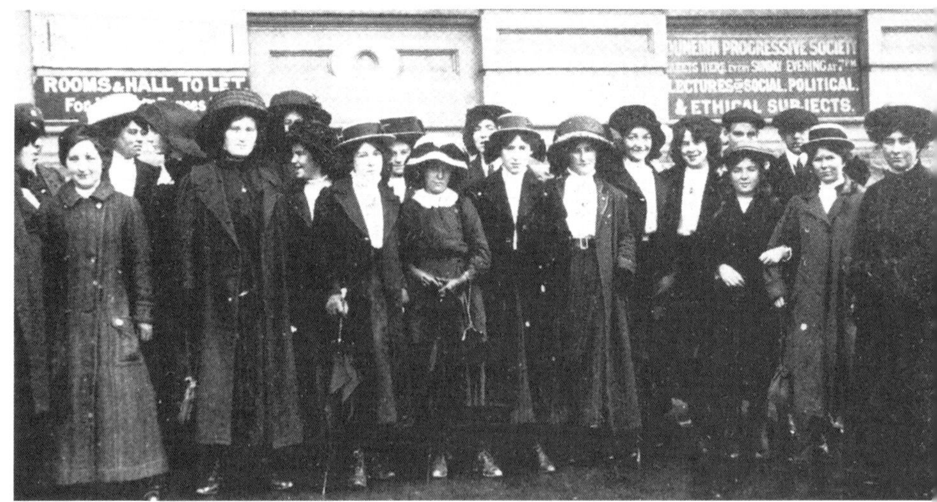

RIGHT **1911 strike at Roslyn Woollen Mill.** In this period of national (and global) industrial unrest, rumours that some of the male supervisory staff at the Roslyn Mill were favouring some girls over others precipitated two strikes in quick succession during 1911. The city's union leaders, all but one of whom were men, and the secretary of the Dunedin Tailoresses' Union, were astonished and taken off guard. The leaders of the factory girls and the mill's manager agreed to submit the issue to arbitration by the Mayor and John T. Paul, a prominent union leader in the city and a Legislative Councillor. After taking extensive evidence, the arbitrators found for the girls. The company happily accepted the recommendation that the offending males be removed to another part of the mill. One engineer was subsequently discharged for immorality. Note how well dressed these factory workers were and how thoroughly they had adopted the conventions governing the presentation of the self in public. They were emphatically affirming that factory work entailed no lack of respectability or refinement; nor, indeed, did striking.

Hocken Collections, *Otago Witness*, 19 July 1911, p. 44; image S13-306b.

the new forms of radicalism. So, too, did the men in the building trades and engineering shops who did not belong to any union – roughly 40 to 50 per cent. Roman Catholic workers, mostly of Irish birth or ancestry, also tended to react negatively to the appeal to desert the Liberals for Labour. Joe Ward, the Liberal Prime Minister (1906–12), was a Catholic of Irish-Australian origin. Besides which, Archbishop Redwood used the Papal anathema against socialism to condemn the local socialists. He strongly backed the Liberals.

During this period of national upheaval, organised Labour lost political ground in Dunedin. Sidey held Dunedin South. Millar's attempts to reform the arbitration system made him so unpopular with organised Labour that he became estranged from his old mates; although their victories were narrow, two youthful Reformers won seats in the House.

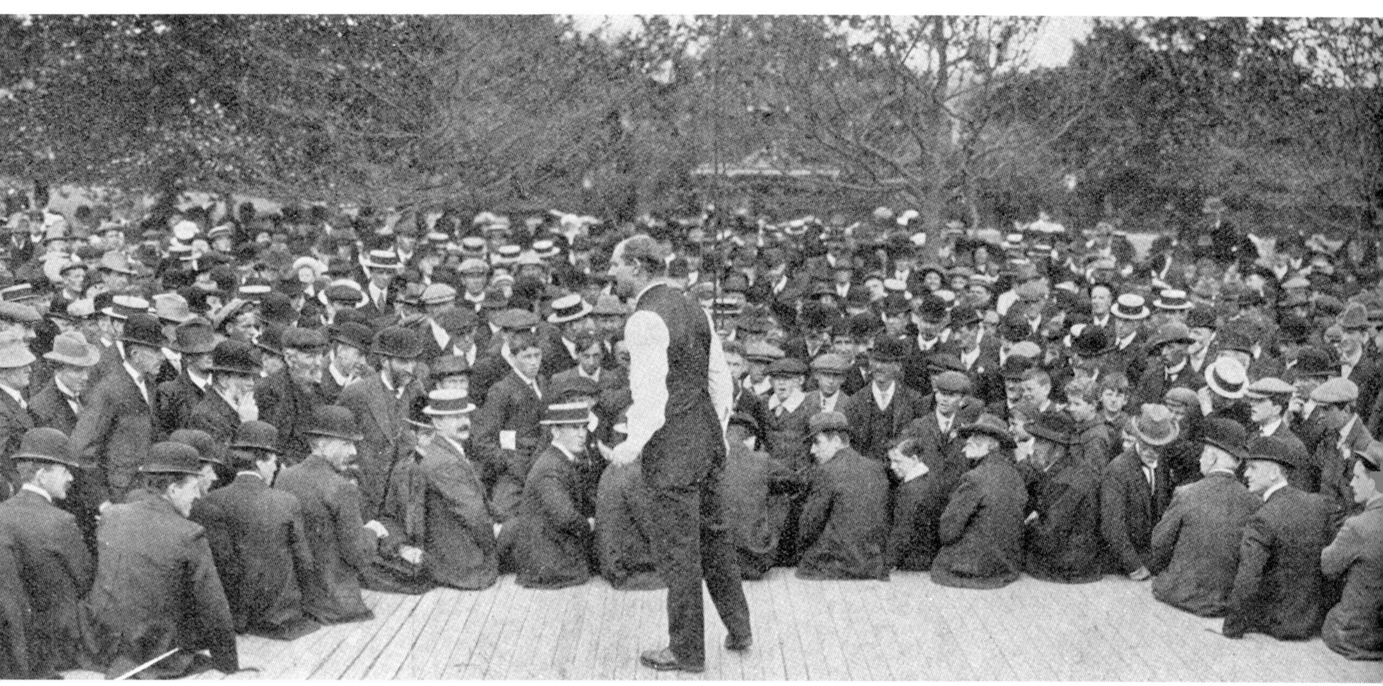

ABOVE **The 1913 Strike.** Although few local unions came out in sympathy with the Watersiders and the Miners, the local branches of both unions, and eventually the Seamen as well, struck. Dunedin was closed down. The Massey government's call for special police brought hundreds from the offices, banks and warehouses of the central city, not to mention hundreds of farmers who camped at Tahuna Park, mobilising unprecedented numbers of working men to support the strikers. The wounds and divisions of 1913 lasted for another generation. Here we see a member of the Strike Committee, Gilchrist, addressing the crowd. Photograph by Guy Morris.

Hocken Collections, *Auckland Weekly News*, 20 November 1913.

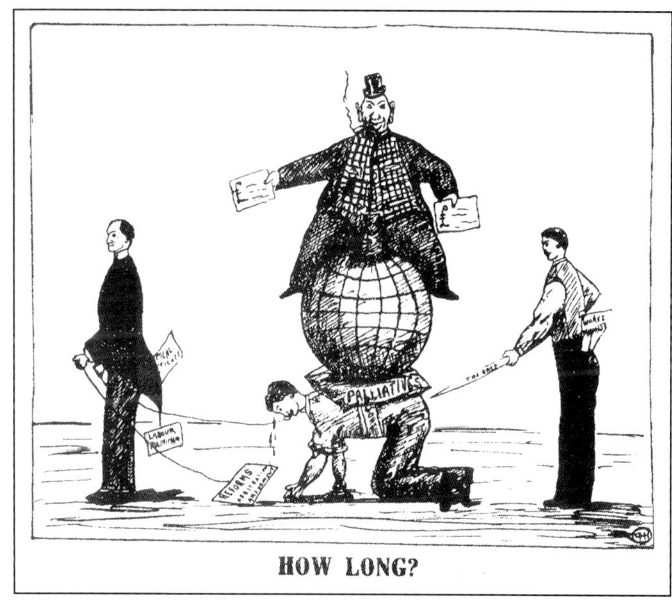

ABOVE **The crude cartoon opposite, from *Industrial Unionist*, 1 May 1913, shows 'Mr Fat' being carried by 'the worker'** who, in turn, is being crushed by the weight of and misled by 'Labour Politician'. Such a vision made social mobility irrelevant. Although this view became more widely held during World War I, thanks to the widespread feeling that capitalists were 'profiteering', no revolutionary movement developed, even after the Bolshevik revolution.

However, the newly formed Labour Party, with which the gaggle of pro-Labour MHRs aligned themselves, not only moved sharply left but increased its support in the mining towns and the working-class electorates in Wellington and Auckland. The intensification of class feeling among working folk drove other working-class folk into the arms of Massey and Reform. Indeed, Reform did better than Labour in winning support from working-class voters in these years.

II

And then World War I came. Within months, inflation had become a massive problem. Union leaders, Labour politicians, socialists and radicals blamed profiteers. 'Mr Fat' became a crude figure of capitalist rapacity. Wealth, not flesh and blood, ought to be conscripted, many argued. But the wartime government, a coalition of Reform and Liberal, instead introduced conscription of men. During Easter 1916 a small number of activists, including Dunedin's leading unionist and Legislative Councillor, J.T. Paul, met in Wellington and formed yet another New Zealand Labour Party. Those present elected Paul as President. The key issue that enabled the once-warring factions to unite was the impending threat of conscription. Paul, a devout Methodist, shared the widespread hostility of his church to any form of state coercion.

In opposing the conscription of men, the fledgling Labour Party risked political suicide before it had time to contest an election, especially as the government had postponed elections until the war was over. In 1914–15, most workers on the Flat, as in Dunedin, strongly supported the war. MacManus volunteered. So did Steve Boreham's sons. Just as most workers despised any man who refused to do a fair day's work for a fair day's pay, so they despised 'shirkers'. The men at Hillside were particularly aggressive on the subject.

An event in Ireland in Easter 1916 slowly had consequences that not only helped Labour survive its opposition to conscription, but brought it the support of the great majority of Irish Catholics. The critical rapprochement occurred in Dunedin. The New Zealand Liberals had long supported Home Rule for Ireland, but had no truck with the new organisation, Sinn Fein, a revolutionary organisation that opposed the war effort and demanded immediate independence. In Easter 1916, even as a handful of activists formed the New Zealand Labour Party, Sinn Fein resorted to armed rebellion in Dublin.

In Dunedin, as throughout Britain and the Empire, initial reactions were critical even among Irish Catholics who supported Home Rule. However, the brutal violence with which the British government suppressed the rebellion caused attitudes to shift both here and throughout the Empire and the United States. Key figures in the Australian Catholic Church quickly condemned the brutal suppression. Some also took a prominent part in opposing both Australian referenda on conscription.

Events in Australia fanned the local fires. A fiery Baptist preacher from Auckland, the Australian-born Howard Elliott, launched an assault on the Catholic church and founded a Protestant Political Association (PPA). Wherever the Orange Lodge existed, the PPA spread (there were 10 lodges in Dunedin, including one in South Dunedin). Members of the most fanatical anti-Catholic denominations, especially the Church of Christ, were also to the fore. Oddly, given that most Catholics still supported Joe Ward's Liberal Party, most supporters of the PPA hated socialists, Catholics, and supporters of the New Zealand Labour Party with even-handed impartiality. On the other side, however, were those who viewed all enemies of Ireland as enemies of Catholicism.

OPPOSITE see caption p. 153.

POBLACHT NA H EIREANN.

THE PROVISIONAL GOVERNMENT OF THE IRISH REPUBLIC TO THE PEOPLE OF IRELAND.

IRISHMEN AND IRISHWOMEN: In the name of God and of the dead generations from which she receives her old tradition of nationhood, Ireland, through us, summons her children to her flag and strikes for her freedom.

Having organised and trained her manhood through her secret revolutionary organisation, the Irish Republican Brotherhood, and through her open military organisations, the Irish Volunteers and the Irish Citizen Army, having patiently perfected her discipline, having resolutely waited for the right moment to reveal itself, she now seizes that moment, and, supported by her exiled children in America and by gallant allies in Europe, but relying in the first on her own strength, she strikes in full confidence of victory.

We declare the right of the people of Ireland to the ownership of Ireland, and to the unfettered control of Irish destinies, to be sovereign and indefeasible. The long usurpation of that right by a foreign people and government has not extinguished the right, nor can it ever be extinguished except by the destruction of the Irish people. In every generation the Irish people have asserted their right to national freedom and sovereignty; six times during the past three hundred years they have asserted it in arms. Standing on that fundamental right and again asserting it in arms in the face of the world, we hereby proclaim the Irish Republic as a Sovereign Independent State, and we pledge our lives and the lives of our comrades-in-arms to the cause of its freedom, of its welfare, and of its exaltation among the nations.

The Irish Republic is entitled to, and hereby claims, the allegiance of every Irishman and Irishwoman. The Republic guarantees religious and civil liberty, equal rights and equal opportunities to all its citizens, and declares its resolve to pursue the happiness and prosperity of the whole nation and of all its parts, cherishing all the children of the nation equally, and oblivious of the differences carefully fostered by an alien government, which have divided a minority from the majority in the past.

Until our arms have brought the opportune moment for the establishment of a permanent National Government, representative of the whole people of Ireland and elected by the suffrages of all her men and women, the Provisional Government, hereby constituted, will administer the civil and military affairs of the Republic in trust for the people.

We place the cause of the Irish Republic under the protection of the Most High God, Whose blessing we invoke upon our arms, and we pray that no one who serves that cause will dishonour it by cowardice, inhumanity, or rapine. In this supreme hour the Irish nation must, by its valour and discipline and by the readiness of its children to sacrifice themselves for the common good, prove itself worthy of the august destiny to which it is called.

Signed on Behalf of the Provisional Government,

THOMAS J. CLARKE.

SEAN Mac DIARMADA. THOMAS MacDONAGH.

P. H. PEARSE. EAMONN CEANNT.

JAMES CONNOLLY. JOSEPH PLUNKETT.

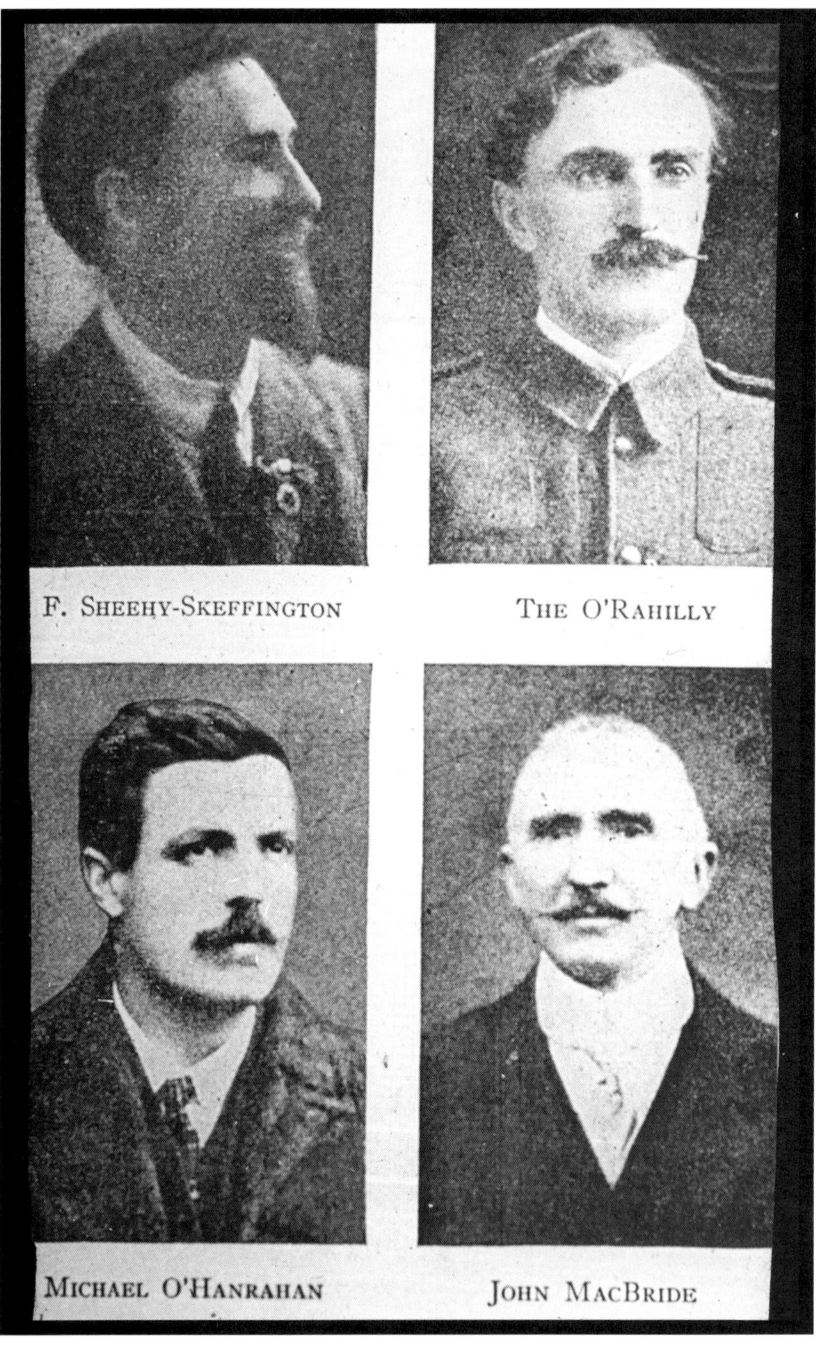

In Dunedin, a group of Irish nationalists and Sinn Fein supporters formed a Maoriland Irish National Association and began publishing *Green Ray*. Arthur McCarthy, long a key figure in the Otago Labour movement, played a critical role in establishing *Green Ray* as the voice of Gaelic Irish republican nationalism. Even the Catholic church swung in behind. In 1917, Father James Kelly was appointed editor of *The New Zealand Tablet*, the official paper of the Dunedin Diocese. Born, raised and educated in Ireland, he was an uncompromising champion of Sinn Fein. In November

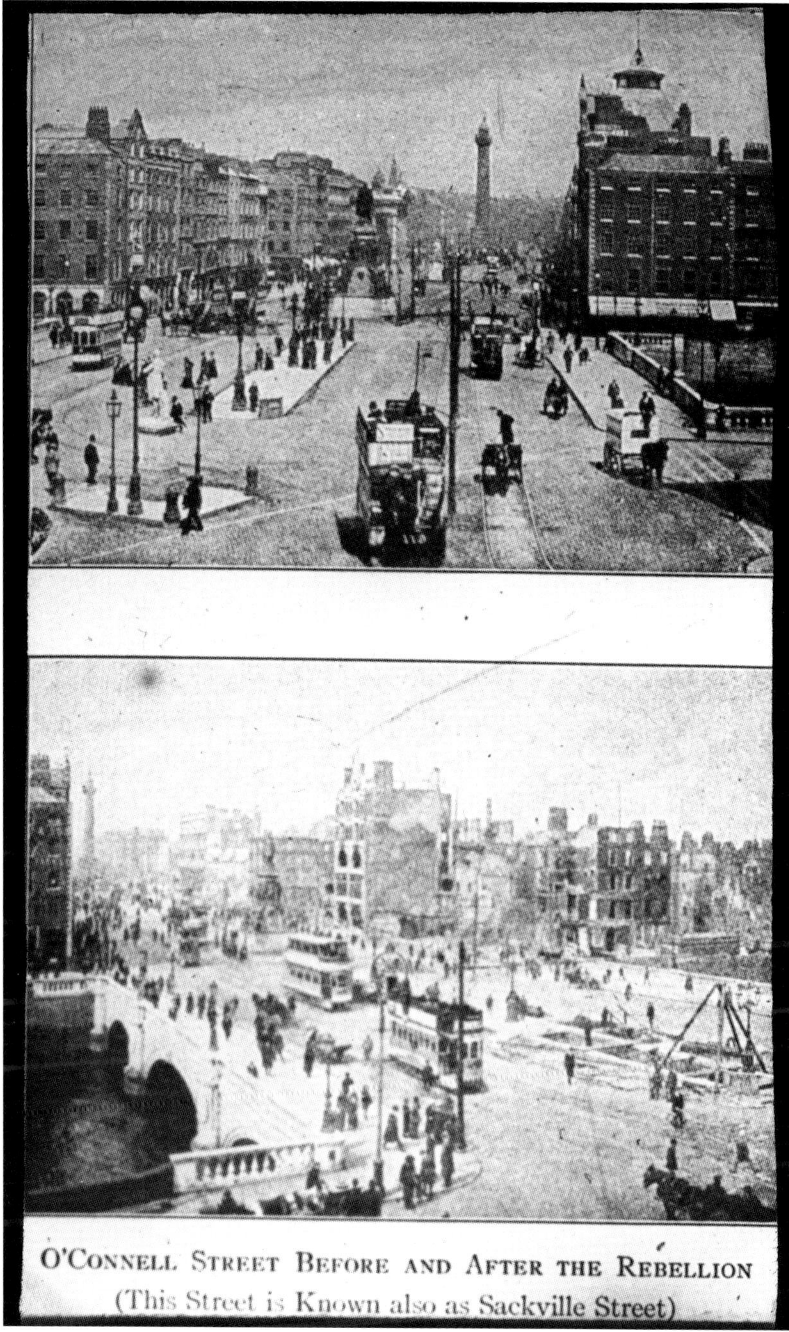

PREVIOUS PAGE AND LEFT
Images from a lantern-slide lecture on Ireland's recent history prepared by the Leader of the Parliamentary Labour Party (1919–33), Harry E. Holland. He stumped the country on the Irish issue and won the loyalty of many who identified with Ireland's campaign for self government, including the Roman Catholic hierarchy. These images show the proclamation of the Provisional government of the Irish Republic (page 151), the portraits of four Irish nationalists, all executed by the British Army, and O'Connell Street, Dublin, before and after the uprising of Easter 1916. One of the nationalists – Francis Sheehy Skeffington – was a famous moderate and pacifist who was brutally killed by British troops even though he had taken no part in the rebellion. Worse still, he was handing out pamphlets warning against looting when arrested and summarily executed. Séan O'Casey, the great dramatist, described Skeffington as the first true Irish socialist martyr. The execution of James Connolly, the famous union leader and revolutionary socialist, whose injuries were so severe that he had to be strapped into an armchair in order to be shot by firing squad, also made a profound impression on Irish everywhere, even though Connolly had been one of the leading rebels.

The four men pictured here were all summarily executed and duly became martyrs to the nationalist cause.

Alexander Turnbull Library, Roy Holland Collection, PA11-019-01, 04 and 03 respectively.

1917, he provoked his enemies to hysteria by referring to Queen Victoria as a 'fat old German woman'. 'Civis' of the *Otago Daily Times* – Alfred Fitchett, once a Methodist minister and now Dean of the Anglican Cathedral – took it upon himself to retaliate, although he made it clear he had no truck with Elliott or the PPA.

To inflame matters further, the government closed *Green Ray* and arrested its editor and its manager. The local Catholic hierarchy and the Labour Party launched a campaign demanding their release. Even Kelly,

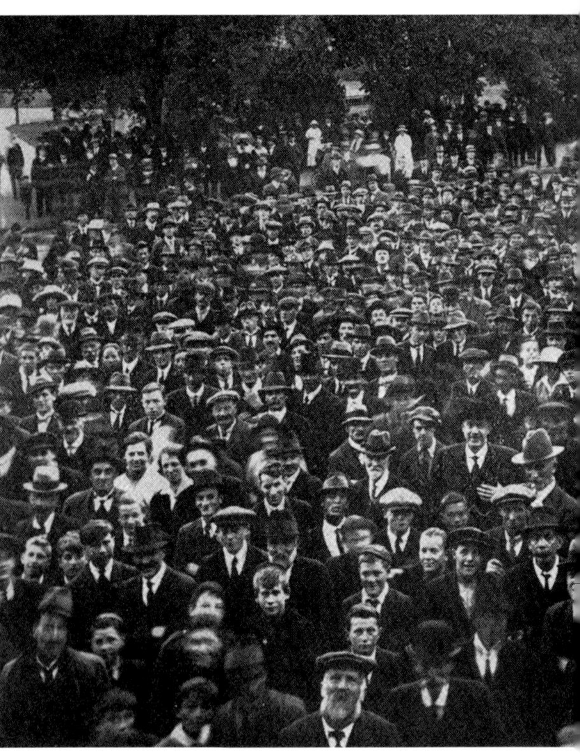

ABOVE **The crowd outside the offices of the Otago Daily Times & Witness Co, lower High Street, election night 1919.** This election aroused extraordinary interest, in part because the war-time National government had postponed the election until after the war; also Ward had taken the Liberals out of the war-time coalition, and frog-marched them leftwards to pre-empt the newly reinvigorated Labour Party. The contest was fierce in Dunedin, where Paul contested Sidey in Dunedin South, Jim Munro contested Sir Charles Statham in Dunedin Central (which included most of Caversham township and Parkside), and in Dunedin West the secretary of the Carpenters' Union, Ed Kellett, ran as Independent Labour against the Labour Party incumbent. Although Labour kept its share of the vote, it failed to win a seat and thus fared worse than in any election since 1890! Vast crowds of men turned out to see the results posted on election night, a popular ritual until World War II.

Hocken Collections, *Otago Witness*, December 1919.

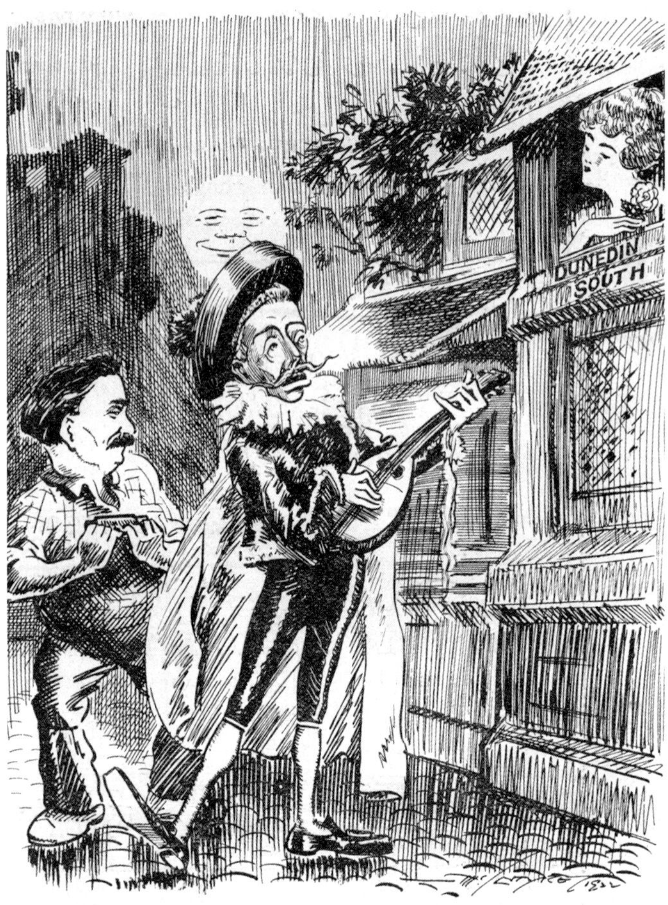

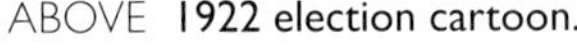

ABOVE **1922 election cartoon.** G.C. McIntyre cashed in on the long-standing reputation of *The Sketcher* and issued two specials for the 1922 election. Here he captures Jack MacManus, secretary of the Labourers' Union and long-time socialist stalwart, and T.K. Sidey competing for the favours of 'Southey' (Dunedin South). Sidey usually voted with Labour on industrial issues, and was especially attentive to the interests of the railway workers; he was just as fierce in his rejection of the Protestant Political Association's demands. Under no circumstances would he have agreed to accept the right of caucus to control his vote, the central requirement of the Labour Party's members. The doggerel in the cartoon attempts to capture one variant of a class-derived accent. Sidey's expensive Laroche car might have done the trick more effectively, but then McIntyre was not pro-Labour.

Hocken Collections.

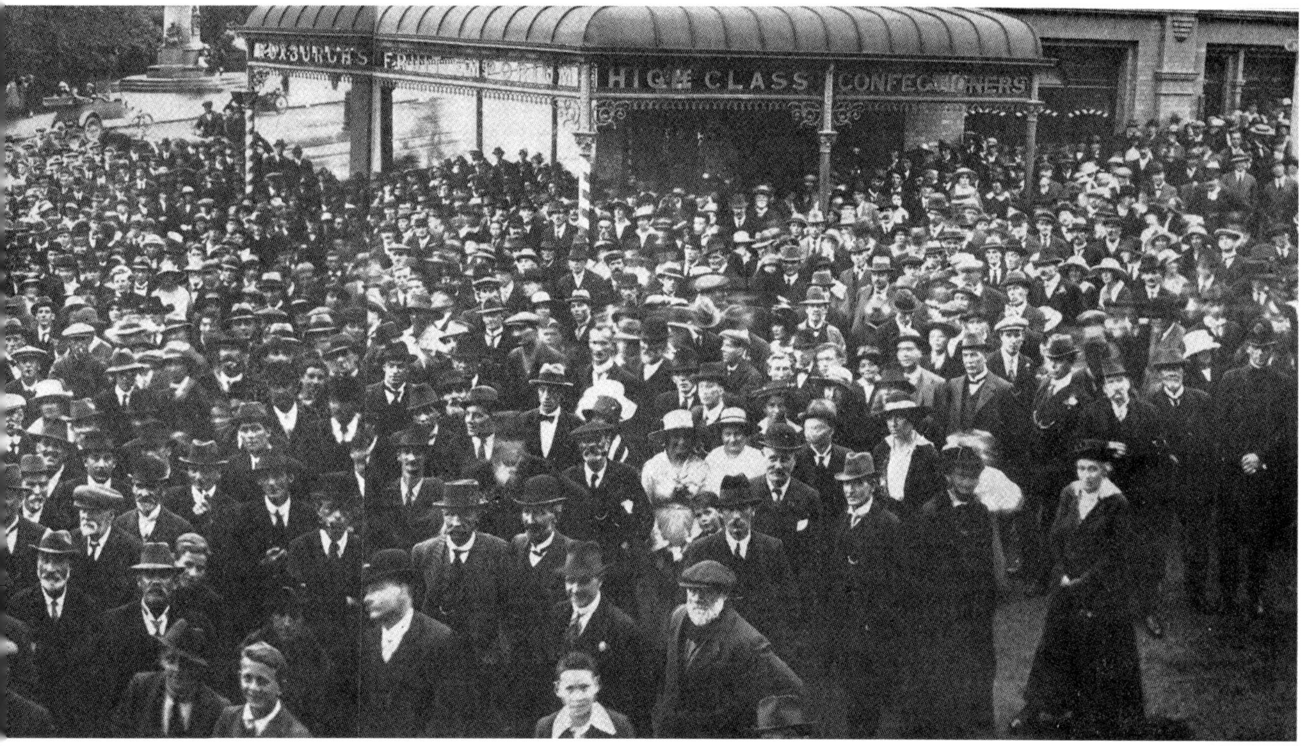

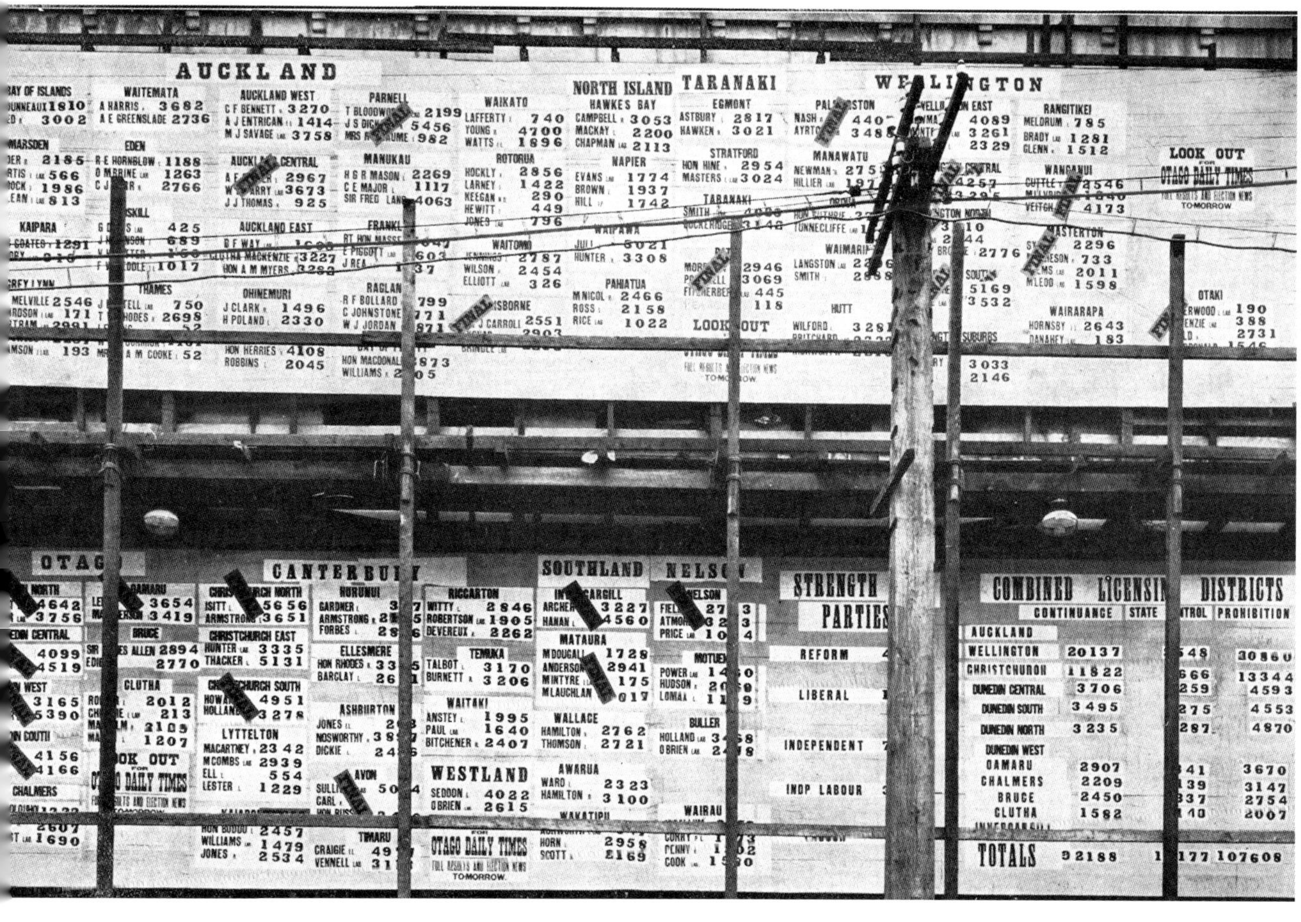
AUCKLAND
NORTH ISLAND
TARANAKI
LOOK OUT
OTAGO DAILY TIMES
SOUTHLAND
WESTLAND
STRENGTH
PARTIES
REFORM
LIBERAL
INDEPENDENT
INDP LABOUR
TOTALS

RIGHT **Class became a political issue,** the word itself becoming a political tool for highlighting inequality. The success of the Bolshevik revolution in 1917, and the vigorous emergence of a local Communist Party preaching the urgent need for a Bolshevik-style revolution locally, set the scene for a new politics of class. G.C. McIntyre's cartoon of 'Mr Gilchrist on the Social Question' captures one of the responses and the importance of both sides of a couple, wife as well as husband, to a family's social standing.

Hocken Collections, *The Sketcher*, December 1922, p. 22.

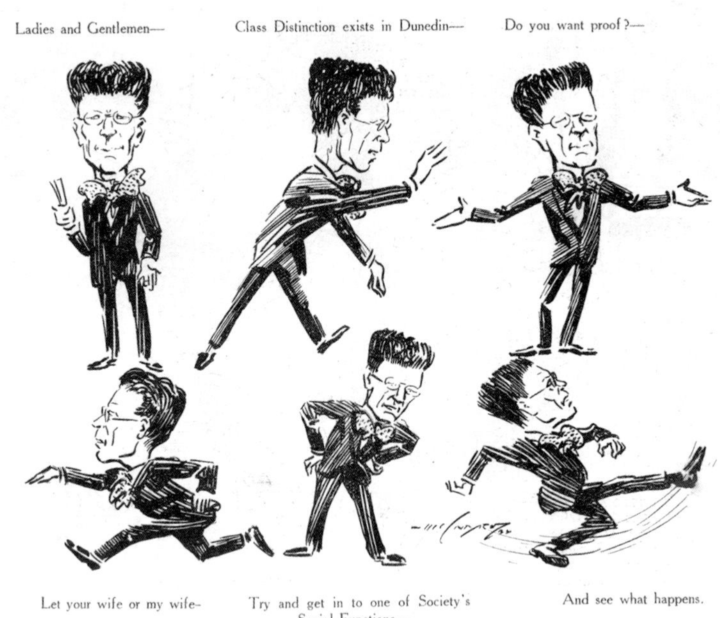

OPPOSITE, BOTTOM **No sooner had the Caversham branch of the Labour Party opened its hall, than it organised a Queen Carnival.** These events had been very popular since World War I, when they were often used as fund-raising ventures. Initially, various companies and organisations put forward a woman, often from a well-known family, but the eventual Queen was not necessarily the most beautiful or even very young. By the end of the 1930s, however, the emphasis on youth and beauty had increased, as can be seen here. What mattered most was the affirmation that working people could organise such events, aspire to be Queen for a day, or one of the Royal retinue. The long campaign of working people to achieve full participation in society, a campaign that began in England and Scotland in the early 1800s, had reached its triumphant conclusion in events like this, as well as in the election of the first Labour government. The late Noel Hudson, who kindly lent me this photograph, is the pageboy to the right of the Queen and came from a prominent Labour family. This, and the other of Savage opening the hall, were given to me by Noel Hudson, son of the MP.

once pro-Liberal rather than pro-Labour, became decidedly more friendly towards Labour. In 1918, local left-wing Catholics and socialists cooperated to produce *Green Ray*'s successor, *The Democrat*.

At this point, Harry Holland set out to translate the rapprochement between the Catholics and the socialists into a national agreement. Holland stumped the country with his lantern slides, wooing the Irish wherever he went with his account of recent Irish history (see pages 151–53). He aligned Labour behind Sinn Fein's demand for independence and attacked Britain's hypocrisy in claiming to stand for small nations while brutally waging war on those Irish who demanded independence. Thanks to the Irish Civil War, most New Zealanders of Irish ancestry, led by their bishops and clergy, swung from the prevaricating liberals to support the only party that championed their cause. In the elections of 1922, the year that Ireland won independence, Labour reaped the benefit of Holland's Irish campaign. The Dunedin accommodation was now national.

III

There has long been a myth, widely accepted in this country, that the labour movement and the Labour Party had their origins on the West Coast. Nothing could be further from the truth. The masters and journeymen who created the Labour Party and swept the city and its suburbs in 1890 created the Liberal–Labour coalition. The Workingman's Political Committee in Dunedin remained the sheet anchor for the idea of a Labour presence in Parliament, even when the unions were in total disarray. Thanks to the Arbitration Act, which Dunedin's unions had demanded since the early 1880s, Dunedin's skilled artisans and mechanics again mobilised and laid

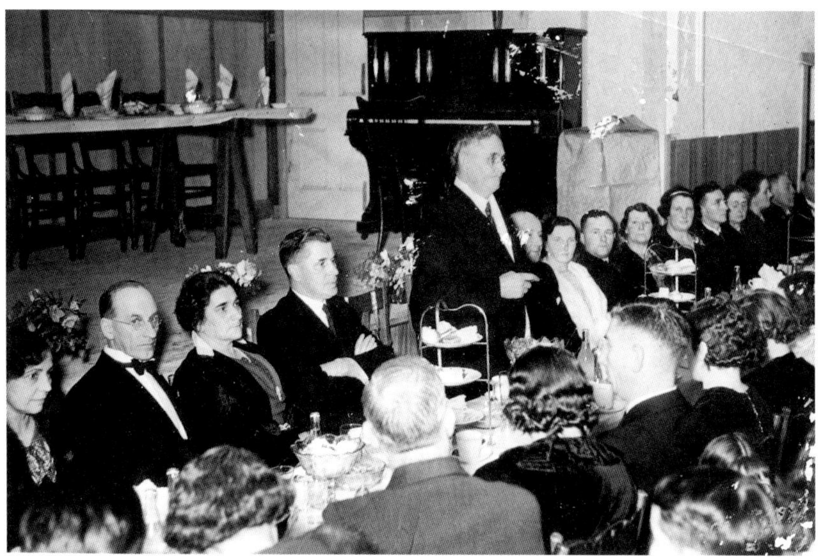

LEFT **The opening of the Harry E. Holland Memorial Hall, Caversham, 1937,** marks a fitting climax to the mobilisation of working men, as does the election both of the first Labour Council and the first Labour government two years earlier in 1935. From left: Mrs and Mr E.T. Cox (Mayor of Dunedin); Mrs and Hon. F. Jones; Rt Hon. M.J. Savage (Prime Minister); Hon. P. Webb; Mrs and Dr D.G. McMillan (he was the MP for Dunedin West); Mrs P. Neilson; Mrs and Mr W.A. Hudson. Peter Neilson, the MP for Dunedin Central (1935–46), which included most of Caversham township, Parkside, and Kensington, is out of sight. He and James Munro, MP for Dunedin North (1922–45), had been pioneers for socialism and partners in a bakery before entering Parliament. Neilson had a nephew who ran a milk treatment station in David Street, which Labour nationalised in 1947, and one of his sisters had married into a prominent Catholic family, the Rossbothams. I am grateful to Neilson's granddaughter, Anne Garden, for the information on Neilson, and to Noel Hudson for the photograph.

the foundation for the unionisation of the unskilled. A few hot-heads and socialists demanded that Labour desert the Lib–Lab government and go their own way. The Red Feds made that inevitable.

Over the next 10 years, in Dunedin, as in the other main towns, the unions of the unskilled and their members created the electoral basis for the formation of the second New Zealand Labour Party. It took Dunedin's skilled unions some time to accept the need for a national organisation. Ironically, given that in Britain Irish Catholics were excluded from skilled trades, the savage reaction of the British government to Sinn Fein not only helped bring most unions of skilled workers into line behind the Labour Party but saw the formation of the historic alliance of the Catholic church and the Labour Party. It took the 'Great Depression', however, to complete Labour's rendezvous with destiny.

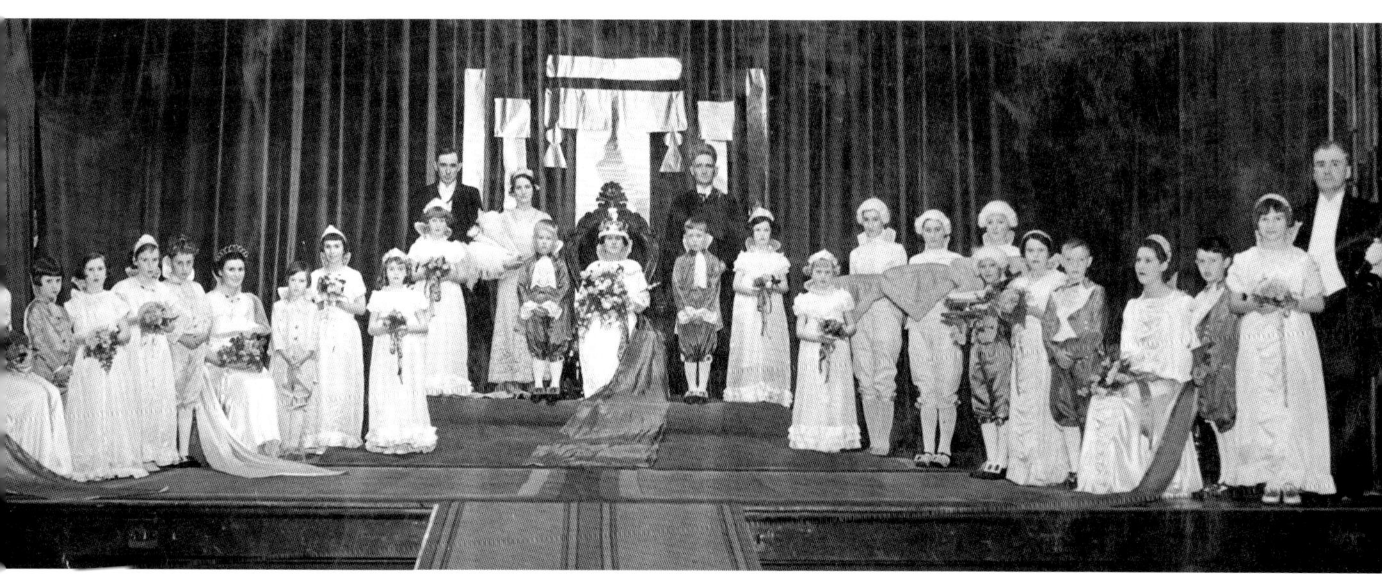

Epilogue

IT IS OFTEN ASSUMED that small local markets sustained the handicrafts and small-scale producers that came to dominate New Zealand. Small local markets were never the whole story, however. Even in the world's largest industrial economies, boasting many factories and mills with as many as 30,000 employees, small-scale firms, often owned and run by a family or partnership, continued and continue to be significant. Except where the market's size made mass production efficient, still unusual in this country, skill remained central to production. This was as true in the primary as in the secondary sector. It was common here, as it was in Britain, even for larger firms to be managed personally, both with regard to governance and to cultural style.

Although the Liberal-Labour governments decided that New Zealand's economic future lay in producing commodities for the British market, and all of their successors agreed until Britain entered Europe in 1973, geographic distance, modest tariffs, and later import controls ensured the survival of the vibrant manufacturing sector that has been the major focus of this book. Comparable nations, notably Australia and the United States, provided much more aggressive protection to their nascent industries and still do. One cannot escape the suspicion that in the last 30 years our own pursuit of deregulation and freer trade, based in part on the assumption that small is ugly and manufacturing dead, has been mistaken, not to say short-sighted, in the extreme.

The values and habits of life that the working men and women of The Flat created, and forged into a political programme, predicated on new meanings for equality and fairness, came to permeate the city and the entire country. A belief in the positive value of work, and the dignity of all labour, lay at the heart of this new culture. These notions were profoundly embedded in the way in which they organised work in their workshops and factories. They were also embedded in the new ideologies among employers that centred on the welfare and well-being of their employees. Husbands may have been slower to extend such rights to their wives, but were often supportive of their daughters, and their daughters, heading off to the city each day for around ten years, acquired knowledge, skills and networks that gave them greater confidence when they married in re-negotiating the patriarchal codes of the Old World. As I said at the end of *Building the New World*, 'We sorely miss the ease with which they confidently translated their faiths into social, economic, and political practice, but the faiths survive even if many of their translations no longer work. We may have lost their world … but we are as we find ourselves because that world once existed.'

Acknowledgements

MOST OF THE WORK for this book, apart from putting it together, has been done over the past 30 years. Some debts have been incurred, however, and I would like to thank in particular Chris Scott at the DCC Archives and Jill Haley at Toitū Otago Settlers Museum. Alison Clark and Mary Lewis at the Hocken Collections have also been most helpful. As ever, I am indebted to David Hood for the maps.

I have benefited from the help of Bill McKinlay in writing captions for the series of photographs on bootmaking. Linda Wigley (of Toitū Otago Settlers Museum) assisted with determining what the photographs of women in clothing factories actually illustrate, and Dr Jane Malthus helped to date various photographs and identify the sewing machines in McKinlay's footwear factory. Jack Rutherford, a grandson of Robert the grocer, also helped, as he has done so often across the years. Sadly Noel Hudson, an old friend, died before this book was finished.

A word about place names. In order to avoid confusion, all street and place names are those used today.

I am also most grateful to John McIndoe, Ross Grimmett, Séan Brosnahan and Anne Garden for their interest and help. Without Wendy Harrex, this book would have been stillborn. I am also grateful to Fiona Moffat and Rachel Scott at Otago University Press for their work on the book.

Select Bibliography

THE ORIGINS of this book lie in the research undertaken for the Caversham Project. Most of the material in the text, as distinct from the captions, is more fully discussed in the Project's major publications: Erik Olssen, *Building the New World: Work, politics and society in Caversham, 1880s–1920s* (Auckland University Press, 1994) (now an e-book); Barbara Brookes, Annabel Cooper and Robin Law (eds), *Sites of Gender: Women, men & modernity in Southern Dunedin, 1880–1939* (Auckland University Press, 2003). Chapter 2 of that book, 'The Landscape of Gender Politics', written by Cooper, Law, myself and Kirsten Thomlinson, remains the best history of the Flat and its politics. *Sites of Gender* is available as a Google book. Finally, *An Accidental Utopia? Social mobility and the foundations of an egalitarian society, 1880–1940* (Otago University Press, 2011), co-written by Clyde Griffen and myself, with indispensable statistical analyses by Frank Jones, provides the fullest account of worklife, marital and inter-generational mobility on the Flat and the most thorough analysis of the relationship between those mobility patterns and the social and political history of the area.

Several local studies have also been invaluable: Geoff Stedman's thesis, 'The South Dunedin Flat: A Study in Urbanisation 1849–1965', 2 vols (MA thesis, University of Otago, 1966), remains the essential starting point. Also useful are Alma Rutherford, *The Edge of the Town: Historic Caversham as seen through its streets and buildings* (John McIndoe, 1978); H.J.A. Aitken, *St Kilda: The first hundred years: a short history of the Borough of St Kilda 1875–1975* (Borough of St Kilda, 1975); and Barbara A. Newton, *Our St Clair – A Resident's History* (Kenore Press, 2003). Father P.R. Mee's various histories are essential for the history of the Roman Catholic church and the Flat's Irish Catholics; see especially *The Turn of the Tide: 'A Historette' to the establishment of St Bernadette's Parish, Forbury, Dunedin* (New Zealand Tablet, 1977) and *St Patrick's School South Dunedin 1878–1978* (1978).

I have also found helpful several works on particular firms, especially: E.M. Seed, 'History of the brick, tile and pottery industries in Otago' (MA thesis, University of Otago, 1954); John H. Angus, *The Ironmasters: The first one hundred years of H.E. Shacklock Limited* (H.E. Shacklock Ltd, 1973); Kathryn G. Lucas, *A New Twist: A centennial history of Donaghy's Industries Limited* (Donaghy's Industries Limited, 1979); G.J. McLean, *Spinning Yarns: A centennial history of Alliance Textiles Ltd and its predecessors 1881–1981* (Alliance Textiles Ltd, 1981), and S.R.H. Jones, *Doing Well and Doing Good: Ross & Glendining – Scottish enterprise in New Zealand* (Otago University Press, 2010).

Index

Page numbers in **bold** refer to illustrations.